Progress in Mathematics 6

Edited by
J. Coates and
S. Helgason

Gérard Lion
Michèle Vergne

The Weil representation, Maslov index and Theta series

Springer Science+Business Media, LLC

Michèle Vergne

Department of Mathematics
Massachusetts Institute of Technology
Cambridge, MA 02139
U.S.A.

Gérard Lion

Université de Paris
X U.E.R. de Sciences Economiques
92001 Nanterre
France

Library of Congress Cataloging in Publication Data

Lion, Gerard, 1949–
The Weil representation, Maslov index, and theta series.

(Progress in mathematics; 6)
Includes bibliographical references.
1. Symplectic groups. 2. Representations of groups. 3. Series, Theta. 4. Forms, Modular. 5. Lifting theory. I. Vergne, Michèle, joint II. Title. III. Title: Maslov index, and theta series. IV. Series: Progress in mathematics (Cambridge); 6.
QA171.L765 512'.22 80-15942

CIP—Kurztitelaufnahme der Deutschen Bibliothek

Lion, Gérard:
The Weil representation, Maslov index and theta series / Gérard Lion : Michèle Vergne.—Boston, Basel, Stuttgart : Birkhäuser, 1980.
(Progress in mathematics : 6)

ISBN 978-0-8176-3007-2 **ISBN 978-1-4684-9154-8 (eBook)**
DOI 10.1007/978-1-4684-9154-8

NE: Vergne, Michèle:

Originally published by Birkhäuser Boston in 1980.

To the Memory

of

Maria Vergne

Preface.

In these notes, the Shale-Weil representation of the symplectic group is discussed, as well as some of its applications to number theory.

The monograph is composed of two parts:

In Part I, written by Gérard Lion and Michèle Vergne, we introduce the Shale-Weil representation and establish a relation between its cocycle and the Maslov index.

In Part II, written by Michèle Vergne, applications of θ-series to liftings of modular forms are given.

Although the results of the first part enlightens the exposition of the classical transformation properties of θ-functions, a reader mainly interested by these applications to liftings could read directly the second part with an eventual glance to earlier paragraphs. The two parts have separate introductions and bibliographical notes.

<u>The authors</u>

Table of Contents

Part I: The Shale-Weil representation and the Maslov Index.

by

Gérard Lion and Michèle Vergne

1.0. Introduction:

Relations between symplectic geometry, Maslov index, representations of the Heisenberg group and the Shale-Weil representation of the symplectic group are discussed in this chapter. We first give in 1.1 the basic definitions and properties of symplectic vector spaces, Lagrangian subspaces, the Heisenberg Lie algebra and the action of the symplectic group of those objects.

The Schrödinger representation of the Heisenberg group N associated to a Lagrangian plane ℓ is constructed in 1.2.

We prove in 1.3 the Stone-von Neumann theorem which asserts that all unitary representations of N whose restriction to the center of N acts by the same non-trivial character are essentially the same: two such irreducible representations are equivalent. Although this uniqueness theorem underlies the construction of the Shale-Weil representation R, we will give however a direct construction of R, independent of the proof of this theorem. Thus results of Section 1.3 will not be needed subsequently.

The Schrödinger representations W_ℓ and $W_{\ell'}$ of the Heisenberg group associated to the Lagrangian planes ℓ and ℓ' are equivalent: we give in 1.4 a canonical choice of an operator $\mathcal{F}_{\ell',\ell}$ such that:

$$W_\ell(n) = \mathcal{F}_{\ell',\ell}^{-1}\; W_{\ell'}(n)\; \mathcal{F}_{\ell',\ell} \quad \text{for every } n \in N.$$

(This operator in appropriate coordinates is a partial Fourier transform.)

Before going further, we have to introduce, in 1.5, the Maslov index of a triple of Lagrangian planes: Let (V,B) be a symplectic vector space, ℓ_1, ℓ_2 and ℓ_3 three Lagrangian planes, then

$$Q_{123}(x_1 \oplus x_2 \oplus x_3) = B(x_1,x_2) + B(x_2,x_3) + B(x_3,x_1)$$

is a quadratic form on $\ell_1 \oplus \ell_2 \oplus \ell_3$, which can be diagonalized with p times the eigenvalue 1 and q times the eigenvalue -1. A modified definition of the Maslov index $\tau(\ell_1,\ell_2,\ell_3)$ due to M. Kashiwara is $\tau(\ell_1,\ell_2,\ell_3) = \text{sign}\, Q_{123} = p - q$. We prove in 1.5 that this Maslov index verifies a fundamental chain property.

The symplectic group G acts on Lagrangian planes, on the Heisenberg Lie algebra $\mathfrak{n}$ and on the Heisenberg Lie group N. If W_ℓ is the Schrödinger representation of N associated to ℓ, the representation $W'(n) = W_\ell(g\cdot n)$ is equivalent to W_ℓ: The natural action $A(g)$ of the symplectic group on functions on N transforms the representation W_ℓ of N into the representation $n \to W_{g\cdot\ell}(g\cdot n)$. The Fourier operator $\mathcal{F}_{\ell,g\cdot\ell}$ intertwines the representation $W_{g\cdot\ell}$ with W_ℓ. Thus the canonical unitary operator $R_\ell(g) = \mathcal{F}_{\ell,g\cdot\ell}\ A(g)$ satisfies the fundamental relation:

$$W_\ell(g\cdot n) = R_\ell(g)\ W_\ell(n)\ R_\ell(g)^{-1}, \text{ for every } n \in N.$$

We prove in (1.6) that

$$\mathcal{F}_{\ell_1,\ell_2}\,\mathcal{F}_{\ell_2,\ell_3}\,\mathcal{F}_{\ell_3,\ell_1} = e^{\frac{i\pi}{4}\tau(\ell_1,\ell_2,\ell_3)}\,\mathrm{Id}$$

and deduce from this formula that

$$R_\ell(g_1g_2) = c_\ell(g_1,g_2)\,R_\ell(g_1)\,R_\ell(g_2)$$

with $\quad c_\ell(g_1,g_2) = e^{-\frac{i\pi}{4}\tau(\ell,g_1\ell,g_1g_2\ell)}$.

It is known that the Shale-Weil projective representation R_ℓ is not equivalent to a true representation of G. We construct in 1.7 a function $s(\tilde{\ell}_1,\tilde{\ell}_2)$, defined on couples $(\tilde{\ell}_1,\tilde{\ell}_2)$ of oriented Lagrangian planes, invariant under the action of the symplectic group. Let us write

$$c(\ell_1,\ell_2,\ell_3) = e^{\frac{i\pi}{4}\tau(\ell_1,\ell_2,\ell_3)},$$

we prove the relation

$$c(\ell_1,\ell_2,\ell_3)^2 = s(\tilde{\ell}_1,\tilde{\ell}_2)\,s(\tilde{\ell}_2,\tilde{\ell}_3)\,s(\tilde{\ell}_3,\tilde{\ell}_1).$$

This leads to the results of Shale and Weil that the projective representation R_ℓ is equivalent to a true representation of the metaplectic group, a double covering of G.

The Section 1.8 is devoted to the construction of the universal covering group G of $SL(2,\mathbb{R})$ by elementary means. Some explicit formulas for the Shale-Weil representation of G are given.

Let Λ be the manifold of all Lagrangian planes of (V,B). In Section 1.9, we use the chain property of τ to construct both universal coverings of Λ and of $Sp(B)$. We here use the fact that the function $(g_1,g_2) \to \tau(\ell\,,g_1\cdot\ell,g_1g_2\ell)$

is a $\mathbb{Z}$-valued cocycle of Sp(B). We relate our construction of the Maslov index to the formulas in coordinates of Leray and Souriau. The results of this section are independent of the rest of the notes. We follow in the Section 1.9 (as in many other parts of these notes) an idea of M. Kashiwara, and we are happy to thank him.

The appendix extends the notions of Part I to a local field k. The Kashiwara index $\tau(\ell_1,\ell_2,\ell_3)$ is then defined to be the class of the form Q_{123} in the Witt group W_k. Thus we obtain a canonical cocyle of the symplectic group $G = Sp(n,k)$ with values in W_k. We define a function $s(\tilde{\ell}_1,\tilde{\ell}_2)$ on couples of oriented Lagrangian planes, invariant by the action of G. We describe then the metaplectic group, using this function s, and prove as in 1.7 that the Weil projective representation of G lifts to a representation of the metaplectic group. The results of this appendix are due to Patrice Perrin. Similar results had been obtained independently by R. Rao.

Bibliographical notes are given at the end of Part I.

The second author gave a seminar in 1978-79 at the Massachusetts Institute of Technology on these subjects and thank the participants for several improvements of the text, especially Martin Andler, Victor Kac, Steve Paneitz, Carolyn Schröeder and Bob Styer. We thank Patrice Perrin for discussions on his results and Masaki Kashiwara for communicating some unpublished texts.

Our thanks go to Sophie Koulouras for the patient typing of the text.

The Shale-Weil Representation
and the Maslov Index

1.1. Symplectic vector spaces and the Heisenberg Lie algebra.

Let V be a finite dimensional real vector space. Let B be a non-degenerate skew symmetric form on V. Hence $\dim V$ is even. Let $\dim V = 2n$. Then we can choose a basis $(P_1, P_2, \cdots, P_n, Q_1, Q_2, \cdots, Q_n)$ of V with the relations:

1.1.1. $$B(P_i,P_j) = 0 \qquad B(Q_i,Q_j) = 0$$
$$B(P_i,Q_j) = \delta_{ij} \qquad B(Q_i,P_j) = -\delta_{ij}.$$

We will call such a basis a symplectic basis of (V,B).

We consider the Lie algebra $\mathfrak{n} = V + \mathbb{R}E$, with the bracket law defined as follows:

$$[X,Y] = B(X,Y)E, \text{ if } X,Y \in V$$

$$[\mathfrak{n},E] = 0, \text{ i.e. } \mathbb{R}E \text{ is the center of } \mathfrak{n}.$$

$\mathfrak{n}$ is called the Heisenberg Lie algebra. If (P_i,Q_j) is a symplectic basis of (V,B), the Lie algebra $\mathfrak{n}$ has as basis (P_i,Q_j,E) with the Heisenberg commutation relations, or "canonical commutation relations:

1.1.2. $$[P_i,Q_j] = \delta_{ij}E$$
$$[P_i,P_j] = 0$$
$$[Q_i,Q_j] = 0.$$

Let A_n be the associative algebra of differential operators with polynomial coefficients on $\mathbb{C}^n$ and the corresponding bracket $[D_1,D_2] = D_1D_2 - D_2D_1$. The algebra A_n is generated by $p_i = \frac{\partial}{\partial x_i}$, $q_j = x_j$ satisfying the canonical relations $[p_i,q_j] = \delta_{ij}$. The corresponding representation $P_i \to p_i$, $Q_j \to iq_j$, $E \to i\mathrm{Id}$ of $\mathfrak{n}$ will be of particular importance to us.

We will use some simple lemmas for symplectic spaces: If L is a subspace of (V,B) we will denote by $L^\perp$ the orthogonal complement of L in V relative to B, i.e.

$$L^\perp = \{x \in V;\ B(x,y) = 0 \quad \forall y \in V\}.$$

We then have:

1.1.3. a) $(\dim L) + \dim(L^\perp) = 2n$

b) $(L^\perp)^\perp = L$

c) $(L_1 + L_2)^\perp = L_1^\perp \cap L_2^\perp$

d) $(L_1 \cap L_2)^\perp = L_1^\perp + L_2^\perp$.

a), b), c) are clear; d) is easily deduced from c) by using the relation $(L^\perp)^\perp = L$.

If a subspace ℓ of V is such that $\ell = \ell^\perp$, ℓ is called a Lagrangian subspace of V. We have then $B(x,y) = 0$ for every $x,y \in \ell$, hence ℓ is totally isotropic with respect to the form B; moreover if x is such that $B(x,\ell) = 0$, then

$x \in \ell$; i.e. ℓ is a maximal totally isotropic subspace of (V,B).

Lemma 1.1.4. Let ℓ be a Lagrangian subspace of (V,B). There exists a Lagrangian subspace ℓ' such that $\ell \oplus \ell' = V$.

Proof: Let ℓ_0' be maximal among the totally isotropic subspaces such that $\ell' \cap \ell = 0$. Hence we have $\ell_0'^{\perp} + \ell^{\perp} = V$. However we have $\ell_0'^{\perp} \subset \ell + \ell_0'$, for otherwise, we could choose $x \in \ell_0'^{\perp}$ and not in $\ell + \ell_0'$ and the subspace ℓ_0' contained in $\ell_0' + \mathbb{R}x$ would not be maximal. Hence we obtain $\ell_0' + \ell = V$, which is the equality required.

1.1.5. Hence given ℓ a Lagrangian subspace, we can find ℓ' such that $\ell \oplus \ell' = V$; the bilinear form B clearly induces a pairing between ℓ and $\ell' = V/\ell$. So we can choose a symplectic basis $(P_1, P_2, \cdots, P_n, Q_1, Q_2, \cdots, Q_n)$ such that $\ell = \sum_{i=1}^{n} \mathbb{R}P_i$, $\ell' = \sum_{i=1}^{n} \mathbb{R}Q_i$.

1.1.6. We define the symplectic group $G = Sp(B)$: By definition $g \in G$ if g is an invertible linear transformation of the vector space V preserving the form B, i.e. for every $x,y \in V$, $B(gx,gy) = B(x,y)$.

1.1.7. Let $V = \mathbb{R}P \oplus \mathbb{R}Q$, with $B(P,Q) = 1$, be the two dimensional canonical symplectic space. Then $g = \begin{pmatrix} a & b \\ c & d \end{pmatrix}$ belongs to $Sp(B)$ if and only if $B(aP + cQ, bP + dQ) = B(P,Q) = 1$, i.e. $ad - bc = 1$. Hence the symplectic group for a two-dimensional symplectic vector space is isomorphic to $SL(2,\mathbb{R})$.

1.1.8. Let V be a $2n$ dimensional symplectic vector space. We choose a decomposition $V = \ell \oplus \ell'$ of V into two complementary Lagrangian spaces. We write x for an element of ℓ and y for an element of ℓ'. We write an element g of $G = Sp(B)$ as $g = \left(\begin{array}{c|c} a & b \\ \hline c & d \end{array}\right)$, with

$$\begin{array}{ll} a: \ell \to \ell, & b: \ell' \to \ell \\ c: \ell \to \ell', & d: \ell' \to \ell' \end{array} .$$

The conditions for g to be in G are:

$$\begin{array}{lll} B(g\cdot x, g\cdot x') = 0 & & x, x' \in \ell \\ B(g\cdot x, g\cdot y) = B(x,y) & & x \in \ell,\ y \in \ell' \\ B(g\cdot y, g\cdot y') = 0 & & y, y' \in \ell' \end{array} .$$

We identify ℓ and ℓ'^* and ℓ' to $(\ell)^*$, via the bilinear map $B(x,y)$, $x \in \ell$, $y \in \ell'$.

We identify ℓ' with ℓ^* via $x \to B(x,y)$ for $y \in \ell'$.

1.1.9. Hence these conditions are equivalent to:

$$\begin{array}{c} {}^t ca = {}^t ac \\ {}^t ad - {}^t cb = id \\ {}^t db = {}^t bd \end{array} .$$

In particular we have

$$g^{-1} = \left(\begin{array}{c|c} {}^t d & -{}^t b \\ \hline -{}^t c & {}^t a \end{array}\right) .$$

As $g^{-1} \in G$ we have also $c^t d = d^t c$, $a^t b = b^t a$, $a^t d - b^t c = id$.

If $x \in \ell$, $y \in \ell'$, and $u: \ell' \to \ell'$, then:

$$B(x,uy) = B({}^t ux,y).$$

1.1.10. From 1.1.5, we see that the group G acts transitively on the couples (ℓ,ℓ') of transverses Lagrangian planes, as they can be transformed to the canonical pair $\ell = \sum_{i=1}^{n} \mathbb{R}P_i$, $\ell' = \sum_{i=1}^{n} \mathbb{R}Q_i$ by a symplectic automorphism.

1.2. The Heisenberg group and the Schrödinger representation.

We consider the Heisenberg group N as being the simply connected Lie group of Lie algebra $\mathfrak{n}$. Via the exponential map exp, N is identified with the $2n + 1$ vector space $V + \mathbb{R}E$ with the multiplication law:

$$\exp(v + tE)\cdot\exp(v' + t'E) = \exp(v + v' + (t + t' + \frac{B(v,v')}{2})E)$$

where $v,v' \in V$, $t,t' \in \mathbb{R}$.

For dv the euclidean measure on the vector space V, $dv\,dt$ is the Haar measure on N. ($dv\,dt$ is invariant by left and right translations.)

The subgroup $\{\exp tE\}$ is the center Z of N.

The group $G = Sp(B)$ acts as a group of automorphisms on N by $g\cdot(\exp v + tE) = \exp(gv + tE)$. In particular G preserves the center Z of N.

1.2.1. Let ℓ be a Lagrangian subspace of (V,B), then $\ell + \mathbb{R}E$ is an abelian subalgebra of $\mathfrak{n}$. We consider the group $L = \exp(\ell + \mathbb{R}E)$ which is an abelian subgroup of N.

1.2.2. Let us consider the function $f(\exp v + tE) = e^{2i\pi t}$ on the group N, f is a function with values in the torus $T = \{z \in \mathbb{C},\ |z| = 1\}$. It is immediate to see that f restricted to L defines a character of L, i.e.

$$f(h_1h_2) = f(h_1)f(h_2) \quad \text{for } h_1,h_2 \in L.$$

(If we order the pairs (H,ψ) of a subgroup H of N together with a character ψ of H via the relation $(H,\psi) < (H_1,\psi_1)$ if $H \subset H_1$ and $\psi_1|H = \psi$, it is easy to see that the pair (L,f) is maximal for this order.)

1.2.3. Let us consider (ℓ,ℓ') as in (1.1.4). Every element n of N can be written uniquely as $n = \exp y' \exp(x + tE)$, with $y' \in \ell'$, $x \in \ell$, $t \in \mathbb{R}$. Hence the coset space N/L can be identified with ℓ'. The euclidean measure dv' on ℓ' defines on N/L a positive measure $d\dot{n}$ invariant by the left action of N on the homogeneous space N/L. We recall that such a measure $d\dot{n}$ is unique up to multiplication by a positive constant.

1.2.4. We consider for a given Lagrangian subspace ℓ the representation $W(\ell)$ induced by the character f of the group L. $W(\ell)$ is the Schrödinger representation of N associated to ℓ. We write

$$W(\ell) = \operatorname{Ind}\uparrow_{L}^{N} f.$$

By definition of induced representation, $W(\ell)$ is realized as follows: The Hilbert space $H(\ell)$ is the completion of the space of continuous functions φ on N such that

1.2.4. a) $\varphi(nh) = f(h)^{-1}\varphi(n)$, for every $n \in N$ and every $h \in L$.

1.2.4. b) The function $n \to |\varphi(n)|$ on N/L is square integrable with respect to the invariant measure $d\dot{n}$ on N/L.

The norm of φ is:

$$\|\varphi\|^2 = \int_{N/L} |\varphi(n)|^2 \, d\dot{n} .$$

The representation $W(\ell)$ is defined to be the representation of N in $H(\ell)$ given by left translations:

i.e. $(W(\ell)(n_0)\varphi)(n) = \varphi(n_0^{-1}n)$ for $\varphi \in H(\ell)$, $n_0 \in N$.

As $\exp tE$ is in the center of the group N, we have:

$$\begin{aligned}(W(\ell)(\exp tE)\varphi)(n) &= \varphi((\exp - tE)n) \\ &= \varphi(n \exp - tE) \\ &= e^{2i\pi t}\varphi(n)\end{aligned}$$

i.e. $W(\ell)(\exp tE) = e^{2i\pi t} Id_{H(\ell)}$,

where $Id_{H(\ell)}$ denotes the identity operator on $H(\ell)$.

1.2.5. Let us consider (ℓ, ℓ') as in 1.1.4. Each element of N is written uniquely as $n = \exp y \cdot \exp(x + tE)$ where $y \in \ell'$, $x \in \ell$. Hence, if $\varphi \in H(\ell)$, the condition 1.2.4 a) is written as $\varphi(\exp y \cdot \exp x \cdot \exp tE) = e^{-2i\pi t}\varphi(\exp y)$, $(y \in \ell'$, $x \in \ell$, $t \in E)$. So φ is completely determined by its restriction to $\exp \ell'$. Hence the application $\varphi \to \varphi(y) = \varphi(\exp y)$ defines an isometry R of $H(\ell)$ with $L^2(\ell')$. The representation $\widehat{W}(n) = RW(n)R^{-1}$ acts on $L^2(\ell')$ by the following formulas:

$$(\widetilde{W}(\exp x)\varphi)(y) = e^{2i\pi B(x,y)}\varphi(y) \qquad x \in \ell,\ y \in \ell'$$

$$(\widetilde{W}(\exp y_0)\varphi)(y) = \varphi(y - y_0) \qquad y, y_0 \in \ell'$$

$$\widetilde{W}(\exp tE) = e^{2i\pi t}\mathrm{Id}.$$

Let us consider $x = \Sigma x_i P_i$, $y = \Sigma y_j Q_j$ as in 1.1.5. Then $L^2(\ell')$ is identified with $L^2(\mathbb{R}^n)$. We consider the space $\mathcal{S}(\mathbb{R}^n)$ of rapidly decreasing functions on $\mathbb{R}^n$. Then it is easy to see that if

$$X \in \mathfrak{n},\ f \in \mathcal{S},\ dW(X)\cdot f = \frac{d}{dt} W(\exp tX)\cdot f\Big|_{t=0}$$

defines a representation of $\mathfrak{n}$, which is the infinitesimal representation associated to $\widetilde{W}$. From the preceding formulas, we see immediately that

$$d\widetilde{W}(P_j) = 2i\pi x_j$$

$$d\widetilde{W}(Q_i) = -\frac{\partial}{\partial x_i}$$

$$d\widetilde{W}(E) = 2i\pi \mathrm{Id}.$$

In a sense to be made precise, this is the unique representation of $\mathfrak{n}$ which can be exponentiated to a unitary representation of N.

1.3. The Weyl transform and the Stone-Von Neumann theorem.

1.3.1. Let G be a topological group. A unitary representation T of G in the Hilbert space H is a homomorphism $g \to T(g)$ of G in the group of unitary operators on H (i.e. $T(g_1\, g_2) = T(g_1) \cdot T(g_2)$. We also require the continuity of the maps from G to H given by $g \to T(g)x$ for every $x \in H$.

There is an obvious notion of equivalence: if (T,H) and (T',H') are two representations of G in H and H', we say that $T \sim T'$ if an isomorphism $I: H \to H'$ exists such that the diagram:

$$\begin{array}{ccc} H & \xrightarrow{\ I\ } & H' \\ \big\downarrow T_1(g) & & \big\downarrow T_2(g) \\ H & \xrightarrow{\ I\ } & H' \end{array}$$

is commutative for every $g \in G$.

If T_1 and T_2 are two representations of G in H_1 and H_2, we can form $T = T_1 \oplus T_2$ represented in $H = H_1 \oplus H_2$ by

$$\left(\begin{array}{c|c} T_1(g) & 0 \\ \hline 0 & T_2(g) \end{array}\right).$$

We say that T is irreducible, if T cannot be written as $T_1 \oplus T_2$, or equivalently if there is no closed subspace H_1 of H stable under all the operators $T(g)$.

Let H be a Hilbert space, we will denote by $\overline{H}$ the same space H, but with the multiplication law $t \cdot x = \overline{t}x$ and the scalar product $\langle x,y\rangle_{\overline{H}} = \overline{\langle x,y\rangle}_H$. An isomorphism of H and $\overline{H}$ is then given by an antilinear map $\sigma: H \to H$, such that $\sigma(\lambda x) = \overline{\lambda}\sigma(x)$, $\|\sigma(x)\| = \|x\|$. If T is a unitary representation of G in H, the same formulas for $T(g)$ define a unitary representation of G in $\overline{H}$ that we denote by $\overline{T}$. If we have defined $\sigma: H \to \overline{H}$ an antilinear isomorphism between H and $\overline{H}$, the representation $\overline{T}$ is equivalent to the representation $\sigma^{-1} \circ T \circ \sigma$ on the Hilbert space H.

Let H_1, H_2 be two Hilbert spaces; we consider the Hilbert space $H_1 \otimes H_2$: $H_1 \otimes H_2$ is the completion of the vector space spanned by finite linear combinations $\sum_{i=1}^{N} v_i \otimes w_i$, $v_i \in H_1$, $w_i \in H_2$ for the natural inner product such that $\langle v_1 \otimes w_1, v_2 \otimes w_2\rangle = \langle v_1,v_2\rangle\langle w_1,w_2\rangle$. If (e_i) is a Hilbert basis of H_1 and (f_j) a Hilbert basis of H_2, then $e_i \otimes f_j$ is a Hilbert basis of $H_1 \otimes H_2$. Also $H_1 \otimes H_2$ is the Hilbert sum of the Hilbert subspaces $(H_1 \otimes \mathbb{C}f_j)$. If we are mainly interested in H_1, we will say that $H_1 \otimes H_2$ is a multiple of the space H_1, with multiplicity equal to $\dim H_2$, i.e. finite or $+\infty$.

Let T_1 be a unitary representation of G in H_1. We consider the representation T of G in $H_1 \otimes H_2$ given by $T(g) = T_1(g) \otimes \mathrm{Id}_{H_2}$, where Id_{H_2} denotes the identity operator on H_2. We will then say that T is a multiple of the representation T_1 (with multiplicity $\dim H_2$): $H_1 \otimes H_2$

can be written as $\bigoplus_j (H_1 \otimes Cf_j)$ and $T = \bigoplus_j T_1(g) \otimes 1_{Cf_j}$.

Let T be a unitary representation of a Lie group, with a left invariant Haar measure dg, on a Hilbert space H. We define the space H^∞ of C^∞ vectors for T by the following condition: $x \in H^\infty$ if the map $g \to T(g)x$ is C^∞.

Let φ be a function with compact support. We can form the operator:

$$T(\varphi) = \int \varphi(g)\ T(g)\ dg, \text{ i.e.}$$

$$\langle T(\varphi)x,y\rangle = \int_G \varphi(g)\ \langle T(g)x,y\rangle\ dg\ .$$

This integral makes sense if φ is in $L^1(G,dg)$, or if x is in H^∞ and $\varphi(g)dg$ a distribution with compact support. For φ_1 and φ_2 continuous functions with compact support, we form the convolution product $\varphi_1 *_G \varphi_2$ defined by:

$$(\varphi_1 * \varphi_2)(g) = \int_G \varphi_1(u)\ \varphi_2(u^{-1}g)\ du\ .$$

We have

$$\int_G (\varphi_1 * \varphi_2)(g)\psi(g)dg = \iint_{G\times G} \varphi_1(g)\varphi_2(h)\psi(gh)dg\ dh$$

if ψ is continuous with compact support. Thus we have defined a structure of algebra on the space of continuous functions. with compact support, as well as on $L^1(G)$ or on the space $\mathcal{E}'(G)$ of distributions with compact support.

The following proposition is immediate to prove:

If φ satisfies 1.3.3 a), the function $\varphi^*(n) = \overline{\varphi(n^{-1})}$ also satisfies 1.3.3 a); thus identifying φ with a function on V, we have $\varphi^*(v) = \overline{\varphi(-v)}$.

If φ satisfies 1.3.3.a) the function $n \to \varphi(n_0^{-1}n)$ as well as the function $n \to \varphi(nn_0)$ satisfies 1.3.3.a) (for $n_0 \in N$); we will denote them by $n_0 *_B \varphi$ and $\varphi *_B n_0^{-1}$, (convolution with Dirac distributions).

If $n_0 = \exp u_0$, identifying φ with a function on V,

$$(n_0 *_B \varphi)(u) = e^{i\pi B(u_0,u)} \varphi(u-u_0)$$

$$(\varphi *_B n_0)(u) = e^{i\pi B(u,u_0)} \varphi(u-u_0) .$$

1.3.6. From Proposition 1.3.2, we deduce:

a) $W_T(\varphi_1 *_B \varphi_2) = W_T(\varphi_1) \circ W_T(\varphi_2)$

b) $W_T(\varphi^*) = W_T(\varphi)^*$

c) $W_T(n_0 *_B \varphi) = T(n_0) \cdot W_T(\varphi)$

d) $W_T(\varphi *_B n_0) = W_T(\varphi) \cdot T(n_0)$.

1.3.7. We consider $T = W(\ell)$; we will now see that the Weyl transform W extends to an isomorphism from the space $L^2(V)$ to the space of Hilbert-Schmidt operators on $H(\ell)$. We recall some facts on Hilbert-Schmidt operators:

Let H be a Hilbert space. We recall that a Hilbert-Schmidt operator $A: H \to H$ on H is an operator such that for some orthonormal basis (e_i) of H,

$$\sum_i \|Ae_i\|^2 = \sum_{i,j} |\langle Ae_i, e_j\rangle|^2 = \sum_{i,j} |\langle e_i, A^* e_j\rangle|^2 = \sum_{i,j} |a_{ij}|^2 < \infty .$$

a mass one Lebesgue measure on $\mathbb{R}E/\mathbb{Z}E$.

A function satisfying 1.3.3 a) is determined by its restriction to exp V. Hence the space of continuous functions on N satisfying 1.3.3. a) is identified with the space of continuous functions on V. We still denote by $\varphi(v) = \varphi(\exp v)$ the restriction of φ to V.

Let T be a unitary representation of N satisfying

$$T(\exp tE) = e^{2i\pi t} Id_H .$$

Then T is a representation of B. We form

1.3.4. $W_T(\varphi) = T(\varphi) = \int_B \varphi(b)T(b)db = \int_V \varphi(v)T(\exp v)dv$

for φ a continuous compactly supported function on V (or B). $W_T(\varphi)$ is the Weyl transform of the functions φ. We will also consider $W_T(\varphi)$ for φ a rapidly decreasing function on V.

Compactly supported continuous functions satisfying 1.3.3. a) form an algebra under the convolution product on B. We have then defined a structure of algebra on the space of continuous functions with compact support on V. We have:

$$\begin{aligned}(\varphi_1 *_B \varphi_2)(\exp v) &= \int_V \varphi_1(\exp u)\, \varphi_2(\exp -u \exp v)du \\ &= \int_V \varphi_1(\exp u)\, \varphi_2(\exp(v-u) \exp - \frac{B(u,v)}{2} E)du \\ &= \int_V \varphi_1(\exp u)\, \varphi_2(\exp(v-u))\, e^{i\pi B(u,v)}du\end{aligned}$$

i.e.

1.3.5. $(\varphi_1 *_B \varphi_2)(v) = \int_V \varphi_1(u)\, \varphi_2(v-u)\, e^{i\pi B(u,v)}du .$

If φ satisfies 1.3.3 a), the function $\varphi^*(n) = \overline{\varphi(n^{-1})}$ also satisfies 1.3.3 a); thus identifying φ with a function on V, we have $\varphi^*(v) = \overline{\varphi(-v)}$.

If φ satisfies 1.3.3.a) the function $n \to \varphi(n_0^{-1}n)$ as well as the function $n \to \varphi(nn_0)$ satisfies 1.3.3.a) (for $n_0 \in N$); we will denote them by $n_0 *_B \varphi$ and $\varphi *_B n_0^{-1}$, (convolution with Dirac distributions).

If $n_0 = \exp u_0$, identifying φ with a function on V,

$$(n_0 *_B \varphi)(u) = e^{i\pi B(u_0,u)}\varphi(u-u_0)$$

$$(\varphi *_B n_0)(u) = e^{i\pi B(u,u_0)}\varphi(u-u_0) .$$

1.3.6. From Proposition 1.3.2, we deduce:

a) $W_T(\varphi_1 *_B \varphi_2) = W_T(\varphi_1) \circ W_T(\varphi_2)$

b) $W_T(\varphi^*) = W_T(\varphi)^*$

c) $W_T(n_0 *_B \varphi) = T(n_0) \cdot W_T(\varphi)$

d) $W_T(\varphi *_B n_0) = W_T(\varphi) \cdot T(n_0)$.

1.3.7. We consider $T = W(\ell)$; we will now see that the Weyl transform W extends to an isomorphism from the space $L^2(V)$ to the space of Hilbert-Schmidt operators on $H(\ell)$. We recall some facts on Hilbert-Schmidt operators:

Let H be a Hilbert space. We recall that a Hilbert-Schmidt operator $A: H \to H$ on H is an operator such that for some orthonormal basis (e_i) of H,

$$\sum_i \|Ae_i\|^2 = \sum_{i,j} |<Ae_i,e_j>|^2 = \sum_{i,j} |<e_i,A^*e_j>|^2 = \sum_{i,j} |a_{ij}|^2 < \infty .$$

This sum doesn't depend on the choice of the orthonormal basis (e_i) and is denoted by $\|A\|^2_{H.S}$. For A and B Hilbert-Schmidt operators, B^*A is of trace class and $\langle A,B\rangle = \mathrm{Tr}B^*A = \sum_i \langle B^*Ae_i,e_i\rangle = \sum_i \langle Ae_i,Be_i\rangle$ defines a scalar product on the space $\mathcal{HS}_2(H)$ of Hilbert-Schmidt operators on H; hence $\mathcal{HS}_2(H)$ is a Hilbert space, having as basis the operators $E_{i,j}(x) = \langle x,e_i\rangle e_j$. Let $x,y \in H$, we define the rank one operator $E_{x,y}(v) = \langle v,y\rangle x$ on H. Clearly $E_{x,y}$ is Hilbert-Schmidt and

$$\langle E_{x_1,y_1},E_{x_2,y_2}\rangle = \langle x_1,x_2\rangle\overline{\langle y_1,y_2\rangle}.$$

Hence as the map $x,y \to E_{x,y}$ is linear in x, antilinear in y we obtain an isometry from $H \otimes \overline{H}$ onto $\mathcal{HS}_2(H)$.

Let us suppose now that $H = L^2(E,dy)$ where (E,dy) is a measure space: We will see that the space of Hilbert-Schmidt operators on $L^2(E,dy)$ is equivalent to the space $L^2(E \times E,dxdy)$ of square integrable functions $K(x,y)$ on $E \times E$ via

$$(K\varphi)(x) = \int K(x,y)\varphi(y)dy$$

Let (e_i) be a Hilbert basis of $L^2(E)$. We denote by

$$(e_i \otimes e_j)(\varphi) = \langle\varphi,e_j\rangle\, e_i.$$

Hence

$$\begin{aligned}((e_i \otimes e_j)(\varphi))(x) &= \langle\varphi,e_j\rangle\, e_i(x)\\ &= (\int \varphi(y)\, \overline{e}_j(y)dy)\, e_i(x)\end{aligned}$$

$$= \int \varphi(y)\, e_i(x)\overline{e}_j(y)dy$$

$$= \int K_{ij}(x,y)\, \varphi(y)dy$$

$$\text{with } K_{ij}(x,y) = e_i(x)\overline{e_j(y)}.$$

The Hilbert-Schmidt operators are of the form $\Sigma\, a_{ij}\, e_i \otimes e_j$ with $\Sigma\, |a_{ij}|^2 < \infty$; hence $K(x,y) = \Sigma\, a_{ij}\, e_i(x)\overline{e_j(y)}$ is such that

$$\int |K(x,y)|^2\, dxdy = \Sigma\, |a_{ij}|^2 < \infty .$$

Thus the operator associated to the element $\Sigma\, a_{ij}\, e_i \otimes e_j$ of $H \otimes \overline{H}$ is the operator $\int K(x,y)\, \varphi(y)dy$ with $K(x,y) = \Sigma\, a_{ij}\, e_i(x)\overline{e_j(y)}$. Let us remark here for later reference: that if $\psi_0 \in L^2(V)$ is of norm 1, the projector $P: H \to \mathbf{C}\psi_0$ given by $P(\varphi) = \langle\varphi, \psi_0\rangle\psi_0$ is given by the kernel $K = \psi_0(x)\overline{\psi_0(y)}$.

1.3.8. We now prove

<u>Proposition</u>: Let $T = W(\ell)$; the Weyl transform $W = W_T$ extends to an isomorphism from $L^2(V)$ to the space of Hilbert-Schmidt operators on $H(\ell)$.

<u>Proof</u>: We have

$$W_T(\varphi) = \int_V \varphi(v)\, T(v)dv .$$

We write $V = \ell' \oplus \ell$ according to the choice of ℓ, ℓ' as in 1.1.4. Hence

$$W_T(\varphi) = \iint \varphi(y+x)\quad T(\exp(y+x))\, dydx, \quad y \in \ell',\ x \in \ell$$

$$= \iint \varphi(y+x)\quad T(\exp y)\quad T(\exp x)e^{-i\pi B(y,x)}\, dxdy.$$

Let us identify $H(\ell)$ with $L^2(\ell')$ as in 1.2.5, and we write $W_T(\varphi)\cdot f$ for $f \in L^2(\ell')$

$$(W_T(\varphi)\cdot f)(\xi) = \iint e^{-i\pi B(y,x)}\varphi(y+x)(\tilde{W}(\exp y)\tilde{W}(\exp x)f)(\xi)\, dxdy$$

$$= \iint e^{-i\pi B(y,x)}\varphi(y+x)e^{2i\pi B(x,\xi-y)}f(\xi-y)\, dxdy$$

Changing y to $\xi - y$, we have

$$= \iint e^{-i\pi B(\xi-y,x)}\varphi(\xi-y+x)e^{2i\pi B(x,y)}f(y)\, dxdy$$

$$= \iint e^{i\pi B(x,\xi+y)}\varphi(\xi-y+x)f(y)\, dxdy.$$

Hence if we define:

$$K_\varphi(\xi,y) = \int e^{i\pi B(x,y+\xi)}\varphi(\xi-y+x)\, dx,$$

we have written $W_T(\varphi)$ as a kernel operator, i.e.

$$(W_T(\varphi)f(\xi) = \int_{\ell'} K_\varphi(\xi,y)f(y)dy.$$

To prove our proposition, we have to see that if $\varphi \in L^2(V)$ the corresponding kernel $K_\varphi \in L^2(\ell' \times \ell')$. The bilinear form $B(x,\xi)(x \in \ell,\ \xi \in \ell')$ defines a nondegenerate bilinear map on $\ell \times \ell'$. Hence the partial Fourier transform

$$(\mathcal{F}_x\varphi)(y,\xi) = \int e^{-2i\pi B(x,\xi)}\varphi(y+x)\, dx$$

is a unitary isomorphism from $L^2(\ell' \oplus \ell)$ onto $L^2(\ell' \times \ell')$. But our kernel is then

$$(K_\varphi)(\xi,y) = (\mathcal{F}_x\varphi)(\xi-y,\ -\tfrac{y+\xi}{2})$$

which is clearly in $L^2(\ell' \times \ell')$. Hence our proposition is proven.

1.3.9. Corollary: If U is a bounded operator on $H(\ell)$ commuting with all the operators $W(\ell)(n)$ for $n \in N$ then U is a scalar operator.

Proof: Let $U: H(\ell) \to H(\ell)$ commuting with all the operators $W(\ell)(n)$ for every n, then U commutes with the operator $W_T(\varphi)$ for φ function on V which is compactly supported and hence by continuity with all the operators $W(\varphi)$ for $\varphi \in L^2(V)$, in particular with all the Hilbert-Schmidt operators. Taking a Hilbert basis (e_i), the relation $U(e_i \otimes e_j) = (e_i \otimes e_j)U$ for every (i,j) implies immediately that $U = \lambda \, Id$ for some $\lambda \in \mathbf{C}$. It follows that there cannot be any closed invariant proper subspace H_1 of $H(\ell)$ (the projector P_{H_1} has to be Id_H by the preceding corollary). Hence the part a) of the theorem is proven.

1.3.10. Let us fix $\ell = \ell_0$, $\ell' = \ell'_0$, $H_0 = H(\ell_0) = L^2(\ell'_0)$, $W_0 = W(\ell_0)$ We remark that, if φ_1, φ_2 belong to the space $\mathcal{S}(V)$ of rapidly decreas functions on V, then $\varphi_1 *_B \varphi_2 \in \mathcal{S}(V)$. Hence we have defined a structure of (noncommutative) algebra on $\mathcal{S}(V)$. In the Weyl transform given by W_0 the corresponding kernel $K_\varphi(\xi,y)$ for $\varphi \in \mathcal{S}(V)$ is a rapidly decreasing function of (ξ,y), and we obtain this way all the operators with rapidly decreasing kernels.

Now we will prove the Stone-Von Neumann theorem. Let us first sketch the idea of the proof: via the Weyl transform W_{T_0}, we have identified the algebra $(\mathcal{S}(V), *_B)$ with a subalgebra $\mathfrak{A}_0$ of operators on H_0. If (T,H) is a representation of N

into a Hilbert space H satisfying $T(\exp tE) = e^{2i\pi t}\,\mathrm{Id}_H$, the Weyl transform W_T defines a homomorphism of $(\mathcal{S}(V), *_B)$ into a subalgebra $\mathcal{O}\!\iota$ of operators on H. Hence we obtain a homomorphism $\varphi: \mathcal{O}\!\iota_0 \to \mathcal{O}\!\iota$ of algebras such that $\varphi(A^*) = \varphi(A)^*$.

Let us recall that if V_0 and V are two finite-dimensional complex Hilbert spaces, and φ a homomorphism from $\mathrm{End}_{\mathbb{C}} V_0$ to $\mathrm{End}_{\mathbb{C}} V$ satisfying $\varphi(A^*) = \varphi(A)^*$, $\varphi(1) = 1$, then there exist V_1 and an isomorphism $I: V_0 \otimes V_1 \to V$ such that for

$$A \in \mathrm{End}_{\mathbb{C}}\, V_0,\ \varphi(A) = I \circ (A \otimes \mathrm{Id}_{V_1}) \circ I^{-1}.$$

We will give here the proof for the finite-dimensional case as our proof in the general case will be similar:
Let $x_1 \in V_0$ with $\|x_1\| = 1$; we consider P_1 the projector $P_1(x) = \langle x, x_1\rangle x_1$ on the one-dimensional space $\mathbb{C}x_1$; we have $P_1^2 = P_1$, $P_1 = P_1^*$.

Let us consider $\varphi(P_1)$. First we remark that $\varphi(P_1) \neq 0$. In fact we can find $(x_1, x_2, \cdots, x_n)$ an orthonormal basis of V_0 starting with x_1. We denote by P_i the projector on $\mathbb{C}x_i$. If g_i is any operator on V such that $g_i(x_1) = x_i$, then $g_i P_1 g_i^* = P_i$. Hence $\sum_i g_i P_1 g_i^* = \mathrm{Id}_{V_0}$. Applying the homomorphism φ, we obtain $\sum_{i=1}^n \varphi(g_i)\varphi(P_1)\varphi(g_i)^* = \mathrm{Id}_V$ so $\varphi(P_1) \neq 0$ and verify $\varphi(P_1)^2 = \varphi(P_1)$, $\varphi(P_1)^* = \varphi(P_1)$, i.e. $\varphi(P_1)$ is a projector on a subspace V_1 of V. From the relation $\sum_{i=1}^n \varphi(g_i)\varphi(P_1)\varphi(g_i)^* = 1$, we see that every element of V is of the form $\sum_{i=1}^n \varphi(g_i) w_i$ with $w_i \in V_1$. We consider the surjective map

$I: V_0 \otimes V_1 \to V$ defined by $I(A\cdot x_1 \otimes w) = \varphi(A)w$. This map is well defined as, if $(A-B)x_1 = 0$, $(A-B)P_1 = 0$, then $(\varphi(A)-\varphi(B))\varphi(P_1) = 0$ meaning that $\varphi(A)w = \varphi(B)w$ for $w \in V_1$. Clearly now $I(Ax \otimes w) = \varphi(A)I(x \otimes w)$ and it is easy to conclude the proof.

Now we come back to our case. Let us choose a function ψ_1 in $\mathcal{S}(\ell')$ such that $\|\psi_1\|_{L^2(\ell')} = 1$. We consider the projector P_1 on $\mathbb{C}\psi_1$, P_1 is given by the kernel $\overline{\psi_1(y)}\,\psi_1(x)$ and hence is of the form $W_0(\varphi_1)$ for $\varphi_1 \in \mathcal{S}(V)$. As P_1 is a projector on the one-dimensional subspace $\mathbb{C}\psi_1$ we have $P_1^2 = P_1$, $P_1^* = P_1$, $P_1 \circ W_0(n) \circ P_1 = \alpha(n)P_1$ with $\alpha(n) = \langle W_0(n)\psi_1, \psi_1\rangle$. Hence from 1.3.6 the function φ_1 satisfies

1.3.11. $\quad \varphi_1 *_B \varphi_1 = \varphi_1, \quad \varphi_1^* = \varphi_1, \varphi_1 *_B n *_B \varphi_1 = \alpha(n)\varphi_1.$

As $\varphi_1 \in \mathcal{S}(V)$, we can calculate $W_T(\varphi_1)$ for any unitary representation (T,H) of N satisfying $T(\exp tE) = e^{2i\pi t}\,\mathrm{Id}_H$.

1.3.12. __Lemma:__ The space H is generated by the elements $T(n)W_T(\varphi_1)\cdot x$ for $n \in N$, $x \in H$.

__Proof:__ Let $y \in H$ be such that $\langle y, T(n)W_T(\varphi_1)\cdot x\rangle = 0$ for every $n \in N$, $x \in H$. We compute for $n_0 = \exp u$,

$$\begin{aligned}\langle y, T(n_0)W_T(\varphi_1)T(n_0)^{-1}.x\rangle = 0 &= \int_V \langle y, T(\exp u)T(\exp v)T(\exp -u)\cdot x\rangle \varphi_1(v)dv \\ &= \int_V \langle y, T(\exp(v+B(u,v)E)).x\rangle \varphi_1(v)dv \\ &= \int_V \langle y, T(\exp v)x\rangle \varphi_1(v)e^{2i\pi B(u,v)}dv = 0.\end{aligned}$$

Our function $\varphi_1(v)\ \langle y, T(\exp v)x\rangle$ is a continuous function in L^2 (as $\varphi_1(v) \in \mathcal{S}(V)$ and $|\langle y, T(\exp v)x\rangle| \leq \|y\|\|x\|$ is bounded). The preceding equality means that the Fourier transform of this function with respect to the bilinear form $B(u,v)$ is identically zero. Hence $\varphi_1(v)\ \langle y, T(\exp v)x\rangle \equiv 0$. As φ_1 is not identically zero, there exist v_0 such that $\langle y, T(\exp v_0)x\rangle = 0$ for every $x \in H$. This implies $\langle y, H\rangle = 0$ hence $y = 0$.

Now from the relations 1.3.11 and lemma 1.3.6, we deduce that $W_T(\varphi_1)$ is a projector on the subspace $H_1 = W_T(\varphi_1)H$ of H. As in the finite-dimensional case, we wish to define $I: H_0 \otimes H_1 \to H$ via the formula $I(W_0(n)\psi_1 \otimes w) = T(n)\cdot w$, with $n \in N$, $w \in H_1$. We first verify

1.3.13. For $w_1 = W_T(\varphi_1)x_1$, $w_2 = W_T(\varphi_1)x_2$, $n_1, n_2 \in N$

$$\langle T(n_1)w_1, T(n_2)w_2\rangle_H = \langle W_0(n_1)\psi_1, W_0(n_2)\psi_1\rangle_{H_0}\ \langle w_1, w_2\rangle_{H_1}.$$

<u>Proof</u>: We have

$$\langle T(n_1)W_T(\varphi_1)x_1,\ T(n_2)W_T(\varphi_1)x_2\rangle_H$$
$$= \langle W_T(\varphi_1)T(n_2)^{-1}T(n_1)W_T(\varphi_1)x_1, x_2\rangle_H .$$

Using relations 1.3.11 and lemma 1.3.6, this equals

$$= \alpha(n_2^{-1}n_1)\langle W_T(\varphi_1)x_1, x_2\rangle_H = \alpha(n_2^{-1}n_1)\langle W_T(\varphi_1)x_1, W_T(\varphi_1)x_2\rangle_H$$

which is the desired equality.

As the representation W_0 is irreducible, the set of linear combinations $\Sigma\, c_i W_0(n_i)\cdot\psi_1$ is a dense subspace of H_0.

It is clear now that we can define as isometry I from $H_0 \otimes H_1 \to H$ via the formula

$$I\left(\sum_{i=1}^{N} W_0(n_i)\psi_1 \otimes w_i\right) = \sum_{i=1}^{N} T(n_i)w_i .$$

This map is well defined as if $\sum_{i=1}^{N} W_0(n_i)\cdot\psi_1 \otimes w_i = 0$ the equality 1.3.13 implies $\|\sum_{i=1}^{N} T(n_i)w_i\|_H = \|\Sigma W_0(n_i)\psi_1 \otimes w_i\|_{H_0\otimes H_1} = 0$.

The operator I is surjective by 1.3.12. Hence I is a unitary isomorphism between $H_0 \otimes H_1$ and H. Clearly $I\,(W_0(n) \otimes Id_{H_1})\,I^{-1} = T(n)$. Hence T is a multiple of the representation W_0.

1.4. Fourier transforms and intertwining operators.

Let ℓ_1 and ℓ_2 be two Lagrangian planes. We can form the unitary representations $W_1 = W(\ell_1)$ and $W_2 = W(\ell_2)$ of N. By the Stone-Von Neumann theorem, we know that they are equivalent, i.e. there exists a unitary operator $\mathcal{F}_{2,1}: H(\ell_1) \to H(\ell_2)$ such that:

1.4.1. $\qquad \mathcal{F}_{2,1}\ W_1(n) = W_2(n)\ \mathcal{F}_{2,1}$, for every $n \in N$.

$\mathcal{F}_{2,1}$ is determined by this relation up to a scalar of modulus one, as follows from 1.3.9.

1.4.2. Let us first compute $\mathcal{F}_{2,1}$ in the case where $\ell_1 = \mathbb{R}P_1 \oplus \cdots \oplus \mathbb{R}P_n$ and $\ell_2 = \mathbb{R}Q_1 \oplus \cdots \oplus \mathbb{R}Q_n$. We adopt the following conventions: $x = (x_1, x_2, \cdots, x_n)$, $y = (y_1, y_2, \cdots, y_n)$, $x \cdot y = \sum_{i=1}^{n} x_i y_i$, $x \cdot P = \sum_{i=1}^{n} x_i P_i$, $y \cdot Q = \sum_{i=1}^{n} y_i Q_i$.

Then W_1 acts on $L^2(dy)$ and W_2 on $L^2(dx)$ by the following formulas:

$$(W_1(\exp x_0 P)\varphi)(y) = e^{2i\pi x_0 \cdot y}\ \varphi(y)$$

$$(W_1(\exp y_0 Q)\varphi)(y) = \varphi(y - y_0)$$

$$(W_2(\exp x_0 P)\varphi)(x) = \varphi(x - x_0)$$

$$(W_2(\exp y_0 Q)\varphi)(x) = e^{-2i\pi x \cdot y_0}\ \varphi(x).$$

We denote by $\mathcal{F}$ the Fourier transform from $L^2(dy) \simeq H(\ell_1)$ to $L^2(dx) \simeq H(\ell_2)$ given by

1.4.3. $$(\mathcal{F}\varphi)(x) = \int e^{-2i\pi x\cdot y}\,\varphi(y)dy.$$

Since $\mathcal{F}$ transforms translation operators into multiplication operators it is immediate that $\mathcal{F}\circ W_1(n) = W_2(n)\circ\mathcal{F}$. Hence $\mathcal{F}_{2,1} = \mathcal{F}$.

1.4.4. Let ℓ_1 and ℓ_2 be two Lagrangian planes. Let $L_1 = \exp(\ell_1 + \mathbb{R}E)$ and $L_2 = \exp(\ell_2 + \mathbb{R}E)$ be the subgroups of N associated to ℓ_1 and ℓ_2 (1.2.1). We consider, as in 1.2.4, the Hilbert spaces $H(\ell_1)$ and $H(\ell_2)$ canonically associated to ℓ_1 and ℓ_2. We wish to find an operator from H_1 to H_2 intertwining the representations $W(\ell_1)$ and $W(\ell_2)$.

The formal construction is simple: We look for an operator commuting with left translations and transforming a function φ semi-invariant under the right action of L_1 (i.e. verifying 1.2.4 a)) into a function φ semi-invariant under L_2. Hence it is natural to "force" φ to be semi-invariant under L_2 by averaging right translates of φ under L_2, taking in account that φ verifies 1.2.4 a) for $h \in L_1 \cap L_2$.

Hence we will define formally:

1.4.5. $$(\mathcal{F}_{\ell_2,\ell_1}\varphi)(n) = \int_{L_2/L_1\cap L_2} \varphi(nh_2)f(h_2)d\dot{h}_2$$

where $d\dot{h}_2$ denotes a positive L_2-invariant measure on the homogeneous space $L_2/L_1\cap L_2$. As $d\dot{h}_2$ is unique up to multiplication by a positive scalar, we remark that $\mathcal{F}_{\ell_2,\ell_1}$ is therefore defined up to multiplication by a <u>positive constant</u>.

Let us compute $\mathcal{F}_{\ell_2,\ell_1}$ for the preceding example. We have

$L_2 \cap L_1 = \{ \exp tE \}$. Hence $(L_2/L_2 \cap L_1, d\dot{h}_2)$ is identified with (ℓ_2, dy), and

$$(\mathcal{F}_{\ell_2,\ell_1}\varphi)(n) = \int_{\ell_2} \varphi(n \exp y\cdot Q)dy.$$

We identify $H(\ell_1)$ with $L^2(dy)$ by $\varphi(y) = \varphi(\exp y\cdot Q)$, $H(\ell_2)$ with $L^2(dx)$ by $\varphi(x) = \varphi(\exp x\cdot P)$. Our operator $\mathcal{F}_{\ell_2,\ell_1}$ becomes

$$\begin{aligned}(\mathcal{F}_{\ell_2,\ell_1}\varphi)(\exp x\cdot P) &= \int_{\ell_2} \varphi(\exp x.P \exp y\cdot Q)dy \\ &= \int_{\ell_2} \varphi(\exp x\cdot P \exp y\cdot Q \exp{-x\cdot P} \\ &\qquad \exp x.P)dy \\ &= \int_{\ell_2} \varphi(\exp y.Q \exp{-x\cdot P} \exp(x.y)E)dy \\ &= \int_{\ell_2} \varphi(\exp yQ)e^{-2i\pi x\cdot y}dy, \\ &\qquad \text{as } \varphi \in H(\ell_1) \\ &= \int_{\ell_2} \varphi(y)e^{-2i\pi x\cdot y}dy = (\mathcal{F}\varphi)(x).\end{aligned}$$

Hence $\mathcal{F}_{2,1}$ defined formally by 1.4.3 is indeed a unitary operator given by the Fourier transform.

The following lemma shows that this is basically the only situation.

1.4.6. <u>Lemma</u>: Let ℓ_1 and ℓ_2 be two Langrangian subspaces of (V,B). There exists a symplectic basis $(P_1, P_2, \cdots, P_n, Q_1, Q_2, \cdots, Q_n)$ of V such that:

$$\ell_1 = \mathbb{R}P_1 \oplus \mathbb{R}P_2 \oplus \cdots \oplus \mathbb{R}P_k \oplus \mathbb{R}P_{k+1} \oplus \cdots \oplus \mathbb{R}P_n$$

$$\ell_2 = \mathbb{R}Q_1 \oplus \mathbb{R}Q_2 \oplus \cdots \oplus \mathbb{R}Q_k \oplus \mathbb{R}P_{k+1} \oplus \cdots \oplus \mathbb{R}P_n.$$

Proof: Let W be maximal in the collection of all totally isotropic subspaces S satisfying $(\ell_1+\ell_2) \cap S = 0$. Hence $W^\perp \subset \ell_1 + \ell_2 + W$ by the maximality condition. But we have $(\ell_1 \cap \ell_2) + W^\perp = V$ as $(\ell_1+\ell_2) \cap W = 0$ (1.1.3), hence $(\ell_1 \cap \ell_2) + \ell_1 + \ell_2 + W = V$, i.e. W is a complementary subspace to $\ell_1 + \ell_2$ in V. The bilinear form B defines a duality between $\ell_1 \cap \ell_2$ and $V/\ell_1+\ell_2 \simeq W$. Hence we can choose a basis $P_{k+1}, \cdots, P_n$ of $\ell_1 \cap \ell_2$ and a basis $Q_{k+1}, \cdots, Q_n$ of W such that $B(P_i,Q_j) = \delta_{ij}$, $B(P_i,P_j) = 0$, $B(Q_i,Q_j) = 0$ for $i,j \geq k+1$. Let us consider the subspace $M = \ell_1 \cap \ell_2 + W$. Clearly $V = M \oplus M^\perp$ as $M \cap M^\perp = 0$. We have

$$\ell_1 = \ell_1 \cap \ell_2 + \ell_1 \cap M^\perp$$

$$\ell_2 = \ell_1 \cap \ell_2 + \ell_2 \cap M^\perp.$$

It is immediate to see that $\ell_1 \cap M^\perp$ and $\ell_2 \cap M^\perp$ are transverse Lagrangian planes in $M^\perp$. We then choose $P_1, P_2, \cdots, P_k$ a basis of $\ell_1 \cap M^\perp$, $Q_1, Q_2, \cdots, Q_k$ a basis of $\ell_2 \cap M^\perp$ by duality, and we obtain the lemma.

Let us consider ℓ_1, ℓ_2 and the symplectic basis of V given by the lemma 1.4.6. We will write $x = (x',x'')$, with $x' = (x_1, x_2, \cdots, x_k)$, $x'' = (x_{k+1}, \cdots, x_n)$ and similar notations. Then $H(\ell_1)$ is identified with $L^2(dy'dy'')$ and

$H(\ell_2)$ is identified with $L^2(dx'dy")$. Our formal operator $\mathcal{F}_{2,1}$ becomes then $\varphi(y',y") \to (\mathcal{F}'\varphi)(x',y")$ where $\mathcal{F}'$ denotes the partial Fourier transform in the first k-variables. If we compute $\mathcal{F}_{1,2}$ we see that $\mathcal{F}_{1,2} = (\mathcal{F}')^{-1}$. Hence we have proven the

1.4.7. <u>Proposition:</u> The operator

$$(\mathcal{F}_{\ell_2,\ell_1}\varphi)(n) = \int_{L_2/L_1\cap L_2} \varphi(nh_2)f(h_2)d\dot{h}_2$$

is an intertwining operator between $W(\ell_1)$ and $W(\ell_2)$. $\mathcal{F}_{\ell_2,\ell_1}$ is canonically defined up to multiplication by a positive constant and $\mathcal{F}_{\ell_2,\ell_1} = (\mathcal{F}_{\ell_1,\ell_2})^{-1}$.

We will now make specific the choice of $d\dot{h}_2$ so that $\mathcal{F}_{\ell_2,\ell_1}$ becomes a unitary operator.

1.4.8. Let us first recall the definition of α-densities on a vector space E. Let E be a k-dimensional vector space over $\mathbb{R}$, $\Lambda^k E$ the space of k-vectors in E. $\Lambda^k E$ is one-dimensional over $\mathbb{R}$. For any real number α, we call a density of order α a map $\rho: \Lambda^k E - \{0\} \to \mathbb{R}$ such that $\rho(\lambda v) = |\lambda|^\alpha \rho(v)$ for each $v \in \Lambda^k E - \{0\}$, $\lambda \in \mathbb{R} - \{0\}$. The space of all densities of order α is one-dimensional over $\mathbb{R}$, and is denoted by $\Omega_\alpha(E)$. If $w \in \Lambda^k E^*$ is a k-form, we denote by $|w|^\alpha$ the α density defined by $|w|^\alpha(v) = |(w,v)|^\alpha$, for $v \in \Lambda^k E - \{0\}$. $\Omega_1(E)$ is referred also as the space of volume forms. We denote by $|\Lambda^k E|$ the space $\Lambda^k E$, modulo the equivalence relation $e \sim -e$.

1.4.9. Let ℓ be a Lagrangian subspace of (V,B) and $e \in \Lambda^n \ell$ be a n-vector. We identify ℓ with $(V/\ell)^*$ via the bilinear map B. Hence each element $|e| \in |\Lambda^n \ell|$ gives a volume form $|e|$ on V/ℓ.

Let $L = \exp(\ell \oplus \mathbb{R}E)$ and $X = N/L$. The tangent space of X at the image $\dot{1}$ of the identity element 1 of N is canonically identified with V/ℓ. It is immediate to see that $|e|$ can be extended as an N-invariant volume-form on X. We denote by $d_{|e|}$ the associated measure on X. We denote by $H(\ell,|e|)$ the space $H(\ell)$ where the choice of the inner product on $H(\ell)$ is now determined by:

$$\|\varphi\|^2 = \int_X |\varphi(n)|^2 \, d_{|e|}\dot{n}.$$

1.4.10. Given (V,B), a symplectic vector space, we have a canonical element $\omega \in \Lambda^{2n} V^*$ defined by :

$$\omega = \underbrace{B \wedge B \wedge \cdots \wedge B}_{n\text{-times}}.$$

If E is a vector space, we denote the top-degree term of the graded vector space ΛE by $\Lambda^{\max} E$.

Let E_1 be a subspace of E, we have then:

$$\Lambda^{\max} E \simeq \Lambda^{\max}(E/E_1) \otimes \Lambda^{\max} E_1 .$$

If $E = E_1 \oplus E_2$,

$$\Lambda^{\max} E \simeq \Lambda^{\max}(E/E_1) \otimes \Lambda^{\max}(E/E_2).$$

1.4.11. Let us consider two Lagrangian planes ℓ_1 and ℓ_2. The bilinear form B defines a canonical symplectic form B' on $\ell_1+\ell_2/\ell_1\cap\ell_2$. The spaces $\ell_1/\ell_1\cap\ell_2$ and $\ell_2/\ell_1\cap\ell_2$ are transverse Lagrangian subspaces in $\ell_1+\ell_2/\ell_1\cap\ell_2$.

We have:

$$\Lambda^{max}(V/\ell_1\cap\ell_2) \simeq \Lambda^{max}(V/\ell_1+\ell_2) \otimes \Lambda^{max}(\ell_1+\ell_2/\ell_1\cap\ell_2)$$

$$\Lambda^{max}(\ell_1+\ell_2/\ell_1\cap\ell_2) \simeq \Lambda^{max}(\ell_1+\ell_2/\ell_1) \otimes \Lambda^{max}(\ell_1+\ell_2/\ell_2).$$

Thus $\Lambda^{max}(V/\ell_1\cap\ell_2) \simeq \Lambda^{max}(V/\ell_1+\ell_2) \otimes \Lambda^{max}(\ell_1+\ell_2/\ell_1) \otimes \Lambda^{max}(\ell_1+\ell_2/\ell_2)$

From this, we deduce:

1.4.12. Lemma:

$$\Omega_1(V/\ell_1\cap\ell_2) \simeq \Omega_{1/2}(\ell_1+\ell_2/\ell_1\cap\ell_2) \otimes \Omega_{1/2}(V/\ell_2) \otimes \Omega_{1/2}(V/\ell_1).$$

Proof: Let $u_1 \wedge u_2 \wedge u_3$ be an element of $\Lambda^{max}(V/\ell_1\cap\ell_2)$, with

$$u_1 \in \Lambda^{max}(V/\ell_1+\ell_2),\ u_2 \in \Lambda^{max}(\ell_1+\ell_2/\ell_1) \text{ and } u_3 \in \Lambda^{max}(\ell_1+\ell_2/\ell_2).$$

Thus $u_1 \wedge u_2 \in \Lambda^{max}(V/\ell_1)$, $u_2 \wedge u_3 \in \Lambda^{max}(\ell_1+\ell_2/\ell_1\cap\ell_2)$ and $u_3 \wedge u_1 \in \Lambda^{max}(V/\ell_2)$. Hence for $\rho_1 \in \Omega_{1/2}(\ell_1+\ell_2/\ell_1\cap\ell_2)$, $\rho_2 \in \Omega_{1/2}(V/\ell_2)$ and $\rho_3 \in \Omega_{1/2}(V/\ell_1)$, we define

$$(\rho_1 \otimes \rho_2 \otimes \rho_3,\ u_1 \wedge u_2 \wedge u_3) = \rho_1(u_2 \wedge u_3)\ \rho_2(u_3 \wedge u_1)\ \rho_3(u_1 \wedge u_2)$$

which is homogeneous of degree 1, thus it is a volume form on $V/\ell_1\cap\ell_2$.

As $\ell_1+\ell_2/\ell_1\cap\ell_2$ is provided with a canonical half-density $|\omega'|^{1/2}$, we get a canonical map from

$\Omega_{1/2}(V/\ell_1) \otimes \Omega_{1/2}(V/\ell_2)$ to $\Omega_1(V/\ell_1 \cap \ell_2)$ by $(\rho_1,\rho_2) \to |\omega'|^{1/2} \otimes \rho_2 \otimes \rho_1$.

1.4.13. Let $e_1 \in \Lambda^n \ell_1$ and $e_2 \in \Lambda^n \ell_2$. We recall that the choice of a volume form $\delta \in \Omega_1(\ell_2/\ell_1 \cap \ell_2)$ defines a canonical operator $\mathcal{F}^{\delta}_{\ell_2,\ell_1} : H(\ell_1,e_1) \to H(\ell_2,e_2)$ by :

$$(\mathcal{F}^{\delta}_{\ell_2,\ell_1}\varphi)(n) = \int_{L_2/L_1 \cap L_2} \varphi(nh_2) f(h_2)\delta\dot{h}_2,$$

where $\delta\dot{h}_2$ denotes the L_2 invariant positive measure on $L_2/L_1 \cap L_2$ associated to δ.

We now give a canonical choice of δ in function of e_1 and e_2. We consider the identifications:

$$\Omega_1(V/\ell_1 \cap \ell_2) \simeq \Omega_{1/2}(V/\ell_1) \otimes \Omega_{1/2}(V/\ell_2) \otimes |\omega'|^{1/2}$$

$$\Omega_1(V/\ell_1 \cap \ell_2) \simeq \Omega_1(V/\ell_2) \otimes \Omega_1(\ell_2/\ell_1 \cap \ell_2).$$

Given $e_1 \in \Lambda^n \ell_1$ and $e_2 \in \Lambda^n \ell_2$, there exist a unique $\delta \in \Omega_1(\ell_1/\ell_1 \cap \ell_2)$ such that

$$|e_1|^{1/2} \otimes |e_2|^{1/2} \otimes |\omega'|^{1/2} = |e_2| \otimes \delta.$$

We denote symbolically $\delta = |e_1|^{1/2} \otimes |e_2|^{-1/2} \otimes |\omega'|^{1/2}$. If $\ell_1 = \ell_2$ and $e_1 = e_2$, then $\delta = 1$. If $\ell_1 \cap \ell_2 = 0$, identifying $\Omega_1(L_2) = \Omega_1(V/L_1)$, we have for $e_1 \wedge e_2 = c\omega$, $\delta = |c|^{-1/2}|e_1|$.

1.4.13. <u>Proposition:</u> Let $e_1 \in \Lambda^n \ell_1$, $e_2 \in \Lambda^n \ell_2$ and

$\delta = |e_1|^{1/2} \otimes |e_2|^{-1/2} \otimes |\omega'|^{1/2}$. Then $\mathcal{F}^{\delta}_{\ell_2,\ell_1}$ is a unitary operator from $H(\ell_1,e_1)$ to $H(\ell_2,e_2)$. With this choice $\mathcal{F}_{\ell_1,\ell_2} = \mathcal{F}^{-1}_{\ell_2,\ell_1}$.

Proof: It follows from our calculation in coordinates in 1.4.2. that our normalized operator is precisely $\mathcal{F}$. The general case is deduced from 1.4.6.

1.5. Maslov index.

In this section, we define the Maslov index and prove its properties following an unpublished text of M. Kashiwara.

Let ℓ_1, ℓ_2, ℓ_3 be 3 Lagrangian planes in V. We consider the 3n-dimensional vector space $\ell_1 \oplus \ell_2 \oplus \ell_3$. We define the Maslov index $\tau(\ell_1, \ell_2, \ell_3)$:

1.5.1. Definition (Kashiwara): $\tau(\ell_1, \ell_2, \ell_3)$ is the signature of the quadratic form $Q(x_1 + x_2 + x_3)$ on the vector space $\ell_1 \oplus \ell_2 \oplus \ell_3$ defined by:

$$Q(x_1+x_2+x_3) = B(x_1,x_2) + B(x_2,x_3) + B(x_3,x_1).$$

The signature of Q is defined as follows: In a certain basis of $\ell_1 \oplus \ell_2 \oplus \ell_3$, the matrix of Q is diagonal and contains p times the coefficient $+1$, q times the coefficient -1; the signature of Q is then $p - q$.

It is clear from the definition, that we have:

1.5.2. For any $g \in Sp(B)$, $\tau(g\ell_1, g\ell_2, g\ell_3) = \tau(\ell_1, \ell_2, \ell_3)$.

1.5.3. $\tau(\ell_1, \ell_2, \ell_3) = -\tau(\ell_2, \ell_1, \ell_3) = -\tau(\ell_1, \ell_3, \ell_2)$.

Let ℓ_1 and ℓ_2 be two transverse Lagrangian planes. We have seen that we can always choose a symplectic basis $P_1, P_2, \ldots, P_n, Q_1, Q_2, \ldots, Q_n$, such that $\ell_1 = \sum_{i=1}^{n} \mathbb{R}P_i$, and $\ell_2 = \Sigma\, \mathbb{R}Q_i$, i.e. for ℓ_1', ℓ_2', two other transverse Lagrangian planes, we can certainly find $g \in Sp(B)$ such that $g\ell_1 = \ell_1'$, $g\ell_2 = \ell_2'$. On the contrary, formula 1.5.2 shows that the symplectic group does not act transitively on triples of Lagrangian planes. In fact we will see that the configuration of three transverse Lagrangian planes is determined by their index.

Let ℓ_1, ℓ_2, ℓ_3 be three Lagrangian planes. We suppose first that ℓ_1 and ℓ_3 are transverse, i.e. $V = \ell_1 \oplus \ell_3$. We denote by p_{13} the projection of V on ℓ_1 perpendicular to ℓ_3, and p_{31} the projection of V on ℓ_3 perpendicular to ℓ_1. We have:

1.5.4. Lemma: If ℓ_1 and ℓ_3 are transverse, then $\tau(\ell_1, \ell_2, \ell_3)$ is the signature of the quadratic form on ℓ_2 which associates to $x \in \ell_2$, $B(p_{13}x, p_{31}x) = B(x, p_{31}x) = B(p_{13}x, x)$.

Proof: We have

$$Q(x_1 + x_2 + x_3) = B(x_1, x_2) + B(x_2, x_3) + B(x_3, x_1)$$
$$= B(x_1, p_{31}x_2) + B(p_{13}x_2, x_3) + B(x_3, x_1) \text{ as } \ell_1 \text{ and } \ell_3 \text{ are Lagrangians,}$$
$$= B(p_{13}x_2, p_{31}x_2) - B(x_1 - p_{13}x_2, x_3 - p_{31}x_2).$$

Let $y_1 = x_1 - p_{13}x_2$, $y_2 = x_2$, $y_3 = x_3 - p_{31}x_2$; with respect to these coordinates $Q(x_1 + x_2 + x_3) = B(p_{13}y_2, p_{31}y_2) - B(y_1, y_3)$, hence $\operatorname{sign} Q = \operatorname{sign} B(p_{13}y_2, p_{31}y_2) - \operatorname{sign} B(y_1, y_3)$. As the signature of the form $B(y_1, y_3)$ is equal to zero, we obtain the lemma.

1.5.5. Remark. The bilinear form $S(x,y) = B(p_{13}x, p_{31}y)$ on ℓ_2 is symmetric, since for $x, y \in \ell_2$, $B(x,y) = 0 = B(p_{13}x + p_{31}x, p_{13}y + p_{31}y) = B(p_{13}x, p_{31}y) + B(p_{31}x, p_{13}y)$, as ℓ_1 and ℓ_3 are Lagrangian.

1.5.6. The kernel of the bilinear form S on ℓ_2 is equal to $\ell_1 \cap \ell_2 + \ell_3 \cap \ell_2$: if $B(p_{13}x, p_{31}y) = 0$ for every $y \in \ell_2$, then $B(p_{13}x,y) = 0$, hence $p_{13}x \in \ell_2 \cap \ell_1$. As $x = p_{31}x + p_{13}x$, we have $p_{31}x \in \ell_2 \cap \ell_3$, and $x \in \ell_1 \cap \ell_2 + \ell_3 \cap \ell_2$.

1.5.7. Corollary: Let (ℓ_1,ℓ_2,ℓ_3) be three mutually transverse Lagrangian planes, i.e. $\ell_1 \cap \ell_2 = 0$, $\ell_2 \cap \ell_3 = 0$, $\ell_1 \cap \ell_3 = 0$. Then there exists a symplectic basis $P_1, \cdots, P_n, Q_1, \cdots, Q_n$ and an integer k, $0 \leq k \leq n$, such that

$$\ell_1 = \mathbb{R}P_1 \oplus \cdots \oplus \mathbb{R}P_n$$

$$\ell_2 = RQ_1 \oplus \cdots \oplus \mathbb{R}Q_n$$

$$\ell_3 = \mathbb{R}(P_1 \oplus \varepsilon_1 Q_1) \oplus \cdots \oplus \mathbb{R}(P_n \oplus \varepsilon_n Q_n)$$

with $\varepsilon_i = +1$ if $i \leq k$, $\varepsilon_i = -1$ if $i > k$.

We then have $\tau(\ell_1,\ell_2,\ell_3) = n - 2k$.

Proof: The symmetric form $S(x,y)$ on ℓ_2 is non-degenerate, we hence can choose a basis $Q_1, Q_2, \cdots, Q_n$ of ℓ_2, such that $S(Q_i,Q_j) = -\varepsilon_j \delta_{ij}$. As B gives a duality between ℓ_1 and ℓ_2 we can choose a basis $P_1, P_2, \cdots, P_n$ of ℓ_1 such that $B(P_i,Q_j) = \delta_{ij}$, i.e. $(P_1, \cdots, P_n, Q_1, \cdots, Q_n)$ is a symplectic basis. Let $Z_i = p_{13}Q_i$, with respect to the decomposition $V = \ell_1 \oplus \ell_3$; then $Q_i - Z_i \in \ell_3$. By definition $S(Q_i,Q_j) = B(p_{13}Q_i,Q_j) = -\varepsilon_j \delta_{i,j}$, for every j. So $p_{13}Q_i = -\varepsilon_i P_i$. A basis of ℓ_3 is then $Q_i + \varepsilon_i P_i = \varepsilon_i(P_i + \varepsilon_i Q_i)$, and the lemma is proven.

1.5.8. Proposition. For $\ell_1,\ell_2,\ell_3,\ell_4$ four Lagrangian planes, τ verifies the chain condition:

$$\tau(\ell_1,\ell_2,\ell_3) = \tau(\ell_1,\ell_2,\ell_4) + \tau(\ell_2,\ell_3,\ell_4) + \tau(\ell_3,\ell_1,\ell_4).$$

We visualize this relation as follows:

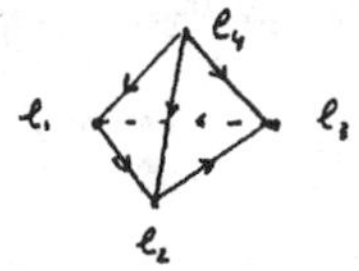

Proof: Let us first suppose that ℓ_4 is such that $\ell_4 \cap \ell_i = 0$ for $i = 1,2,3$. We have that $\tau(\ell_1,\ell_2,\ell_3)$ is the signature of the form $Q(x_1,x_2,x_3)$ on $\ell_1 \oplus \ell_2 \oplus \ell_3$. The second member is the signature of the quadratic form Q' on $\ell_1 \oplus \ell_2 \oplus \ell_3$ given by

$$Q'(y_1,y_2,y_3) = B(p_{14}y_2,y_2) + B(p_{24}y_3,y_3) + B(p_{34}y_1,y_1).$$

The transformations:

$$\begin{aligned} x_1 &= y_1 + p_{14}y_2 & \qquad y_1 &= \tfrac{1}{2}(x_1-p_{14}x_2+p_{14}x_3) \\ x_2 &= y_2 + p_{24}y_3 \quad \text{and} & y_2 &= \tfrac{1}{2}(x_2-p_{24}x_3+p_{24}x_1) \\ x_3 &= y_3 + p_{34}y_1 & y_3 &= \tfrac{1}{2}(x_3-p_{34}x_1+p_{34}x_2) \end{aligned}$$

are reciprocal (as $p_{14}p_{34}y_1 = y_1$ and similar relations). Let us prove that these transformations give the equivalence of Q and Q'.

We have:

$$B(x_1,x_2) = B(p_{14}y_2,y_2) + B(y_1,y_2) + B(y_1,p_{24}y_3) + B(p_{14}y_2,p_{24}y_3).$$

By cyclic permutation, we have to show that

$$B(y_1,y_2) + B(y_2,p_{34}y_1) + B(p_{34}y_1,p_{14}y_2) = 0.$$

Let us write $y_2 = p_{14}y_2 + p_{41}y_2$, this is

$$B(y_1,p_{41}y_2) + B(p_{41}y_2,p_{34}y_1), \text{ as } B(y_1,p_{14}y_2) = 0,$$

$$= B(y_1,p_{41}y_2) + B(p_{41}y_2,y_1) = 0, \text{ as } B(p_{41}y_2,p_{43}y_1) = 0.$$

This proves the proposition for this case.

Now let us take a Lagrangian plane m transverse to all the ℓ_j, $j = 1,2,3,4$, and let us express $\tau(\ell_i,\ell_j,\ell_k)$ as a function of $\tau(\ell_i,\ell_j,m)$. Then the result follows immediately.

1.5.9. Let ρ be an isotropic subspace of V, i.e. $B(\rho,\rho) = 0$. Then B defines a non-degenerate symplectic form on $\rho^\perp/\rho$. For W a subspace of V, we define

$$W^\rho = (W \cap \rho^\perp) + \rho = (W+\rho) \cap \rho^\perp \subset \rho^\perp.$$

We have $(W^\perp)^\rho = (W^\rho)^\perp$ as $(W^\rho)^\perp = ((W \cap \rho^\perp)+\rho)^\perp = \rho^\perp \cap (W^\perp+\rho) = (W^\perp)^\rho$
Hence if W is a Lagrangian plane of V, W^ρ/ρ is a Lagrangian plane in $\rho^\perp/\rho$.

We will prove:

1.5.10. <u>Proposition</u>: Let $\rho \subset (\ell_1 \cap \ell_2) + (\ell_2 \cap \ell_3) + (\ell_3 \cap \ell_1)$ then $\tau(\ell_1,\ell_2,\ell_3) = \tau(\ell_1^\rho,\ell_2^\rho,\ell_3^\rho)$.

We first prove the

1.5.11. Lemma: Let ℓ_1, ℓ_2 two Lagrangian subspaces. Then if $\ell = (\ell \cap \ell_1) + (\ell \cap \ell_2)$, we have $\tau(\ell_1, \ell, \ell_2) = 0$.

Proof: Let us choose $Y_1 \subset \ell \cap \ell_1$, $Y_2 \subset \ell \cap \ell_2$ such that $\ell = Y_1 \oplus Y_2$. For an element $x = (x_1, u+v, x_2)$ of the space $\ell_1 \oplus Y_1 \oplus Y_2 \oplus \ell_2$ with $x_1 \in \ell_1$, $u \in \ell \cap \ell_1$, $v \in \ell \cap \ell_2$, $x_2 \in \ell_2$, we have $Q(x) = B(x_1, v) + B(u, x_2) + B(x_2, x_1)$ $= B(x_2 - v, x_1 - u)$ as $B(u,v) = 0$. Hence the signature of Q is equal to the signature of the quadratic form $B(y_2, y_1)$ on $\ell_2 \oplus \ell_1$ which is obviously zero.

Now let us prove proposition 1.5.10. Let ρ be contained in $\ell_{123} = \ell_1 \cap \ell_2 + \ell_2 \cap \ell_3 + \ell_3 \cap \ell_1$. In particular, as ℓ_{123} is isotropic, $\ell_1 \cap \ell_2$, $\ell_2 \cap \ell_3$ and $\ell_3 \cap \ell_1$ are contained in $\rho^\perp$. We have $\ell_1^\rho = (\ell_1 \cap \rho^\perp) + \rho$ by definition. If $u = u_{12} + u_{23} + u_{31} \in \rho$ with $u_{12} \in \ell_1 \cap \ell_2$, $u_{23} \in \ell_2 \cap \ell_3$, $u_{31} \in \ell_3 \cap \ell_1$, we have $u_{23} = u - u_{12} - u_{31} \in \rho + (\ell_1 \cap \rho^\perp) = \ell_1^\rho$. Hence $u_{23} \in \ell_1^\rho \cap (\ell_2 \cap \rho^\perp)$. So $\ell_1^\rho = (\ell_1^\rho \cap \ell_1) + (\ell_2 \cap \ell_1^\rho) =$ $(\ell_1^\rho \cap \ell_1) + (\ell_2^\rho \cap \ell_1^\rho)$.

We conclude by the preceding formula that

$$\tau(\ell_1, \ell_1^\rho, \ell_2) = \tau(\ell_1, \ell_1^\rho, \ell_2^\rho) = 0.$$

From proposition 1.5.8, using the chain rule, it follows that $\tau(\ell_1, \ell_2, \ell_3) = \tau(\ell_1^\rho, \ell_2^\rho, \ell_3^\rho)$, as seen from the diagram:

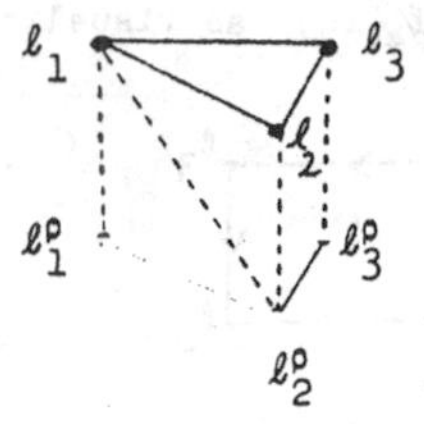

1.5.12. We define, for a sequence $(\ell_1, \ell_2, \cdots, \ell_k)$ of Lagrangian spaces in (V,B), the Maslov index $\tau(\ell_1, \ell_2, \cdots, \ell_k)$ for $k \geq 4$, by:

$$\begin{aligned}\tau(\ell_1,\ell_2,\cdots,\ell_k) &= \tau(\ell_1,\ell_2,\ell_3) + \tau(\ell_1,\ell_3,\ell_4) + \cdots + \tau(\ell_1,\ell_{k-1},\ell_k)\\ &= \tau(\ell_1,\ell_2,\ell) + \tau(\ell_2,\ell_3,\ell) + \cdots + \tau(\ell_{k-1},\ell_k,\ell) + \tau(\ell_k,\ell_1,\ell),\end{aligned}$$

where ℓ is an arbitrary Lagrangian space. (The equality follows from 1.5.8.) We have:

1.5.13. <u>Proposition</u>:

a) The index $\tau(\ell_1,\ell_2,\cdots,\ell_k)$ is invariant under the action of the symplectic group, and its value is unchanged under circular permutation.

b) For any Lagrangian planes ℓ_1,ℓ_2,ℓ_3, ℓ_1',ℓ_2',ℓ_3', we have:

$$\begin{aligned}\tau(\ell_1',\ell_2',\ell_3') &= \tau(\ell_1,\ell_2,\ell_3) + \tau(\ell_1',\ell_2',\ell_2,\ell_1)\\ &\quad + \tau(\ell_2',\ell_3',\ell_3,\ell_2) + \tau(\ell_3',\ell_1',\ell_1,\ell_3)\end{aligned}$$

as visualized by the graphic:

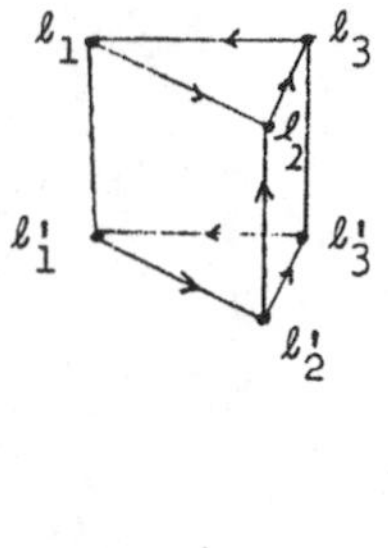

c) $\tau(\ell_1,\ell_2,\ell_3,\ell_4) = -\tau(\ell_2,\ell_1,\ell_4,\ell_3)$ **as visualized by**

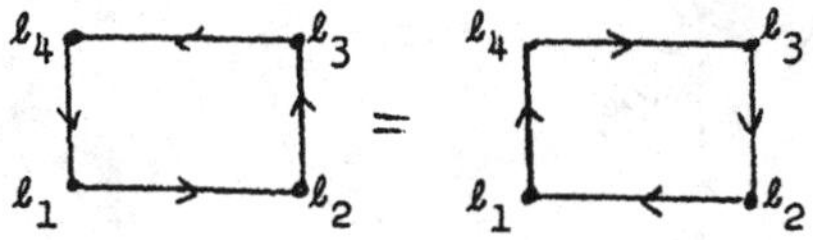

1.6. The cocycle of the Shale-Weil representation and the Maslov index.

Let ℓ_1, ℓ_2, ℓ_3 be three Lagrangian planes in the symplectic vector space (V,B). We consider the canonical unitary intertwining operator (we leave implicit the choice of e_1, e_2, e_3):

$$\mathcal{F}_{i,j}: H(\ell_j) \longrightarrow H(\ell_i)$$

which intertwines $W(\ell_j)$ and $W(\ell_i)$ defined in 1.4.8. It is clear that the operator $\mathcal{F}_{1,3}\ \mathcal{F}_{3,2}\ \mathcal{F}_{2,1}$ is proportional to the identity operator on $H(\ell_1)$ as this operator intertwines the irreducible representation $W(\ell_1)$ with itself. Hence there is a scalar of modulus one $a(\ell_1,\ell_2,\ell_3)$ such that:

$$\mathcal{F}_{\ell_1,\ell_3}\ \mathcal{F}_{\ell_3,\ell_2}\ \mathcal{F}_{\ell_2,\ell_1} = a(\ell_1,\ell_2,\ell_3)\ \mathrm{Id}.$$

(It is easy to see that $a(\ell_1,\ell_2,\ell_3)$ does not depend of e_1, e_2, e_3.)

1.6.1. Theorem: Let ℓ_1, ℓ_2, and ℓ_3 be three Lagrangian planes. Then

$$\mathcal{F}_{1,3}\ \mathcal{F}_{3,2}\ \mathcal{F}_{2,1} = e^{-\frac{i\pi}{4}\tau(\ell_1,\ell_2,\ell_3)}\ \mathrm{Id}_{H_1}.$$

Proof: Let us first compute this for the three dimensional Heisenberg algebra: Let $\ell_1 = \mathbb{R}P$, $\ell_2 = \mathbb{R}Q$, $\ell_3 = \mathbb{R}(P+Q)$. We have $\tau(\ell_1,\ell_2,\ell_3) = -\tau(\ell_1,\ell_3,\ell_2) = -1$ as follows from 1.5.4. Hence we have to prove that in this case

$$\mathcal{F}_{1,3}\ \mathcal{F}_{3,2}\ \mathcal{F}_{2,1} = e^{\frac{i\pi}{4}}\ \mathrm{Id}.$$

Let us identify $H(\ell_1)$, $H(\ell_2)$ and $H(\ell_3)$ with $L^2(\mathbb{R})$ by $(R_1\varphi)(x) = \varphi(\exp xQ)$, $(R_2\varphi)(x) = \varphi(\exp xP)$ and $(R_3\varphi)(x) = \varphi(\exp xP)$. Then on $H = L^2(\mathbb{R})$, we obtain

1) $$(\mathcal{F}_{2,1}\varphi)(\exp xP) = \int \varphi(\exp xP \exp \xi Q)\, d\xi$$
$$= \int e^{-2i\pi x\xi}\, \varphi(\xi)\, d\xi$$

i.e. $\mathcal{F}_{2,1} = \mathcal{F}$

2) $$(\mathcal{F}_{3,2}\varphi)(\exp xP) = \int \varphi(\exp xP \exp \xi(P+Q))\, d\xi.$$

We have: $\exp xP \exp \xi(P+Q) = \exp(x+\xi)P \exp \xi Q \exp \frac{-\xi^2}{2} E$.

Hence

$$(\mathcal{F}_{3,2}\varphi)(x) = \int \varphi(x+\xi)\, e^{i\pi\xi^2}\, d\xi$$
$$= e^{i\pi x^2} \int \varphi(\xi) e^{-2i\pi x\xi}\, e^{\,i\pi\xi^2}\, d\xi, \text{ changing } \xi \text{ to } \xi-x$$

i.e. $\mathcal{F}_{3,2} = e^{i\pi x^2} \mathcal{F} e^{i\pi x^2}$

where $e^{i\pi x^2}$ denotes the multiplication operator.

3) $$(\mathcal{F}_{1,3}\varphi)(\exp xQ) = \int \varphi(\exp xQ \exp \xi P)\, d\xi.$$

We have:

$$\exp xQ \exp \xi P = \exp(\xi-x)P \exp x(P+Q) \exp(-x\xi + \frac{x^2}{2})\, E.$$

Hence

$$(\mathcal{F}_{1,3}\varphi)(x) = \int \varphi(\xi-x)\, e^{2i\pi x\xi}\, e^{-i\pi x^2}\, d\xi$$

$$= e^{i\pi x^2} \int \varpi(\xi)\, e^{2i\pi x\xi}\, d\xi, \text{ changing } \xi \text{ to } \xi + x$$

i.e. $$\mathcal{F}_{1,3} = e^{i\pi x^2} \mathcal{F}^{-1}.$$

Let us consider the function $\psi_z(x) = e^{i\pi z x^2}$ for $\operatorname{Im} z > 0$. This function is in $L^2(\mathbb{R})$.

1.6.2. Lemma.

$$\int e^{-2i\pi x\xi}\, e^{i\pi z\xi^2}\, d\xi = (\tfrac{z}{i})^{-1/2}\, e^{-i\pi z^{-1} x^2}$$

where the determination of the function $z \to (\frac{z}{i})^{-1/2}$ on the simply-connected domain $\operatorname{Im} z > 0$ is 1 for $z = i$.

Proof: We define for $z \in D = \{z;\ \operatorname{Im} z > 0\}$, $x \in \mathbb{C}$

$$\delta(z,x) = \int e^{-2i\pi x\xi}\, e^{i\pi z\xi^2}\, d\xi.$$

It is easy to see that $\delta(z,x)$ is a holomorphic function of (z,x) on $D \times \mathbb{C}$, and for $x \in \mathbb{R}$, $\delta(z,x) = (\mathcal{F}\psi_z)(x)$. It is hence sufficient to prove $\delta(z,x) = (\frac{z}{i})^{-1/2}\, e^{-i\pi z^{-1} x^2}$ for $z = iy$, $y > 0$, $x = iu$. We have then

$$\delta(iy,iu) = \int e^{2\pi u\xi}\, e^{-\pi y\xi^2}\, d\xi ,$$

we change ξ to $y^{-1/2}\xi$

$$= y^{-1/2} \int e^{2\pi u y^{-1/2}\xi}\, e^{-\pi\xi^2}\, d\xi$$

$$= y^{-1/2} \int e^{-\pi(\xi - y^{-1/2}u)^2}\, e^{\pi y^{-1}u^2}\, d\xi$$

$$= y^{-1/2}\, e^{\pi y^{-1}u^2} \int e^{-\pi\xi^2}\, d\xi$$

$$= y^{-1/2}\, e^{\pi y^{-1}u^2} ,$$

which is the desired formula.

We will denote $\mathcal{F}\varphi$ by $\hat{\varphi}$. From the lemma 1.6.1 we deduce:

1.6.3. Lemma. Let φ be a function in $\mathcal{S}(\mathbb{R})$, then:

$$\int \hat{\varphi}(\xi)\, e^{i\pi\xi^2}\, d\xi = e^{\frac{i\pi}{4}} \int \varphi(\xi)\, e^{-i\pi\xi^2}\, d\xi.$$

Proof: Let $z \in D$. We consider the function $\int \hat{\varphi}(\xi)\, e^{i\pi z\xi^2}\, d\xi$. This is clearly a continuous function of z for z in $\bar{D}$; we have for $z \in D$ by the Plancherel formula, and lemma 1.6.1

$$\int \hat{\varphi}(\xi)\, e^{i\pi z\xi^2}\, d\xi = \int \varphi(\xi)(e^{i\pi z\xi^2})^\wedge\, d\xi$$

$$= (\tfrac{z}{i})^{-1/2} \int \varphi(\xi)\, e^{-i\pi z^{-1}\xi^2}\, d\xi.$$

If we now take the limit of this continuous function when z tends to 1 we obtain the lemma.

We now calculate $\mathcal{F}_{3,2}\, \mathcal{F}_{2,1}$; we have from the preceding calculations, for $\varphi \in \mathcal{S}(\mathbb{R})$

$$(\mathcal{F}_{3,2}\, \mathcal{F}_{2,1}\varphi)(x) = \int \hat{\varphi}(x+\xi)\, e^{i\pi\xi^2}\, d\xi.$$

But $\hat{\varphi}(x+\xi) = (e^{-2i\pi\xi u}\varphi(u))^\wedge$. Hence

$$(\mathcal{F}_{3,2}\, \mathcal{F}_{2,1}\varphi)(x) = e^{\frac{i\pi}{4}} \int e^{-2i\pi\xi u}\, \varphi(u) e^{-i\pi u^2}\, du$$

i.e. $$\mathcal{F}_{3,2}\, \mathcal{F}_{2,1} = e^{\frac{i\pi}{4}}\, \mathcal{F}\, e^{-i\pi x^2}.$$

Hence

$$\mathcal{F}_{1,3}\, \mathcal{F}_{3,2}\, \mathcal{F}_{2,1} = e^{\frac{i\pi}{4}}\, \mathrm{Id} = e^{\frac{-i\pi}{4}\tau(\ell_1,\ell_2,\ell_3)}\, \mathrm{Id}.$$

The case when $\tau(\ell_1,\ell_2,\ell_3) = 1$ is proven in the same way.

Hence the formula of the Theorem 1.6.1 is proven for the three dimensional Heisenberg Lie algebra. (If $\ell_1 = \ell_2$, then $\mathcal{F}_{2,1} = \mathrm{Id}$, $\mathcal{F}_{1,3} = \mathcal{F}_{3,2}^{-1}$ (1.4.9b) and $\tau(\ell_1,\ell_2,\ell_3) = 0$).

Now we will prove Theorem 1.6.1 by induction on the dimension of V. The case where ℓ_1,ℓ_2,ℓ_3 are transverse follows then from 1.5.7, as we can then decompose the transformations $\mathcal{F}_{\ell_1,\ell_3}\,\mathcal{F}_{\ell_3,\ell_2}\,\mathcal{F}_{\ell_2,\ell_1}$ into products of transformations involving only the variables in the symplectic subspaces $\mathbb{R}P_i + \mathbb{R}Q_i$ without mutual interferences.

We denote by $a(\ell_1,\ell_2,\ell_3)$ the element of T such that $\mathcal{F}_{\ell_1,\ell_3}\,\mathcal{F}_{\ell_3,\ell_2}\,\mathcal{F}_{\ell_2,\ell_1} = a(\ell_1,\ell_2,\ell_3)\ \mathrm{Id}$.

Let us suppose now that, for example, ℓ_1 and ℓ_2 are not transverse and let $\rho = \ell_1 \cap \ell_2$. From 1.5.9 we have $\tau(\ell_1,\ell_2,\ell_3) = \tau(\ell_1,\ell_2,\ell_3^\rho)$. We will show that:

$$\mathcal{F}_{\ell_3,\ell_2} = \mathcal{F}_{\ell_3,\ell_3^\rho}\,\mathcal{F}_{\ell_3^\rho,\ell_2}$$

$$\mathcal{F}_{\ell_1,\ell_3} = \mathcal{F}_{\ell_1,\ell_3^\rho}\,\mathcal{F}_{\ell_3^\rho,\ell_3}\ .$$

$$\begin{array}{ccccccc} \ell_1 & \longrightarrow & \ell_2 & \longrightarrow & \ell_3 & \longrightarrow & \ell_1 \\ & & & \searrow & \uparrow & \nearrow & \\ & & & & \ell_3^\rho & & \end{array}\ .$$

Hence we will have

$$\mathcal{F}_{1,3}\,\mathcal{F}_{3,2}\,\mathcal{F}_{2,1} = \mathcal{F}_{\ell_1,\ell_3^\rho}\,\mathcal{F}_{\ell_3^\rho,\ell_3}\,\mathcal{F}_{\ell_3,\ell_3^\rho}\,\mathcal{F}_{\ell_3^\rho,\ell_2}\,\mathcal{F}_{\ell_2,\ell_1}$$

$$= \mathcal{F}_{\ell_1,\ell_3^\rho}\,\mathcal{F}_{\ell_3^\rho,\ell_2}\,\mathcal{F}_{\ell_2,\ell_1}. \text{ (By (1.4.9b).)}$$

But the three Lagrangian planes $\ell_1,\ell_2,\ell_3^\rho$ are contained in $\rho^\perp = \ell_1 + \ell_2 \neq V$. If we consider $V' = \rho^\perp/\rho$, $\ell_1' = \ell_1/\rho$, $\ell_2' = \ell_2/\rho$, $\ell_3' = \ell_3^\rho/\rho$, ℓ_1', ℓ_2' and ℓ_3' are 3 Lagrangian planes in the symplectic vector space V' of dimension strictly smaller. It is easily checked that $a(\ell_1',\ell_2',\ell_3') = a(\ell_1,\ell_2,\ell_3^\rho)$ and $\tau(\ell_1',\ell_2',\ell_3') = \tau(\ell_1,\ell_2,\ell_3^\rho)$ by 1.5.9.

Hence we only need to prove:

1.6.4 Lemma. Let ℓ_1, ℓ_2, and ℓ three Lagrangian planes such that $\ell = (\ell \cap \ell_1) + (\ell \cap \ell_2)$, then

$$\mathcal{F}_{\ell_2,\ell_1} = \mathcal{F}_{\ell_2,\ell}\,\mathcal{F}_{\ell,\ell_1}.$$

Then our result will follow, as for example for $\ell_3^\rho = (\ell_3 \cap \rho^\perp) + \rho$, ρ being contained in $\ell_3^\rho \cap \ell_1$, $\ell_3^\rho = \ell_3^\rho \cap \ell_3 + \ell_3^\rho \cap \ell_1$.

Proof: Let us compute $(\mathcal{F}_{\ell_2,\ell}\,\mathcal{F}_{\ell,\ell_1}\varphi)(g)$

$$= \int_{\ell_2/\ell\cap\ell_2} (\mathcal{F}_{\ell,\ell_1}\varphi)(g \exp x)\, d\dot{x}$$

$$= \int_{x\in\ell_2/\ell\cap\ell_2} \int_{y\in\ell/\ell_1\cap\ell} \varphi(g \exp x \exp y)\, d\dot{x}\, d\dot{y}.$$

We have $\ell = \ell_1 \cap \ell + \ell_2 \cap \ell$, hence $\ell_1 \cap \ell_2$ is contained in $\ell^\perp = \ell$. Let us consider Y_1 a complementary subspace for

$\ell \cap \ell_2 \cap \ell_1 = \ell_1 \cap \ell_2$ in $\ell \cap \ell_2$; then $\ell = \ell_1 \cap \ell + Y_1$. Let Y_2 be a complementary subspace for $\ell \cap \ell_2$ in ℓ_2, then $\ell_2 = Y_2 + \ell \cap \ell_2 = Y_2 + Y_1 + \ell_1 \cap \ell_2$. Hence $Y_1 \oplus Y_2$ is a complementary subspace for $\ell_1 \cap \ell_2$ in ℓ_2. Our integral $\mathcal{F}_{\ell_2,\ell}\, \mathcal{F}_{\ell,\ell_1}\varphi$ using the choice of this complementary subspace is then:

$$\int_{Y_2} \int_{Y_1} \varphi(g \exp y_2 \exp y_1)\, dy_1\, dy_2.$$

But as $Y_1, Y_2 \subset \ell_2$, $B(Y_1,Y_2) = 0$ hence

$$\begin{aligned}(\mathcal{F}_{\ell_2,\ell}\, \mathcal{F}_{\ell,\ell_1}\varphi)(g) &= \int_{Y_1+Y_2} \varphi(g \exp (y_1+y_2))\, dy_1\, dy_2 \\ &= \int_{\ell_2/\ell_1\cap\ell_2} \varphi(g \exp u)\, d\dot{u} \\ &= (\mathcal{F}_{\ell_2,\ell_1}\varphi)(g) \qquad \text{q.e.d.}\end{aligned}$$

We now define the Segal-Shale-Weil projective representation of the symplectic group $G = Sp(B)$.

Let G be the symplectic group of the vector space (V,B) (1.1.6.). The group G acts on N leaving the center $\exp \mathbb{R}E$ of N fixed. Let (W,H) be "the" irreducible representation of N into the Hilbert space H such that $W(\exp tE) = e^{2i\pi t}\, Id_H$. If we consider the representation (W^g,H) of N in H defined by $W^g(n) = W(g.n)$, then from the Stone-Von Neumann theorem, the representation W^g is equivalent to W. Hence there exists a unitary operator $R(g)$ on H such that:

1.6.5. $$R(g)\, W(n)\, R(g)^{-1} = W(g\cdot n)$$

for every $n \in N$. This determines $R(g)$ up to a scalar of modulus one.

Let us consider a function $g \to R(g)$ from G into the unitary operator $U(H)$ of H such that for every $g \in G$, $n \in N$ the relation 1.6.5. holds. Then from 1.3.9. we have that for g_1, $g_2 \in G$

1.6.6. $$R(g_1g_2) = c(g_1,g_2)\, R(g_1)\, R(g_2)$$

with $c(g_1,g_2)$ a scalar of modulus one. (The operator $R(g_1)\, R(g_2)\, R(g_1g_2)^{-1}$ is a unitary operator commuting with the representation W). From the relation

$$R((g_1g_2)\cdot g_3) = R(g_1\cdot(g_2g_3))$$

we deduce that for g_1, g_2, $g_3 \in G$, we have:

1.6.7. $$c(g_1g_2,g_3)\, c(g_1,g_2) = c(g_1,g_2g_3)\, c(g_2,g_3).$$

We say that $g \to R(g)$ is a projective representation of the group G and $c(g_1,g_2)$ is the associated cocycle. We will now define a canonical choice $g \to R(g)$ satisfying 1.6.5, and compute the associated cocycle.

1.6.8. Let us consider ℓ a Lagrangian subspace of (V,B) with a given form e. The group G operates on functions on N by $(A(g)\varphi)(n) = \varphi(g^{-1}.n)$. If φ is a function in $H(\ell,e)$

(satisfying 1.2.4. a),b)) it is clear that $A(g)\varphi \in H(g\cdot\ell, g\cdot e)$. Hence we may consider $A(g)$ as a unitary operator from $H(\ell,e)$ to $H(g.\ell, g\cdot e)$. It is immediate to verify that $A(g)$ is a unitary operator satisfying

$$A(g)\, W_\ell(n)\, A(g)^{-1} = W_{g\ell}(g\cdot n).$$

Let $(\ell_1,e_1),(\ell_2,e_2)$ be two Lagrangian subspaces and $\mathcal{F}^{\delta}_{\ell_2,\ell_1}$ our canonical unitary operator (1.4.9). The diagram

$$\begin{array}{ccc} H(\ell_1,e_1) & \xrightarrow{\ \mathcal{F}^{\delta}_{\ell_2,\ell_1}\ } & H(\ell_2,e_2) \\ \big\downarrow {\scriptstyle A(g)} & & \big\downarrow {\scriptstyle A(g)} \\ H(g\ell_1, ge_1) & \xrightarrow{\ \mathcal{F}^{g\cdot\delta}_{g\ell_2,g\ell_1}\ } & H(g\ell_2, ge_2) \end{array}$$

is commutative:

We now define for every $g \in G$

1.6.9. $$R_\ell(g) = \mathcal{F}^{\delta}_{\ell,g\cdot\ell}\cdot A(g)$$

where $\delta = |g\cdot e|^{1/2} \otimes |e|^{-1/2} \otimes |\omega'|^{1/2} = \delta(g)$. Clearly δ doesn't depend of e (only of relative size of $g.e$, with respect to e). Hence $R_\ell(g)$ depends only on ℓ.

The function $g \to R_\ell(g)$ is a function on G with values in the unitary operators on $H = H(\ell,e)$ verifying:

1.6.10. $$R_\ell(g)\, W_\ell(n) \circ R_\ell(g)^{-1} = W_\ell(g\cdot n)$$

for every $g \in G$, and $n \in N$.

We will often write $R(g) = R_\ell(g)$, if the choice of ℓ is without ambiguity. We have:

$$R_\ell(g_1)\ R_\ell(g_2) = \mathcal{F}_{\ell,g_1\ell}\cdot A(g_1)\,\mathcal{F}_{\ell,g_2\ell}\cdot A(g_2)$$

$$= \mathcal{F}_{\ell,g_1\ell}\cdot\mathcal{F}_{g_1\ell,g_1g_2\ell}\cdot A(g_1)\cdot A(g_2) \quad \text{(from 1.6.8.)}$$

$$= e^{\frac{i\pi}{4}\tau(\ell,g_1\ell,g_1g_2\ell)}\,\mathcal{F}_{\ell,g_1g_2\ell}\ A(g_1g_2)$$

as: $A(g_1)\ A(g_2) = A(g_1g_2)$

$$\mathcal{F}_{\ell,g_1\ell}\cdot\mathcal{F}_{g_1\ell,g_1g_2\ell}\cdot\mathcal{F}_{g_1g_2\ell,\ell} = e^{-\frac{i\pi}{4}\tau(\ell,g_1g_2\ell,g_1\ell)}\ \text{Id} \quad (1.6.1.).$$

Hence we have:

1.6.11. <u>Theorem</u>:

$$R_\ell(g_1g_2) = c_\ell(g_1,g_2)\ R_\ell(g_1)\ R_\ell(g_2)$$

with:

$$c_\ell(g_1,g_2) = e^{-\frac{i\pi}{4}\tau(\ell,g_1\ell,g_1g_2\ell)}$$

1.6.12. For a projective representation R of a group G with cocycle c, we consider the Mackey obstruction group G_c. The group G_c is the set $G_c = G \times T$ with the law of multiplication being given by $(g_1,t_1)\cdot(g_2,t_2) = (g_1g_2, t_1t_2\ c(g_1,g_2)^{-1})$. (The cocycle relation 1.6.7. is in fact equivalent to the associativity of the above given law.) If we consider the function $R(g,t) = tR(g)$, R is now a unitary representation of the group G_c, as

$$R((g_1,t_1)\cdot(g_2,t_2)) = R(g_1g_2,t_1t_2\; c(g_1,g_2)^{-1})$$

$$= t_1t_2\; c(g_1,g_2)^{-1}\; R(g_1g_2)$$

$$= t_1\; R(g_1)\; t_2\; R(g_2).$$

Hence we can think of the projective representation R of G as a true representation of the Mackey group G_c.

In our case, our cocycle is given by the formula

$$c_\ell(g_1,g_2) = e^{-\frac{i\pi}{4}\,\tau(\ell,g_1\ell,g_1g_2\ell)} .$$

We consider the $\mathbb{Z}$-valued function $\tau_\ell(g_1,g_2) = \tau(\ell,g_1\ell,g_1g_2\ell)$. This function satisfies the following:

1.6.13. <u>Lemma</u>:

$$\tau_\ell(g_1g_2,g_3) + \tau_\ell(g_1,g_2) = \tau_\ell(g_1,g_2g_3) + \tau_\ell(g_2,g_3).$$

Proof: Applying the cochain relation 1.5.8. to $\ell,g_1\ell,g_1g_2\ell,g_1g_2g_3\ell$, we have

$$\tau(\ell,g_1\ell,g_1g_2\ell) = \tau(\ell,g_1\ell,g_1g_2g_3\ell) + \tau(g_1\ell,g_1g_2\ell,g_1g_2g_3)$$

$$+ \tau(g_1g_2\ell,\ell,g_1g_2g_3\ell)$$

which is the relation we want, if we remark that

$$\tau(g_1\ell,g_1g_2\ell,g_1g_2g_3\ell) = \tau(\ell,g_2\ell,g_2g_3\ell) = \tau_\ell(g_2,g_3) .$$

1.6.14. Hence we can define the group $\widetilde{G}_\ell = G \times \mathbb{Z}$ with the following associative law

$$(g_1,n_1)\cdot(g_2,n_2) = (g_1g_2, n_1+n_2+\tau(\ell, g_1\ell, g_1g_2\ell)).$$

We will study in the appendix the group $\widehat{G}_\ell$ in detail and prove the the connected component of the identity of $\widetilde{G}_\ell$ is the universal cover group of G, i.e. the calculation of our cocycle in the form $e^{\frac{i\pi}{4}\tau}$ allows us not only to construct the metaplectic group, but also the universal covering group of G.

The map $(g,n) \to (g, e^{\frac{i\pi}{4}n})$ is a homomorphism of $\widetilde{G}_\ell = G \times \mathbb{Z}$ into the Mackey group G_c. Hence we define the Shale-Weil representation $\widehat{R}_\ell$ on $\widehat{G}_\ell = G \times \mathbb{Z}$ by the formula

1.6.15. $$\widehat{R}_\ell(g,n) = e^{\frac{i\pi}{4}n} R_\ell(g).$$

This is a true representation of $\widehat{G}_\ell$.

1.6.16. Let us suppose we have chosen another Lagrangian plane ℓ' and define the projective representation $R_{\ell'}$. We have $R_{\ell'}(g) = b(g)\, R_\ell(g)$, with $b(g) = b_{\ell,\ell'}(g)$ a scalar of modulus one. From the definition of $R_{\ell'}(g) = \mathcal{F}_{\ell',g\ell'}\, A(g)$ we see that $b(g)$ is the scalar such that the following diagram is commutative:

$$\begin{array}{ccccc}
H(\ell') & \xrightarrow{A(g)} & H(g\cdot\ell') & \xrightarrow{\mathcal{F}_{\ell',g\ell'}} & H(\ell') \\
\downarrow{\scriptstyle \mathcal{F}_{\ell,\ell'}} & & \vdots{\scriptstyle \mathcal{F}_{g\ell,g\ell'}} & & \downarrow{\scriptstyle \mathcal{F}_{\ell,\ell'}} \\
H(\ell) & \xrightarrow{A(g')} & H(g\cdot\ell) & \xrightarrow{b(g)\mathcal{F}_{\ell,g\ell}} & H(\ell)
\end{array} .$$

As the first square of the diagram with the arrow in dashes is clearly commutative, b(g) is given by the commutativity of the second square, i.e.

$$\begin{aligned} b(g) &= \mathcal{F}_{\ell,\ell'}\,\mathcal{F}_{\ell',g\ell'}\,\mathcal{F}_{g\ell',g\ell}\,\mathcal{F}_{g\ell,\ell} \\ &= \mathcal{F}_{\ell,\ell'}\,\mathcal{F}_{\ell',g\ell'}\,\mathcal{F}_{g\ell',\ell}\,\mathcal{F}_{\ell,g\ell'}\,\mathcal{F}_{g\ell',g\ell}\,\mathcal{F}_{g\ell,\ell} \\ &= e^{-\frac{i\pi}{4}\,\tau(\ell,g\ell',\ell')+\tau(\ell,g\ell,g\ell')}\,\mathrm{Id} \qquad \text{by } 1.6.1 \\ &= e^{-\frac{i\pi}{4}\,\tau(\ell,g\ell,g\ell',\ell')}\,\mathrm{Id} \qquad (1.5.12) \end{aligned}$$

i.e.

1.6.17. $$b_{\ell,\ell'}(g) = e^{-\frac{i\pi}{4}\,\tau(\ell,g\ell,g\ell',\ell')}\,.$$

It is clear that we have

$$c_\ell(g_1,g_2) = c_{\ell'}(g_1,g_2)b(g_1)b(g_2)b(g_1g_2)^{-1}$$

as follows from

$$R_{\ell'}(g_1g_2) = b(g_1g_2)R_\ell(g_1g_2)\,.$$

In fact we have:

$$\begin{aligned} &\tau(\ell,g_1\ell,g_1g_2\ell) = \\ &\tau(\ell',g_1\ell',g_1g_2\ell') + \tau(\ell,g_1\ell,g_1\ell',\ell') + \tau(\ell,g_2\ell,g_2\ell',\ell') \\ &\qquad - \tau(\ell,g_1g_2\ell,g_1g_2\ell',\ell') \end{aligned}$$

as follows from the (1.5.13,1.5.14) and the relation $\tau(\ell, g_2\ell, g_2\ell', \ell') = \tau(g_1\ell, g_1g_2\ell, g_1g_2\ell', g_1\ell')$

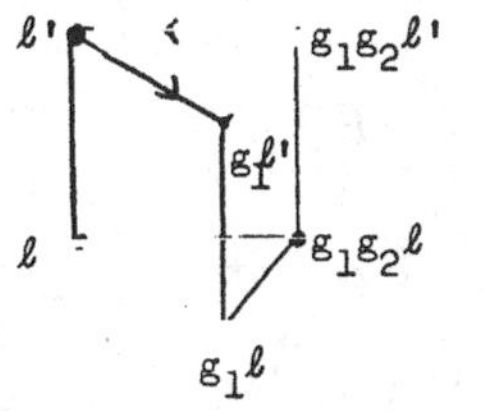

1.6.18. In particular the groups $\tilde{G}_\ell$ and $\tilde{G}_{\ell'}$ are isomorphic via the map from $\tilde{G}_\ell$ to $\tilde{G}_{\ell'}$ given by:

$$(g,n) \to (g, n + \tau(\ell, g\ell, g\ell', \ell')).$$

1.6.19. We now will give the formulas for the canonical projective representation $R_\ell(g)$ of G. We choose a decomposition $V = \ell \oplus \ell'$ of V into two complementary Lagrangian spaces. We write x for an element of ℓ and y for an element of ℓ'. The space ℓ' is isomorphic to ℓ^* via $x \to B(x,y)$ for $y \in \ell'$.

We identify $H(\ell)$ with $L^2(\ell')$ by $\varphi(y) = \varphi(\exp y)$. By definition $R_\ell(g) = \mathcal{F}^{\delta}_{\ell, g\ell}\ A(g)$,

i.e. $$(R_\ell(g)\varphi)(y) = \int_{\ell/g\ell\cap\ell} (A(g)\varphi)(\exp y \exp x)\ \delta\dot{x}$$

$$= \int_{\ell/g\ell\cap\ell} \varphi(\exp g^{-1}y \exp g^{-1}x)\ \delta\dot{x}\ .$$

We have: $\ell \cap g\ell = \{x \in \ell;\ g^{-1}x \in \ell\}$, i.e. $\ell \cap g\ell = \ker {}^tc$. We have: $g^{-1}x = {}^tdx - {}^tcx$, with ${}^tdx \in \ell$, ${}^tcx \in \ell'$, so

$$\exp(g^{-1}x) = \exp(-{}^t cx)\exp({}^t dx)\exp\frac{B({}^t cx, {}^t dx)}{2} E.$$

As $\varphi \in H(\ell)$, and ${}^t dx \in \ell$

$$\begin{aligned}\varphi(\exp g^{-1}y \exp g^{-1}x) &= \varphi(\exp g^{-1}y \exp -{}^t cx)e^{-i\pi B(d{}^t cx,x)}\\ &= \varphi(\exp -{}^t cx \exp g^{-1}y)e^{2i\pi B(g^{-1}y,{}^t cx)}e^{-i\pi B(d{}^t cx,x)}\\ &= \varphi(\exp -{}^t cx \exp g^{-1}y)e^{-2i\pi B({}^t by,{}^t cx)}e^{-i\pi B(d{}^t cx,x)}.\end{aligned}$$

Finally writing $g^{-1}y = -{}^t by + {}^t ay$, we obtain

$$(R_\ell(g)\varphi)(y) = \int_{x\in\ell/\ker{}^t c} \varphi({}^t ay - {}^t cx)e^{i\pi B(a{}^t by,y)}e^{-2i\pi B({}^t by,{}^t cx)}e^{-i\pi B(d{}^t cx,x)}\delta x$$

Changing x to $-x$, we obtain finally:

1.6.20.

$$(R_\ell(g)\varphi)(y) = \int_{x\in\ell/\ker{}^t c} \varphi({}^t ay + {}^t cx)e^{i\pi B(a{}^t by,y)}e^{2i\pi B({}^t by,{}^t cx)}e^{-i\pi B(d{}^t cx,x)}\delta x.$$

The choice of δ is as follows: there is u unique in $|\Lambda^{max}\ell/\ell\cap g\cdot\ell|$ such that: for $v \in \Lambda^{max}(g\cdot\ell/\ell\cap g\cdot\ell)$ and $\alpha \in \Lambda^{max}(\ell\cap g\cdot\ell)$, $v\wedge\alpha = g\cdot(u\wedge\alpha)$ and $|u\wedge v| = |\omega'|$, where ω' is the canonical form on the symplectic space $\ell+g\cdot\ell/\ell\cap g\cdot\ell$. We choose δ such that $|(\delta,u)| = 1$.

1.6.21. Let us consider some special cases of this formula.

1) Let $g(a) = \left(\begin{array}{c|c} a & 0 \\ \hline 0 & {}^t a^{-1}\end{array}\right)$, with $a \in GL(\ell)$, then

$$(R_\ell(g(a))\varphi)(y) = |\det a|^{1/2}\,\varphi({}^t ay).$$

2) Let $u(x) = \left(\begin{array}{c|c} 1 & x \\ \hline 0 & 1\end{array}\right)$, with $x = {}^t x$, then

$$(R_\ell(u(x))\varphi)(y) = e^{i\pi B(xy,y)}\varphi(y)\,.$$

3) Let $g = \left(\frac{a\,|\,b}{c\,|\,d}\right)$, with c invertible, then

$$(R_\ell(g)\varphi)(y) = \int_{\ell'} \varphi(y')e^{i\pi(B(c^{-1}dy',y')+B(ac^{-1}y,y)-2B(c^{-1}y,y'))}\,\delta y'\,.$$

If a given symplectic basis of V is chosen, with $\ell = \oplus \mathbb{R}P_i$ and $\ell' = \oplus \mathbb{R}Q_i$, then $\delta y' = |\det c|^{-1/2}\,dy$.

We also give a formula for the cocycle $c_\ell(g_1,g_2)$ on the open set $\{(g_1,g_2)\}$ of $G \times G$ where

$$g_1 = \begin{pmatrix} a_1 & b_1 \\ c_1 & d_1 \end{pmatrix},\quad g_2 = \begin{pmatrix} a_2 & b_2 \\ c_2 & d_2 \end{pmatrix},\quad \text{with } c_1, c_2 \text{ invertible.}$$

Let us write

$$g_1g_2 = \begin{pmatrix} a_3 & b_3 \\ c_3 & d_3 \end{pmatrix}.$$

We have then:

1.6.22. $$c_\ell(g_1,g_2) = e^{-\frac{i\pi}{4}\operatorname{sign}(c_1^{-1}c_3c_2^{-1})}\,.$$

<u>Proof</u>: We have $c_3 = c_1a_2 + d_1c_2$. Thus $c_1^{-1}c_3c_2^{-1} = a_2c_2^{-1} + c_1^{-1}d_1$ is a symmetric matrix (1.1.9). We now compute

$$\tau(\ell, g_1\ell, g_1g_2\ell) = \tau(g_1^{-1}\ell, \ell, g_2\ell)\,,$$

using 1.5.4.

Let us write $x \in \ell$, on the form $x = z_1 + z_3$ with z_1 in $g_1^{-1}\ell$ and z_3 in $g_2\ell$, i.e.

$x = {}^t d_1u - {}^t c_1u + a_2v + c_2v$, with $u,v \in \ell$. We then have:

${}^t c_1u = c_2v$. τ is the signature of the quadratic form Q' defined by $v \to B({}^t d_1u - {}^t c_1u,\ a_2v + c_2v)$. We choose

$y = c_2 v$ as variable. Then

$$Q'(y) = B({}^t d_1 {}^t c_1^{-1} y, y) - B(y, a_2 c_2^{-1} y) = B(c_1^{-1} d_1 y + a_2 c_2^{-1} y, y)$$
$$= B(c_1^{-1} c_3 c_2^{-1} y, y)$$

and the formula follows.

1.7. Oriented Lagrangian planes and the metaplectic group.

Let us consider $c(\ell_1,\ell_2,\ell_3) = e^{-\frac{i\pi}{4}\tau(\ell_1,\ell_2,\ell_3)}$ We will now show that there exists a function $s(\mathcal{l}_1,\mathcal{l}_2)$ defined on couples of oriented Lagrangian planes, invariant under the symplectic group, such that

$$c(\ell_1,\ell_2,\ell_3)^2 = s(\mathcal{l}_1,\mathcal{l}_2)^{-1}\, s(\mathcal{l}_2,\mathcal{l}_3)^{-1}\, s(\mathcal{l}_3,\mathcal{l}_1)^{-1}.$$

We will use this fact to prove that the Shale-Weil projective representation is a representation of the two-sheeted covering group G_2 of $G = Sp(B)$.

1.7.1. Definition. An oriented vector space of dimension n is a couple (L,e), where L is a real vector space of dimension n, and e an orientation of L, i.e. a connected component of $\Lambda^n L - \{0\}$.

If (L,e) is an oriented vector space, we define the oriented dual vector space $(L,e)^* = (L^*,e^*)$ of (L,e) by choosing e^* such that $\langle x,y\rangle > 0$ for $x \in e^*$, $y \in e$. We will sometimes write e for any element $x \in e \subset \Lambda^n V$, where there will be no confusion.

If (V_1,e_1) and (V_2,e_2) are two oriented vector spaces and A a linear invertible map from V_1 to V_2, we define the sign of the determinant of A denoted by $\varepsilon(A) = \pm 1$, by the condition

$$(\Lambda^n A)e_1 = c\,\varepsilon(A)e_2 \quad \text{with} \quad c > 0.$$

Let us remark that if (V_1,e_1), (V_2,e_2), (V_3,e_3) are three oriented vector spaces and $A_1: V_1 \to V_2$, $A_2: V_2 \to V_3$ are invertible linear maps, then $\mathcal{E}(A_2A_1) = \mathcal{E}(A_2)\mathcal{E}(A_1)$. Also $\mathcal{E}(A) = \mathcal{E}(A^{-1})$.

If $(V_2,e_2) = (V_1^*,e_1^*)$ and A is a linear invertible map from V_1 to V_1^*, $\mathcal{E}(A)$ is defined without ambiguity independently of the orientation of V_1, as it is easily seen by taking the opposite orientation in V_1.

Now if $A = A_3A_2A_1: V \to V^*$ with $A: V \xrightarrow{A_1} V_1 \xrightarrow{A_2} V_2 \xrightarrow{A_3} V^*$ then $\mathcal{E}(A)$ does not depend on the orientation on (V,V_1,V_2), if we choose on V^* the dual orientation of the one on V.

1.7.2. Let ℓ and m be two Lagrangian planes of the symplectic space (V,B). We define $g_{m,\ell}: \ell \to m^*$ by $\langle g_{m,\ell}(x),y\rangle = B(x,y)$. The kernel of $g_{m,\ell}$ is $\ell \cap m$, so if ℓ and m are transverse, $g_{m,\ell}$ is invertible.

Let (ℓ_1,e_1) and (ℓ_2,e_2) two oriented Lagrangian planes, which are transverse (i.e. $\ell_1 \cap \ell_2 = 0$), then $g_{\ell_2,\ell_1} = g_{2,1}: (\ell_1,e_1) \to (\ell_2,e_2)^*$ is invertible, and we define

1.7.3. $$\mathcal{E}((\ell_1,e_1),(\ell_2,e_2)) = \mathcal{E}(g_{2,1}).$$

This depends only on the relative orientation of (ℓ_1,e_1) and (ℓ_2,e_2). More generally if ℓ_1 and ℓ_2 are not transverse, we define $\mathcal{E}((\ell_1,e_1),(\ell_2,e_2))$ as follows: Let e be an orientation of $\rho = \ell_1 \cap \ell_2$. Then e defines an orientation $\tilde{e}_i$, $i = 1,2$ on

ℓ_i/ρ by $\tilde{e}_i \wedge e = e_i$; ℓ_1/ρ and ℓ_2/ρ are two transverse Lagrangian planes of $\ell_1 + \ell_2/\rho = \rho^{\perp}/\rho$. We define

$$\varepsilon((\ell_1,e_1),(\ell_2,e_2)) = \varepsilon((\ell_1/\rho,\tilde{e}_1),(\ell_2/\rho,\tilde{e}_2)).$$

It is easy to see that this does not depend on the choice of the orientation e of ρ, as if we change e to $-e$, both orientations $\tilde{e}_1,\tilde{e}_2$ change simultaneously.

If $\ell_1 = \ell_2$, we define

$$\begin{aligned}\varepsilon((\ell_1,e_1),\ (\ell_2,e_2)) &= 1 \ \text{ if } \ e_1 = e_2\\ &= -1 \ \text{ if } \ e_1 \neq e_2 ,\end{aligned}$$

which can be thought as a special case of the preceding formula.

We remark that:

$$\varepsilon((\ell_1,e_1),(\ell_2,e_2)) = (-1)^{n-\dim(\ell_1\cap\ell_2)}\varepsilon((\ell_2,e_2),(\ell_1,e_1))$$

as ${}^t g_{1,2} = -g_{2,1}$.

1.7.4. <u>Definition</u>. Let (ℓ_1,e_1) and (ℓ_2,e_2) be two oriented Lagrangian planes. We define:

$$s((\ell_1,e_1),(\ell_2,e_2)) = i^{(n-\dim(\ell_1\cap\ell_2))}\varepsilon((\ell_1,e_1),(\ell_2,e_2)).$$

Hence we have $s((\ell_1,e_1)),(\ell_2,e_2))\cdot s((\ell_2,e_2),(\ell_1,e_1)) = 1$.

Let Λ be the manifold of all Lagrangian planes of V and $\tilde{\Lambda}$ the manifold of oriented Lagrangian planes. The map $p: (\ell,e) \to \ell$, realizes $\tilde{\Lambda}$ as a two-sheeted covering of Λ. The symplectic group acts on the space of oriented Lagrangian planes. We will write $\tilde{\ell}$ for a Lagrangian oriented plane (ℓ, e). Clearly we have

1.7.5. $$s(g\tilde{\ell}_1, g\tilde{\ell}_2) = s(\tilde{\ell}_1, \tilde{\ell}_2).$$

Now we prove

1.7.6. <u>Theorem</u>. Let $\tilde{\ell}_1, \tilde{\ell}_2, \tilde{\ell}_3 \in \tilde{\Lambda}$, then

$$e^{\frac{i\pi}{2}\tau(p(\tilde{\ell}_1), p(\tilde{\ell}_2), p(\tilde{\ell}_3))} = s(\tilde{\ell}_1, \tilde{\ell}_2)\, s(\tilde{\ell}_2, \tilde{\ell}_3)\, s(\tilde{\ell}_3, \tilde{\ell}_1).$$

(We remark that the second member depends only on $p(\tilde{\ell}_1)$, $p(\tilde{\ell}_2)$ and $p(\tilde{\ell}_3)$.)

<u>Proof</u>: We will prove the theorem by induction on the dimension of V.

Let us first prove this theorem, when $\ell_i = p(\tilde{\ell}_i)$ are mutually transverse. We recall (1.5.4) that $\tau(\ell_1, \ell_2, \ell_3)$ is the signature of the quadratic form on ℓ_2 given by $Q'(x) = B(p_{13}x, p_{31}x)$. Let $S(x,y) = B(p_{13}x, p_{31}y)$ be the associate symmetric form. We define

$$a_{132}: \ell_2 \to \ell_3^* \to \ell_1 \to \ell_2^*$$

by $a_{132} = g_{21} \circ (g_{31})^{-1} g_{32}$. We see that $S(x,y) = \langle a_{132}(x), y\rangle$, since if $x = x_1 + x_3$, $x_1 \in \ell_1$, $x_3 \in \ell_3$:

$$S(x,y) = B(p_{13}x, y) = B(x_1, y),$$

$$g_{32}x = g_{31}x_1,$$

since for $u \in \ell_3$, $B(x,u) = B(x_1,u)$. Hence $(g_{31})^{-1}g_{32}x = x_1$ and $\langle a_{132}x,y\rangle = B(x_1,y)$. As sign $Q' = p - q = n - 2q$, where p is a number of positive signs in Q', and q the number of negative signs, we have:

$$e^{\frac{i\pi}{2}\tau(\ell_1,\ell_2,\ell_3)} = e^{\frac{i\pi}{2}\text{sign } Q'}, \quad \mathcal{E}(a_{123}) = (-1)^q.$$

Hence

$$\begin{aligned} e^{\frac{i\pi}{2}\tau(\ell_1,\ell_2,\ell_3)} &= i^{(n-2q)} = i^n(-1)^q = i^n\mathcal{E}(a_{132}) \\ &= i^n\mathcal{E}(g_{21})\mathcal{E}(g_{31})\mathcal{E}(g_{32}) \\ &= i^n\varepsilon(\tilde{\ell}_1,\tilde{\ell}_2)\varepsilon(\tilde{\ell}_2,\tilde{\ell}_3)(-1)^n\varepsilon(\tilde{\ell}_3,\tilde{\ell}_1) \\ &= s(\tilde{\ell}_1,\tilde{\ell}_2)\, s(\tilde{\ell}_2,\tilde{\ell}_3)\, s(\tilde{\ell}_3,\tilde{\ell}_1). \end{aligned}$$

Let us now suppose that $\ell_3 \cap \ell_2 = 0$ and $\ell_3 \cap \ell_1 = 0$. Let $\rho = \ell_1 \cap \ell_2$, then $\ell_3 \cap \rho = 0$. We consider the planes $\ell_1' = \ell_1/\rho$, $\ell_2' = \ell_2/\rho$, $\ell_3' = (\ell_3 \cap \rho^\perp) + \rho/\rho \simeq (\ell_3 \cap \rho^\perp)$.

As $\ell_1 \subset \rho^\perp$ and $\ell_1 + \ell_3 = V$ we see that $\rho^\perp = \ell_1 + (\ell_3 \cap \rho^\perp)$. Hence the three Lagrangian planes ℓ_1',ℓ_2',ℓ_3' are transverse in $\rho^\perp/\rho$.

Let us choose an orientation e on ρ and an orientation e_3' on ℓ_3'. We take the orientation e_1', e_2' on ℓ_1',ℓ_2' such that $e_i = e_i' \wedge e$, and consider the corresponding oriented planes $\tilde{\ell}_1', \tilde{\ell}_2', \tilde{\ell}_3'$.

We have $\tau(\ell_1,\ell_2,\ell_3) = \tau(\ell_1^\rho,\ell_2^\rho,\ell_3^\rho) = \tau(\ell_1',\ell_2',\ell_3')$. So

$$e^{\frac{i\pi}{2}\tau(\ell_1,\ell_2,\ell_3)} = s(\tilde{\ell}_1',\tilde{\ell}_2')\, s(\tilde{\ell}_2',\tilde{\ell}_3')\, s(\tilde{\ell}_3',\tilde{\ell}_1').$$

We have $s(\tilde{\ell}_1', \tilde{\ell}_2') = s(\tilde{\ell}_1, \tilde{\ell}_2)$, as the dimension of a Lagrangian plane in $\rho^{\perp}/\rho$ is $n - \dim \rho = n'$. We have only to show now that $s(\ell_2', \ell_3')\, s(\ell_3', \ell_1') = s(\ell_2, \ell_3)\, s(\ell_3, \ell_1)$.

We consider the maps $g_{\ell_2,\ell_3}, g_{\ell_1,\ell_3}$ which are invertible, as $\ell_3 \cap \ell_2,\ \ell_3 \cap \ell_1 = 0$. Let us form

$$F = (g_{\ell_3,\ell_1})^{-1}\, g_{\ell_3,\ell_2} : \ell_2 \longrightarrow \ell_3^* \longrightarrow \ell_1$$

$$F' = (g_{\ell_3',\ell_1'})^{-1}\, g_{\ell_3',\ell_2'} : \ell_2' \longrightarrow (\ell_3')^* \longrightarrow \ell_1' .$$

Clearly F is the identity on $\rho = \ell_1 \cap \ell_2$. We consider the diagram

$$\begin{array}{ccccc} \ell_2 & \xrightarrow{\ g_{\ell_3,\ell_2}\ } & \ell_3^* & \xrightarrow{\ (g_{\ell_3,\ell_1})^{-1}\ } & \ell_1 \\ \downarrow p & & \downarrow r & & \downarrow p \\ \ell_2/\rho & \xrightarrow{\ g_{\ell_3',\ell_2'}\ } & (\ell_3 \cap \rho^{\perp})^* & \xrightarrow{\ (g_{\ell_3',\ell_1'})^{-1}\ } & \ell_1/\rho \end{array}$$

where p is the canonical projection, and r is the restriction map from ℓ_3^* to $(\ell_3 \cap \rho^{\perp})^*$. This diagram is commutative (for $x \in \ell_2$, $y \in \ell_3 \cap \rho^{\perp}$, $\langle g_{\ell_2,\ell_3}(x), y\rangle = B(x,y) = B(px,y)$). Hence F' is the map derived from F by quotienting by ρ; as F is the identity on ρ, we see that $\mathcal{E}(F) = \mathcal{E}(F')$, where the sign of F, F' are relative to the orientations (e_2, e_1), (e_2', e_1'). Hence we have:

$$\mathcal{E}(F) = \mathcal{E}(\tilde{\ell}_2, \tilde{\ell}_3)\, \mathcal{E}(\tilde{\ell}_1, \tilde{\ell}_3) = (-1)^n\, \mathcal{E}(\tilde{\ell}_2, \tilde{\ell}_3)\, \mathcal{E}(\tilde{\ell}_3, \tilde{\ell}_1)$$

$$= s(\tilde{\ell}_2, \tilde{\ell}_3)\, s(\tilde{\ell}_3, \tilde{\ell}_1).$$

Similarly $\mathcal{E}(F') = s(\tilde{\ell}_2',\tilde{\ell}_3')\cdot s(\tilde{\ell}_3',\tilde{\ell}_1')$, and we obtain the desired equality.

Now if $\tilde{\ell}_1,\tilde{\ell}_2,\tilde{\ell}_3$ are arbitrary, we choose a Lagrangian plane $\tilde{m}$ transverse to $\tilde{\ell}_1,\tilde{\ell}_2,\tilde{\ell}_3$. We have

$$\begin{aligned}\tau(p(\tilde{\ell}_1),p(\tilde{\ell}_2),p(\tilde{\ell}_3)) &= \tau(p(\tilde{m}),p(\tilde{\ell}_1),p(\tilde{\ell}_2)) \\ &+ \tau(p(\tilde{m}),p(\tilde{\ell}_2),p(\tilde{\ell}_3)) \\ &+ \tau(p(\tilde{m}),p(\tilde{\ell}_3),p(\tilde{\ell}_1)),\end{aligned}$$

and

$$\begin{aligned}e^{\frac{i\pi}{2}\tau(\ell_1,\ell_2,\ell_3)} &= s(\tilde{m},\tilde{\ell}_1)\ s(\tilde{\ell}_1,\tilde{\ell}_2)\ s(\tilde{\ell}_2,\tilde{m})\ s(\tilde{m},\tilde{\ell}_2)\ s(\tilde{\ell}_2,\tilde{\ell}_3) \\ &\cdot s(\tilde{\ell}_3,\tilde{m})\ s(\tilde{m},\tilde{\ell}_3)\ s(\tilde{\ell}_3,\tilde{\ell}_1)\ s(\tilde{\ell}_1,\tilde{m}) \\ &= s(\tilde{\ell}_1,\tilde{\ell}_2)\ s(\tilde{\ell}_2,\tilde{\ell}_3)\ s(\tilde{\ell}_3,\tilde{\ell}_1)\end{aligned}$$

as $s(\tilde{\ell}_i,\tilde{m})\ s(\tilde{m},\tilde{\ell}_i) = 1$ for $i = 1,2,3$.

1.7.7. Let us consider the projective representation $g \to R_\ell(g)$ of the symplectic group G and its associated cocycle:

$$c_\ell(g_1,g_2) = e^{-\frac{i\pi}{4}\tau(\ell,g_1\ell,g_1g_2\ell)}.$$

Let us choose an orientation ℓ^+ on ℓ. The group G acts on oriented Lagrangian planes: we define $s_\ell(g) = s(\ell^+,g\cdot\ell^+)$ ($s_\ell(g)$ does not depend on the choice of the orientation $\ell^{\pm}$ on ℓ). Theorem 1.7.6 is equivalent to the formula:

1.7.8. $$c^2(g_1,g_2) = s(g_1)^{-1}\ s(g_2)^{-1}\ s(g_1g_2).$$

1.7.9. A projective representation P of a group G with

cocycle p is equivalent to a "true" representation if we can modify the operators $P(g)$ as follows: $P'(g) = \alpha(g)P(g)$ with $\alpha(g) \in T$ such that indeed the operators $P'(g)$ now satisfy the relation: $P'(g_1g_2) = P'(g_1)\, P'(g_2)$, i.e.

$$p(g_1,g_2) = \frac{\alpha(g_1)\alpha(g_2)}{\alpha(g_1g_2)} .$$

The formula 1.7.8 can be written in a symbolic way

$$c(g_1,g_2) = \frac{s(g_1)^{-1/2}\, s(g_2)^{-1/2}}{s(g_1g_2)^{-1/2}} .$$

Hence our projective representation R can be made into a "true" representation of a two-fold covering group G_2 of G by choosing $R_2(g) = s(g)^{"-1/2"}R(g)$.

Let us now make precise definitions. We consider the Mackey group G_c(1.6.12) and the subset $G_2 \subset G_c = G \times T$ defined by

1.7.10. $$G_2 = \{(g,t), \text{ with } t^2 = s(g)^{-1}\} .$$

The formula 1.7.8 implies that G_2 is a subgroup of G_c. Hence the representation $\tilde{R}|G_2$, $\tilde{R}(g,t) = t\, R(g)$ is a "true" representation of G_2.

It is clear that the map $G_2 \to G$ defined by $(g,t) \to g$ is a homomorphism from the group G_2 to G, and each fiber consists of two points. Hence we have lifted R to a true representation $\tilde{R}$ of a double covering G_2 of G. The group G_2 is called the metaplectic group.

1.7.11. Let us consider the group $\tilde{G}_\ell = G \times \mathbb{Z}$ and the function $s(g,n) = e^{\frac{i\pi}{2}n} s(g)$ with values in $\mathbb{Z}/4\mathbb{Z}$.

1.7.12. Lemma: $s(g,n)$ is a character of the group $\tilde{G}_\ell$.

Proof: This is equivalent to the relation 1.7.8.

1.8. The universal covering group of $SL(2,\mathbb{R})$.

In this section, we will describe the universal covering group of $SL(2,\mathbb{R})$ using the Maslov index. This construction is valid in the general case, and we will describe in detail the universal covering of the manifold of Lagrangian planes and the universal covering group of $Sp(n,\mathbb{R})$ in Appendix A.

However, in the applications to theta series of Chapter II, we will use only a small part of this general description. Therefore we will prove in this section only the results that we need.

1.8.1. Let $V = \mathbb{R}P \oplus \mathbb{R}Q$ be the two-dimensional canonical symplectic space. We identify $\mathbb{R}P \oplus \mathbb{R}Q$ to $\mathbb{C}$ by $xP + yQ \to x + iy$. The symplectic form B gives the natural orientation on $\mathbb{C}$.

A Lagrangian plane is therefore defined by its angle θ mod π with the real axis.

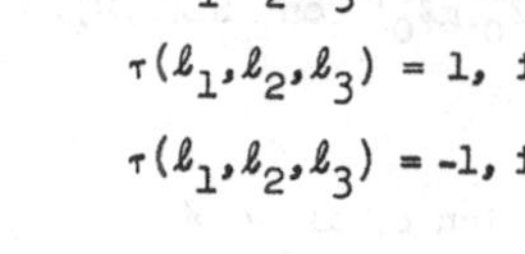

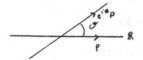

Hence we identify Λ with the torus T by: if $\lambda_\theta = \mathbb{R}e^{i\theta}P$, $u(\lambda_\theta) = e^{2i\theta} \in T$.

Let (ℓ_1,ℓ_2,ℓ_3) be 3 lines. We have:

$\tau(\ell_1,\ell_2,\ell_3) = 0$, if ℓ_1,ℓ_2,ℓ_3 are not all distincts,

$\tau(\ell_1,\ell_2,\ell_3) = 1$, if ℓ_2 is in the interior of the angle ℓ_1,ℓ_3

$\tau(\ell_1,\ell_2,\ell_3) = -1$, if ℓ_2 is in the exterior of the angle ℓ_1,ℓ_3

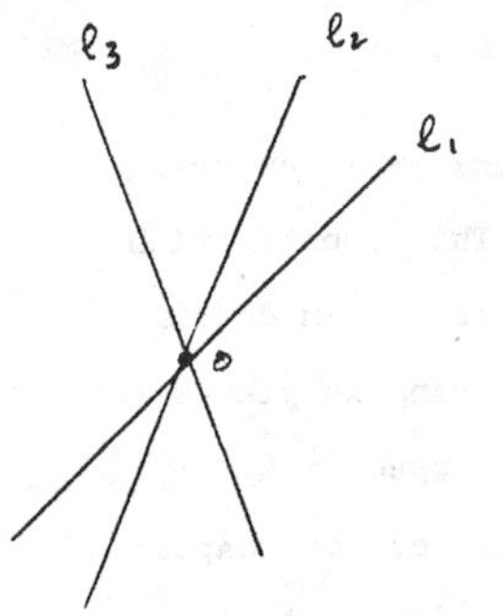

In our preceding identification of Λ with T, $\tau(\ell_1,\ell_2,\ell_3) = 1$ if u_2 is between the points u_1,u_3 when we move around the circle in the canonical way, $\tau(\ell_1,\ell_2,\ell_3) = -1$ if u_2 is between u_3 and u_1.

Let $G = SL(2,\mathbb{R})$. As in 1.6.14, we define the group $\widetilde{G}_\ell$. As a set $\widetilde{G}_\ell = G \times \mathbb{Z}$, the multiplicative law being given by:

$$(g_1,n_1)\cdot(g_2,n_2) = (g_1g_2, n_1+n_2+\tau(\ell, g_1\ell, g_1g_2\ell)) .$$

1.8.2. We consider the oriented plane $\ell_0^+ = \vec{\mathbb{R}P}$.

1.8.3. We define $B_0 = \{g \in G,\ g\ell_0^+ = \ell_0^+\}$. Clearly

$$B_0 = \left\{ \begin{pmatrix} a & b \\ 0 & a^{-1} \end{pmatrix};\ a > 0,\ b \in \mathbb{R} \right\}.$$

1.8.4. We consider the function $s(g) = s(\ell_0^+, g\ell_0^+)$ on $SL(2;\mathbb{R})$. We have:

$$s\begin{pmatrix} a & b \\ 0 & a^{-1} \end{pmatrix} = \text{sign } a\ , \quad s\begin{pmatrix} a & b \\ c & d \end{pmatrix} = i \text{ sign } c, \text{ if } c \neq 0.$$

The function $\tilde{s}(g,n) = e^{\frac{i\pi}{2}n}\, s(g)$ is a character of the group $\widetilde{G}$. We define $\widetilde{G}_0 = \operatorname{Ker} \tilde{s}$. In particular $(e,n) \in \widetilde{G}_0$ is equivalent to $n \in 4\mathbb{Z}$.

1.8.5. We consider the subgroup K of $SL(2;\mathbb{R})$ given by the transformations $u(\theta)z = e^{i\theta}z$ of $\mathbb{C}$, i.e. $u(\theta)$ in real coordinates is the rotation:

$$u(\theta) = \begin{pmatrix} \cos\theta & -\sin\theta \\ \sin\theta & \cos\theta \end{pmatrix}.$$

The map $\theta \to u(\theta)$ is a homomorphism of $\mathbb{R}$ into $SL(2,\mathbb{R})$. As the subgroup of rotations acts simply transitively on the oriented lines $\vec{\ell}$, we see that:

1.8.6. Each element g of $SL(2;\mathbb{R})$ can be written uniquely as $g = u(\theta)b$ with $b \in B_0$,

$$\text{i.e.} \qquad g = \begin{pmatrix} \cos\theta & -\sin\theta \\ \sin\theta & \cos\theta \end{pmatrix} \begin{pmatrix} a & x \\ 0 & a^{-1} \end{pmatrix}, \quad a > 0 .$$

Hence $SL(2,\mathbb{R})$ as a topological space is isomorphic to $T \times \mathbb{R}^+ \times \mathbb{R}$.

1.8.7. We define the function $\mu\colon \mathbb{R} \to \mathbb{Z}$ by

$$\mu(\theta) = 2k, \quad \text{if} \quad \theta = k\pi$$

$$\mu(\theta) = 2k+1 \quad \text{if} \quad k\pi < \theta < (k+1)\pi$$

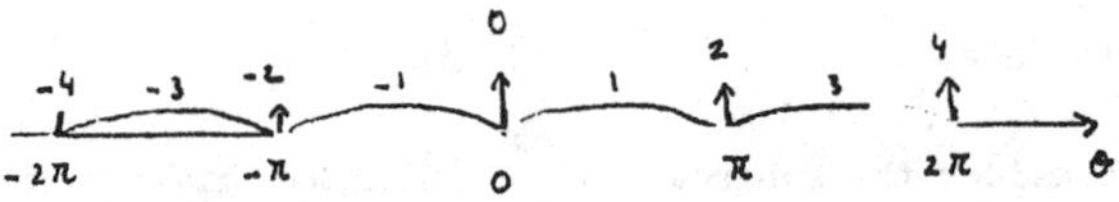

We have: $\mu(-\theta) = -\mu(\theta)$

$$\mu(\theta+k\pi) = \mu(\theta) + 2k \ .$$

1.8.8. <u>Proposition</u>: The map $\theta \to g(\theta) = (u(\theta),-\mu(\theta))$ is an injective homomorphism of $\mathbb{R}$ into $\widetilde{G}_0$.

<u>Proof</u>: We have to prove:

$$\tau(P,e^{i\theta_1}P,e^{i(\theta_1+\theta_2)}P) = \mu(\theta_1) + \mu(\theta_2) - \mu(\theta_1+\theta_2).$$

From the relations

$$\mu(\theta_1+k_1\pi) = \mu(\theta_1) + 2k_1$$

$$\mu(\theta_2+k_2\pi) = \mu(\theta_2) + 2k_2$$

we can assume that $0 \leq \theta_1 < \pi$ and $0 \leq \theta_2 < \pi$. We consider separately the cases:

a) θ_1 or $\theta_2 = 0$

b) $0 < \theta_1 < \pi$ $\quad$ $0 < \theta_1 + \theta_2 < \pi$

$0 < \theta_2 < \pi$

c) $0 < \theta_1 < \pi$ $\quad \theta_1 + \theta_2 = \pi$

$0 < \theta_2 < \pi$

d) $0 < \theta_1 < \pi$ $\quad \pi < \theta_1 + \theta_2 < 2\pi$

$0 < \theta_2 < \pi$

and the relation is easily verified; for example in case b) $\mu(\theta_1) = \mu(\theta_2) = 1 = \mu(\theta_1+\theta_2)$, and $\tau(P, e^{i\theta_1}P, e^{i(\theta_1+\theta_2)}P) = 1$

Now we have:

$$s(u(\theta)) = (-1)^k \quad \text{if} \quad \theta = k\pi$$

$$s(u(\theta)) = i(-1)^k \quad \text{if} \quad k\pi < \theta < (k+1)\pi.$$

Hence $s(u(\theta)) = e^{\frac{i\pi}{2}\mu(\theta)}$ and $g(\theta) = (u(\theta), -\mu(\theta))$ belongs to $\tilde{G}_0$.

1.8.9. For $b_0 \in B_0$, we define $g(b_0) = (b_0, 0)$. As B_0 leaves stable ℓ_0^+, it is clear that the map $b_0 \to (b_0, 0)$ is a homomorphism of B_0 onto its image in $\tilde{G}_0$.

We now are ready to prove:

1.8.10. Proposition: Each element $\tilde{g}$ of $\tilde{G}_0$ is written uniquely as $g(b_0)g(\theta)$ for $\theta \in \mathbb{R}$, $b_0 \in B_0$.

Proof: As each element of G can be written as $b_0 u(\theta)$, we may assume $\pi(\hat{g}) = e$; hence $g = (e, 4n) = g(2n\pi)$. That the decomposition is unique follows from 1.8.8, 1.8.6.

Remark: We will see in Appendix A that we can define a "natural" topology on $\widehat{G}$, using the Maslov index, which makes it into a Lie group. For this topology the map $\pi: (g,n) \to g$ is a covering map, and the map from $\mathbb{R} \times B_0$ into $\widehat{G}_0$ given by $g(b_0)\, g(\theta)$ is a diffeomorphism. Hence $\widehat{G}_0$ is the universal covering group of G.

1.8.11. We will now give another realization of the universal covering group of $G = SL(2,\mathbb{R})$, using the description of $SL(2,\mathbb{R})$ as a group of automorphisms of the upper half-plane.

1.8.12. Let $P^+ = \{z = x + iy;\ z \in \mathbb{C},\ y > 0\}$ be the upper half-plane. The group $SL(2,\mathbb{R})$ acts on P^+ by

$$g = \begin{pmatrix} a & b \\ c & d \end{pmatrix},\ g\cdot z = \frac{az+b}{cz+d},$$

as

$$\mathrm{Im}(g\cdot z) = \mathrm{Im}\left(\frac{az+b}{cz+d}\right) = \frac{1}{2i}\left(\frac{az+b}{cz+d} - \frac{a\bar{z}+b}{c\bar{z}+d}\right)$$

$$= \frac{1}{2i}\,\frac{(az+b)(c\bar{z}+d) - (a\bar{z}+b)(cz+d)}{(cz+d)(c\bar{z}+d)}$$

$$= \frac{1}{2i}\,\frac{z-\bar{z}}{|cz+d|^2}, \text{ i.e.}$$

1.8.13. $$\mathrm{Im}(g\cdot z) = |cz+d|^{-2}\,(\mathrm{Im}\, z).$$

In particular $\mathrm{Im}(g\cdot z) > 0$ if $\mathrm{Im}\, z > 0$.

The map $g \to \Phi_g$, where $\Phi_g(z) = (az+b)(cz+d)^{-1}$ is hence an homomorphism of G into the group of invertible holomorphic transformation of P^+.

(Every biholomorphic transformation of the domain P^+ is a fractional linear transformation of the form $(az+b)(cz+d)^{-1}$ with $\begin{pmatrix} a & b \\ c & d \end{pmatrix} \in SL(2;\mathbb{R})$.)

1.8.14. For $g = \begin{pmatrix} a & b \\ c & d \end{pmatrix}$ and $z \in P^+$, we define $j(g,z) = cz+d$. It is easy to see that $j(g,z) \neq 0$ on P^+ and

1.8.15. $$j(g_1 g_2, z) = j(g_1, g_2 \cdot z) j(g_2, z) .$$

Let us consider $z_0 = i$ as basis point of P^+. We denote by

$$K = \{u(\theta) = \begin{pmatrix} \cos\theta & -\sin\theta \\ \sin\theta & \cos\theta \end{pmatrix}\} .$$

Then K is the stabilizer of the point z_0.

1.8.16. <u>Lemma</u>: The group B_0 acts simply transitively on P^+.

<u>Proof</u>: For $z \in P^+, z = x + iy$, there exist a unique element $b(z) = \begin{pmatrix} a & u \\ 0 & a^{-1} \end{pmatrix}$ of B_0 such that $b(z).i = z = a^2 i + au$, namely

1.8.17.

$$b(z) = \begin{pmatrix} y^{1/2} & y^{-1/2}x \\ 0 & y^{1/2} \end{pmatrix} = \begin{pmatrix} 1 & x \\ 0 & 1 \end{pmatrix}\begin{pmatrix} y^{1/2} & 0 \\ 0 & y^{-1/2} \end{pmatrix} .$$

1.8.18. We will now describe the universal covering group of $SL(2,\mathbb{R})$ using the function j. For $g \in G$ and $z \in P^+$, $j(g,z) = cz + d$ is never zero, hence we can find a determination

φ of $\log(cz+d)$ on the simply connected domain P^+. Let φ be an holomorphic function on P^+ such that $e^{\varphi(z)} = cz + d$, then φ is entirely determined by its value at the point $z = i$ of P^+. Two such determinations φ or φ' of $\log(cz+d)$ differs by $2ik\pi$.

1.8.19. We consider the following group

$$\mathcal{G} = \{(g,\varphi_g);\ e^{\varphi_g(z)} = j(g,z)\}.$$

The multiplicative law being given by

$$(g_1,\varphi_{g_1})\cdot(g_2,\varphi_{g_2}) = (g_1g_2,\varphi') ,$$

where

$$\varphi'(z) = \varphi_{g_1}(g_2\cdot z) + \varphi_{g_2}(z).$$

(We remark that the relation $j(g_1g_2,z) = j(g_1,g_2\cdot z)j(g_2,z)$ implies that $e^{\varphi'(z)} = j(g_1g_2,z)$). We have $(g,\varphi)^{-1} = (g^{-1},-\varphi(g^{-1}z))$.

Let us consider the projection $\mathcal{G} \xrightarrow{\pi} G$ given by $(g,\varphi) \to g$. It is clear that the fiber of this map is isomorphic to $\mathbb{Z}$.

Let $z \neq 0$, $z \in \mathbb{C}$, then z can be written in a unique way $z = e^{x+iy} = e^x e^{iy}$ with $x \in \mathbb{R}$, and y being defined mod 2π. We have then $\log z = x + iy$. Hence to choose $y = \operatorname{Im}\log z$ is equivalent to choose a determination of $\arg z$. We denote by $\operatorname{Arg} z$ the principal determination of $\arg z$, i.e. $-\pi < \operatorname{Arg} z \leq \pi$, and $\operatorname{Log} z = x + i \operatorname{Arg} z$ the corresponding principal determination.

For each element $g \in G$, we denote by $\widetilde{g}$ the particular element of $\mathcal{G}$ above g such that, if $\tilde{g} = (g,\varphi)$, $\varphi(i) = \text{Log}(ci+d)$.

1.8.20. Let $b \in B_0$, i.e. $b = \begin{pmatrix} a & x \\ 0 & a^{-1}\end{pmatrix}$ with $a > 0$. It is clear that $\delta(b) = (b, \text{Log } a^{-1}) = \tilde{b}$ (with $\text{Log } a \in \mathbb{R}$) is an element of $\mathcal{G}$, moreover the map $b \to \delta(b)$ is a group isomorphism of B_0 into its image in $\mathcal{G}$.

Let $u(\theta) \in K$, i.e. $u(\theta) = \begin{pmatrix} \cos\theta & -\sin\theta \\ \sin\theta & \cos\theta \end{pmatrix}$. Then $u(\theta)\cdot i = i$ and $j(u(\theta),i) = e^{i\theta}$. We define the map $\delta\colon \mathbb{R} \to \mathcal{G}$ by $\delta(\theta) = (u(\theta),\varphi_\theta)$, where $\varphi_\theta(i) = i\theta$. It is immediate to verify that δ is an homomorphism of $\mathbb{R}$ into $\mathcal{G}$.

1.8.21. <u>Lemma</u>: Each element of $\mathcal{G}$ can be written uniquely as $\delta(b)\ \delta(\theta)$ with $\theta \in \mathbb{R}$, $b \in B_0$.

<u>Proof</u>: This is immediate.

Let μ be the function defined in 1.8.7.

1.8.22. <u>Proposition</u>: The map $I\colon \mathcal{G} \to \widehat{G}_0$ defined by $I(g,\varphi) = (g, -\mu(\text{Im } \varphi(i)))$ is a group isomorphism from $\mathcal{G}$ to $\widehat{G}_0$.

<u>Proof</u>: Let $(g,\varphi) = \delta(b)\delta(\theta)$ with $b = \begin{pmatrix} a & x \\ 0 & a^{-1}\end{pmatrix}$, $(a > 0)$, an element of $\mathcal{G}$. As $u(\theta)\cdot i = i$, $\varphi(i) = \text{Log } a^{-1} + i\theta$; hence $\text{Im } \varphi(i) = \theta$ and $I(g,\varphi) = g(b)g(\theta)$. So it follows from 1.8.10, 1.8.21 that I is a bijection. We have to prove that I is a group isomorphism. Let us remark first that the function $(g,\varphi) \to \mu(\text{Im } \varphi(i))$ on the group $\mathcal{G}$ is invariant by left and right translations by elements $g(b)$, with $b \in B_0$. For left

translation, this is clear, as if $g(b)(g,\varphi) = (g',\varphi')$, $\varphi'(i) = \mathrm{Log}\ a^{-1} + \varphi(i)$ and $\mathrm{Log}\ a^{-1} \in \mathbb{R}$. For right translations, for $(g',\varphi') = (g,\varphi)\ g(b)$, we have

$$\varphi'(i) = \varphi(b\cdot i) + \mathrm{Log}\ a^{-1}.$$

But for z varying in the upper half space,

$cz + d$ stays in the upper half space if $c > 0$

$cz + d$ stays in the lower half space if $c < 0$

$cz + d$ is constant if $c = 0$.

Hence when we follow $\arg(cz+d) = \theta'$ by continuity from the value $\theta_0 = \arg(ci+d)$, the elements θ and θ' stays in the same open interval $]k\pi,(k+1)\pi[$ or stay equal if $\theta = k\pi$. Hence $\mu(\theta) = \mu(\theta')$.

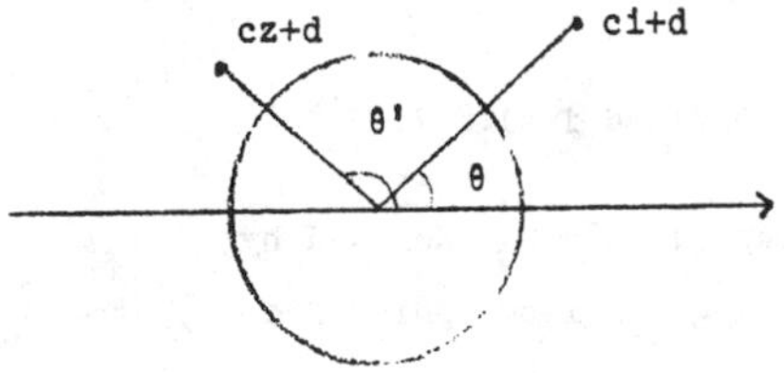

Now we have to prove for $(g_1,\varphi_1)\cdot(g_2,\varphi_2) = (g,\varphi)$, $\mu(\mathrm{Im}(\varphi_1(i)) + \mu(\mathrm{Im}(\varphi_2(i))) - \mu(\mathrm{Im}\ \varphi(i)) = \tau(\ell, g_1\ell, g_1g_2\ell)$. Translating (g_1,φ_1) by an element $g(b_1)$ from the left side, (g_2,φ_2) from an element $g(b_2)$ from the right side, we can assume $(g_1,\varphi_1) = \delta(\theta_1)$, $(g_2,\varphi_2) = \delta(\theta_2)$. The equality to be proven is then $\mu(\theta_1) + \mu(\theta_2) - \mu(\theta_1+\theta_2) = \tau(\ell, e^{i\theta_1}\ell, e^{i(\theta_1+\theta_2)}\ell)$ which follows from 1.8.8.

1.8.23. We recall that we have defined, through the principal determination of $\log(ci+d)$ a particular section $g \to \hat{g}$ of G into $\mathcal{G}$.

1.8.24. Lemma: $I(\hat{g}) = (g,m)$ where

if $g = \begin{pmatrix} a & b \\ c & d \end{pmatrix}$ with $c > 0$ $m = -1$

$g = \begin{pmatrix} a & b \\ c & d \end{pmatrix}$ with $c < 0$ $m = 1$

$g = \begin{pmatrix} a & b \\ 0 & a^{-1} \end{pmatrix}$ with $a > 0$ $m = 0$

$g = \begin{pmatrix} a & b \\ 0 & a^{-1} \end{pmatrix}$ with $a < 0$ $m = -2$.

1.8.25. We define the metaplectic group G_2 as being given by $G_2 = \{(g,d)$, where d is a holomorphic function on P^+ such that $d(z)^2 = cz + d\}$. The composition law being given by: $(g_1,d_1)\cdot(g_2,d_2) = (g_1g_2,d)$, where $d(z) = d_1(g_2z)d_2(z)$.

The map $(g,d) \to g$ from G_2 to $G = SL(2,\mathbb{R})$ is then a covering map with fiber $(g,\pm id)$. The map $(g,\varphi) \to (g,e^{\varphi(z)/2})$ from $\mathcal{G}$ to G_2 is a homomorphism from $\mathcal{G}$ to G_2 with kernel $\{\delta(4k\pi)\}$.

1.9. The universal covering group of the symplectic group.

Let Λ be the manifold of all Lagrangian planes. The group $G = Sp(B)$ acts on Λ. We have constructed in 1.5 a $\mathbb{Z}$-valued function $\tau(\ell_1,\ell_2,\ell_3)$ on triples of Lagrangian planes. This function τ is invariant under the action of G and satisfies the chain condition (1.5.8)

$$\tau(\ell_1,\ell_2,\ell_3) = \tau(\ell_1,\ell_2,\ell_4) + \tau(\ell_2,\ell_3,\ell_4) + \tau(\ell_3,\ell_1,\ell_4).$$

Let ℓ_0 be a fixed element of Λ. The function τ leads naturally to a $\mathbb{Z}$-valued cocycle of the group G by $(g_1,g_2) \to \tau(\ell_0,g_1\ell_0,g_1g_2\ell_0)$. We denote by $\tilde{G}_0 = \tilde{G}_{\ell_0}$ the corresponding extension of G (1.6.14).

Let $\tilde{\Lambda} = \Lambda \times \mathbb{Z} = \{(\ell,\mu);\ \ell \in \Lambda,\ \mu \in \mathbb{Z}\}$. The formula:

$$(g,n)\cdot(\ell,\mu) = (g\cdot\ell,\ n + \mu + \tau(\ell_0,g\cdot\ell_0,g\cdot\ell))$$

defines an action of $\tilde{G}_0$ on $\tilde{\Lambda}$.

Let us consider the function:

$$m_0((\ell_1,u_1),(\ell_2,u_2)) = u_1 - u_2 + \tau(\ell_0,\ell_1,\ell_2).$$

The chain property of τ and its G-invariance implies that m_0 is a $\tilde{G}_0$-invariant function on $\tilde{\Lambda} \times \tilde{\Lambda}$. Thus, the formal properties of τ allows us to construct a covering manifold $\tilde{\Lambda}$ of Λ, a covering group $\tilde{G}_0$ of G and a $\tilde{G}_0$-invariant function on $\tilde{\Lambda} \times \tilde{\Lambda}$ such that

$$\tau(\ell_1,\ell_2,\ell_3) = m(\tilde{\ell}_1,\tilde{\ell}_2) + m(\tilde{\ell}_2,\tilde{\ell}_3) + m(\tilde{\ell}_3,\tilde{\ell}_1).$$

This function m_0, immediately deduced from τ, is usually called the Maslov index. In this section, we relate $(\tilde{\Lambda},\tilde{G}_0)$ to the universal coverings of (Λ,G) and calculate "in coordinates" the function m_0.

1.9.1. We consider the manifold Λ of all Lagrangian planes of (V,B) as a closed submanifold of the Grassmanian of all the n-dimensional planes.

Let (ℓ_0,ℓ_1) be two given transverse Lagrangian planes. Let Λ_ℓ be the open subset of Λ of all the Lagrangian planes m transverses to ℓ. We parametrize Λ_{ℓ_0} as follows: If $m \in \Lambda_{\ell_0}$, there exists a map $\alpha\colon \ell_1 \to \ell_0$ such that $m = \{x + \alpha x;\ x \in \ell_1\}$. The condition for m to be Lagrangian becomes $B(\alpha x,x') = B(\alpha x',x)$, i.e. the bilinear form on ℓ_1 given by $S_\alpha(x,x') \to B(\alpha x,x')$ is symmetric. We have hence identified Λ_{ℓ_0} to the vector space of all symmetric bilinear forms on ℓ_1. In particular Λ_{ℓ_0} is a simply connected neighborhood of ℓ_1.

We note here that $\Lambda_{\ell_0} \cap \Lambda_{\ell_1}$ is then identified with the symmetric forms S_α, such that $\det S_\alpha \neq 0$, thus $\Lambda_{\ell_0} \cap \Lambda_{\ell_1}$ has $n + 1$ connected components, corresponding to the non-degenerate symmetric forms of signature (p,q).

Let (P_i,Q_i) be a symplectic basis of V, and for $k = 0, \cdots, n$

$$\ell_0 = \mathbb{R}P_1 \oplus \cdots \oplus \mathbb{R}P_n$$

$$\ell_1 = \mathbb{R}Q_1 \oplus \cdots \oplus \mathbb{R}Q_n$$

$$\ell_k = \sum_{i=1}^{n} \mathbb{R}(P_i + \varepsilon_i Q_i) \qquad \varepsilon_i = +1 \text{ if } i \leq k$$
$$= -1 \text{ if } i > k .$$

Then each connected component of $\Lambda_{\ell_0} \cap \Lambda_{\ell_1}$ contains one ℓ_k. We parametrize Λ_{ℓ_0} by the matrices $y: \ell_1 \to \ell_0$ symmetric with respect to the basis $(Q_1 \cdots Q_n, P_1 \cdots P_n)$, i.e. we write an element m of Λ_{ℓ_0} as $m = \{x + yx;\ x \in \ell_1\}$.

1.9.2. Lemma: Let $U = \{(m,m') \in \Lambda_{\ell_0} \times \Lambda_{\ell_0};\ m \cap m' = 0\}$ and let U_k be the connected component of U containing (ℓ_1,ℓ_k); then $U = \cup U_k$.

Proof: Let $(m,m') \in U$. We then have to show that there exists a continuous path $(m(t),m'(t))$ contained in U and k, $0 \leq k \leq n$, with $m(0) = \ell_1$, $m'(0) = \ell_k$, $m(1) = m$, $m'(1) = m'$. We write

$$m = \{x + yx;\ x \in \ell_1;\ y: \ell_1 \to \ell_0\}$$

$$m' = \{x + y'x;\ x \in \ell_1;\ y': \ell_1 \to \ell_0\}.$$

The condition $m \cap m' = 0$ is clearly that $(y - y')$ is invertible. Therefore, for any symmetric matrix u the translation $y \to y + u$ on Λ_{ℓ_0} conserves the couples of transverse planes. We can suppose then that $y = 0$, $\det(y') \neq 0$; By a preceding remark, we can then deform (ℓ_1,ℓ') to (ℓ_1,ℓ_k)

for some k.

1.9.3. Proposition:

a) When (ℓ_1,ℓ_2,ℓ_3) moves continuously in such a manner that $\dim(\ell_1 \cap \ell_2)$, $\dim(\ell_2 \cap \ell_3)$, $\dim(\ell_3 \cap \ell_1)$ remains constant, then $\tau(\ell_1,\ell_2,\ell_3)$ remains constant.

b) $\tau(\ell_1,\ell_2,\ell_3) \equiv n + \dim(\ell_1 \cap \ell_2) + \dim(\ell_2 \cap \ell_3) + \dim(\ell_3 \cap \ell_1)$ modulo 2.

Proof: It is enough to show that the rank of the quadratic form $Q(x_1,x_2,x_3) = B(x_1,x_2) + B(x_2,x_3) + B(x_3,x_1)$ does not change when the ℓ_i's move continuously.

Let us compute the kernel I of Q; we have $(x_1,x_2,x_3) \in I$, if and only if $B(x_1,y_2) + B(y_1,x_2) + B(x_2,y_3) + B(y_2,x_3) + B(x_3,y_1) + B(y_3,x_1) = 0$ for every $y_1,y_2,y_3 \in \ell_1,\ell_2,\ell_3$. But $B(x_1 - x_3,y_2) + B(x_2 - x_1,y_3) + B(x_3 - x_2,y_1) = 0$ for any $y_1 \in \ell_1$, $y_2 \in \ell_2$, $y_3 \in \ell_3$ implies $x_1 - x_3 \in \ell_2$, $x_2 - x_1 \in \ell_3$, $x_3 - x_2 \in \ell_1$.

Let us consider the change of variable $y_1 = x_2 + x_3 - x_1$, $y_2 = x_3 + x_1 - x_2$, $y_3 = x_1 + x_2 - x_3$, then we have $y_1 \in \ell_2 \cap \ell_3$, $y_2 \in \ell_3 \cap \ell_1$, $y_3 \in \ell_1 \cap \ell_2$, and $x_1 = \frac{y_2 + y_3}{2}$, $x_2 = \frac{y_3 + y_1}{2}$, $x_3 = \frac{y_1 + y_2}{2}$; hence by this transformation, the kernel is isomorphic to $(\ell_1 \cap \ell_2) \oplus (\ell_2 \cap \ell_3) \oplus (\ell_3 \cap \ell_1)$ and this proves the assertion a).

For b) we have $\tau(\ell_1,\ell_2,\ell_3) = p - q$, where $p + q$ = rank of $Q = 3n - \dim(\ell_1 \cap \ell_2) - \dim(\ell_2 \cap \ell_3) - \dim(\ell_3 \cap \ell_1)$, so

$\tau(\ell_1,\ell_2,\ell_3) \equiv n + \dim(\ell_1 \cap \ell_2) + \dim(\ell_2 \cap \ell_3) + \dim(\ell_3 \cap \ell_1)$ mod 2.

1.9.4. We define:

$$\tilde{\Lambda} = \Lambda \times \mathbb{Z} = \{(\ell,u),\ \ell \in \Lambda,\ u \in \mathbb{Z}\}.$$

Let ℓ_0 be a fixed Lagrangian plane, and let (ℓ_1,u_1) a point of $\tilde{\Lambda}$. We define the following system of neighborhoods of (ℓ_1,u_1): Let ℓ_2 be a Lagrangian plane transverse to ℓ_1, and Ω a neighborhood of ℓ_1 in Λ; we define

$$U(\ell_1,u_1;\Omega,\ell_2) = \{(\ell,u),\ \ell \in \Omega,\ u \in \mathbb{Z},\ \text{with}\ \ u = u_1 + \tau(\ell,\ell_0,\ell_1,\ell_2)\}$$

(where $\tau(\ell,\ell_0,\ell_1,\ell_2)$ is definedin 1.5.12).

1.9.5. Proposition: The $U(\ell_1,u_1;\Omega,\ell_2)$ form a neighborhood basis for a topology on $\tilde{\Lambda}$.

Proof: If $\ell = \ell_1$, $\tau(\ell,\ell_0,\ell_1,\ell_2) = 0$, hence $(\ell_1,u_1) \in U((\ell_1,u_1);(\Omega,\ell_2))$. Let us prove that $\{U(\ell_1,u_1;\Omega,\ell_2)\}$ form a system of neighborhoods. Let ℓ_2' be another Lagrangian plane such that $\ell_1 \cap \ell_2' = 0$. We have

$$\begin{aligned}\tau(\ell,\ell_0,\ell_1,\ell_2) - \tau(\ell,\ell_0,\ell_1,\ell_2') &= \tau(\ell_1,\ell_2,\ell_2') + \tau(\ell_2,\ell,\ell_2')\\ &= \tau(\ell_1,\ell_2,\ell_2') - \tau(\ell,\ell_2,\ell_2').\end{aligned}$$

We have $\ell_1 \cap \ell_2 = 0$, $\ell_1 \cap \ell_2' = 0$, hence ℓ_1 belongs to the open set $\Lambda_{\ell_2} \cap \Lambda_{\ell_2'}$.

Let Ω_1 and Ω_2 be two neighborhoods of ℓ_1 in Λ, and

$\Omega \subset \Omega_1 \cap \Omega_2$ a neighborhood of ℓ_1 contained in the connected component of $\Lambda_{\ell_2} \cap \Lambda_{\ell_2'}$. If $\ell \in \Omega$, then ℓ can be deformed continuously to ℓ_1, remaining transverse to ℓ_2 and ℓ_2'. Hence by 1.9.3, if $\ell \in \Omega$,

$$\tau(\ell,\ell_2,\ell_2') = \tau(\ell_1,\ell_2,\ell_2').$$

This proves that

$$U(\ell_1,\mu_1;\Omega,\ell_2) = U(\ell_1,\mu_1;\Omega,\ell_2') \subset U(\ell_1,\Omega_1;\mu_1,\ell_2) \cap U(\ell_1,\mu_1;\Omega_2,\ell_2').$$

To conclude the proof, it suffices to show that, given some $U = U(\ell_1,\mu_1;\Omega,\ell_2)$, if Ω is sufficiently small, then U is also a neighborhood of any point (ℓ',μ') in U. We choose Ω as being open and contained in Λ_{ℓ_2}. Thus ℓ' and ℓ_2 are also transverses. Now we can easily see that

$$U(\ell',\mu';\Omega,\ell_2) = U(\ell_1,\mu_1;\Omega,\ell_2)$$

as the equality to be verified is $\tau(\ell'',\ell_0,\ell',\ell_2) + \tau(\ell',\ell_0,\ell_1,\ell_2) = \tau(\ell'',\ell_0,\ell_1,\ell_2)$, which follows from the chain condition (1.5.8).

1.9.6. We denote by π the map $\tilde{\Lambda} \to \Lambda$ given by $(\ell,\mu) \to \ell$. It is clearly a continuous map, and the fibers of this map are discrete, so $\tilde{\Lambda}$ is a covering of Λ.

<u>Remark</u>: a) Let $\ell_1 \cap \ell_0 = \{0\}$, i.e. ℓ_1 is in the simply connected set Λ_{ℓ_0}; the neigborhood $U(\ell_1,\mu_1;\Omega,\ell_0)$ of ℓ_1,μ_1 in $\tilde{\Lambda}$ is simply $\Omega \times \mu_1$ as $\tau(\ell,\ell_0,\ell_1,\ell_0) = 0$. Therefore $\pi^{-1}(\Lambda_{\ell_0})$ is isomorphic as a topological space to $\Lambda_{\ell_0} \times \mathbb{Z}$.

b) If $\ell_1 = \ell_0$, then for ℓ_2 such that $\ell_2 \cap \ell_0 = 0$. We have $\tau(\ell,\ell_0,\ell_1,\ell_2) = \tau(\ell_2,\ell,\ell_0)$. The neighborhood $U((\ell_0,\mu_0);(\Omega,\ell_2))$ consists hence of $\{(\ell,\mu)\}$ for $\ell \in \Omega$, and $\mu = \mu_0 - \tau(\ell_0,\ell,\ell_2)$.

1.9.7. Example: Let us consider $V = \mathbb{C}$ with the alternate bilinear form $B(u,v) = -\mathrm{Im}\, u\bar{v}$. Here the Lagrangian planes are just the one dimensional subspaces. We identify the one dimensional subspace $\mathbb{R}e^{i\theta}$ with the element $e^{2i\theta}$, hence Λ is identified with the circle $T = \{u; |u| = 1\}$.

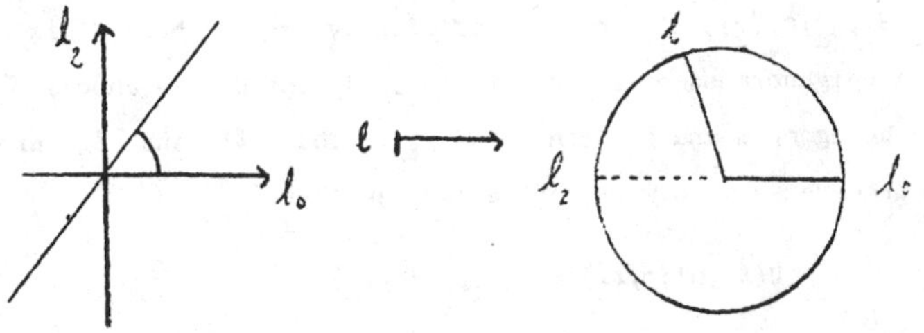

Let us define $\ell_0 = \mathbb{R}$, $\ell_2 = i\mathbb{R}$; in the preceding identification of Λ with T, we have the following numbers for $\tau(\ell_0,\ell,\ell_2)$:

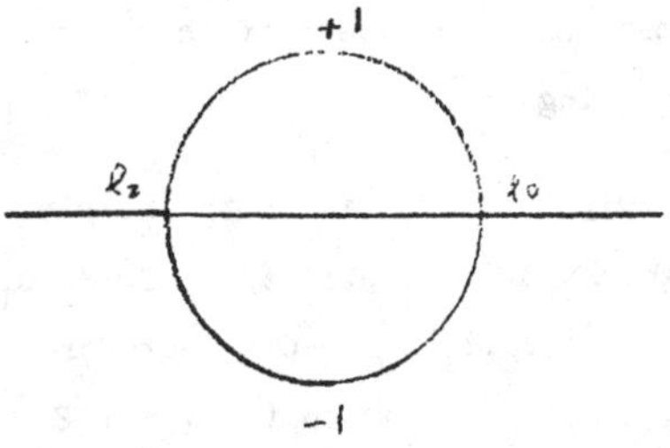

We draw now the manifold Λ with the neighborhood $U((\ell_0,0),(\cap,\ell_2))$

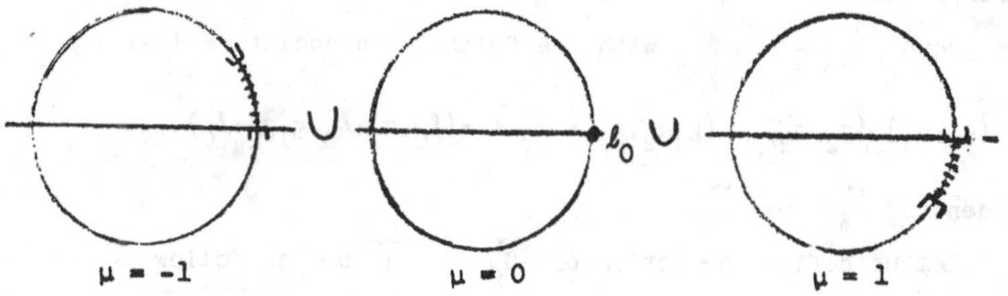

At a point (ℓ,μ) with $\ell \neq \ell_0$, the topology is the usual topology of the circle. Let U be a neighborhood of ℓ_0 in T, then a neighborhood of $(\ell_0,0)$ in Λ, above U, is:

$$\{(\ell_0,0)\} \cup \{(\ell,1);\ \ell \in U \cap (\text{Im } z > 0)\} \cup \{(\ell,-1);\ \ell \in U \cap (\text{Im } z < 0)\}.$$

Hence we see that the connected component of $(\ell_0,0)$ is as follows:

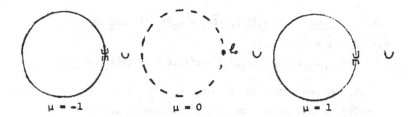

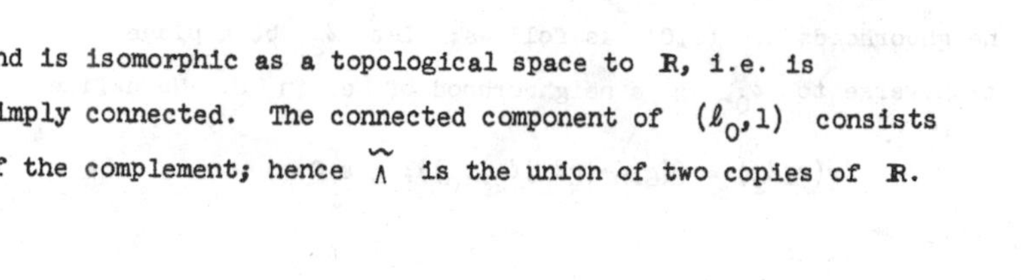

and is isomorphic as a topological space to $\mathbb{R}$, i.e. is simply connected. The connected component of $(\ell_0,1)$ consists of the complement; hence $\widetilde{\Lambda}$ is the union of two copies of $\mathbb{R}$.

1.9.8. Let $G = Sp(B)$ be the symplectic group. Let ℓ_0 be a fixed element of Λ. We recall that we have defined in 1.6.14 the group $\widetilde{G}_{\ell_0} = G \times \mathbb{Z}$ with the following associative law:

$$(g_1,n_1)\cdot(g_2,n_2) = (g_1g_2,n_1 + n_2 + \tau(\ell_0,g_1\cdot\ell_0,g_1g_2\cdot\ell_0)).$$

We denote $\widetilde{G}_{\ell_0}$ by $\widetilde{G}_0$.

Let us define the action of $\widetilde{G}_0$ on $\widetilde{\Lambda}$ by the following formula:

$$(g_1,\mu_1)\cdot(\ell,\mu) = (g_1\ell,\mu_1 + \mu + \tau(\ell_0,g_1\ell_0,g_1\ell)).$$

It is easily verified that $\tilde{g}_1\cdot(\tilde{g}_2\cdot\tilde{\ell}) = (\tilde{g}_1\cdot\tilde{g}_2)\cdot\tilde{\ell}$ for any $\tilde{g}_1 \in \widetilde{G}_0$, $\tilde{g}_2 \in \widetilde{G}_0$, $\tilde{\ell} \in \widetilde{\Lambda}$.

1.9.9. Let us consider the function $m_0\colon \widetilde{\Lambda} \times \widetilde{\Lambda} \to \mathbb{Z}$ defined by $m_0((\ell,\mu),(\ell',\mu')) = \mu - \mu' + \tau(\ell,\ell',\ell_0)$. The following lemma is clear:

1.9.10. <u>Lemma</u>: a) $m_0(\tilde{g}\cdot\tilde{\ell},\tilde{g}\cdot\tilde{\ell}') = m_0(\tilde{\ell},\tilde{\ell}')$ for any $\tilde{\ell},\tilde{\ell}' \in \widetilde{\Lambda}$; $\tilde{g} \in \widetilde{G}_0$,

b) $m_0(\tilde{\ell}_1,\tilde{\ell}_2) + m_0(\tilde{\ell}_2,\tilde{\ell}_3) + m(\tilde{\ell}_3,\tilde{\ell}_1) = \tau(\pi(\ell_1),\pi(\ell_2),\pi(\ell_3))$.

1.9.11. We now define the topology of $\widetilde{G}_0$. Let $e \in G$ be the identity element of G. We give a fundamental system of neighborhoods of $(e,0)$ as follows: Let ℓ_2 be a plane transverse to ℓ_0, Ω a neighborhood of e in G. We define

$$V(\Omega,\ell_2) = \{(g,-\tau(g\ell_0,\ell_2,\ell_0));\ g \in \Omega\}.$$

Let $\mathfrak{A} = \{v(\Omega,\ell_2)\}$. We have to verify the axioms:

a) If $A_1, A_2 \in \mathfrak{A}$, there exist $A \in \mathfrak{A}$ such that $A \subset A_1 \cap A_2$,

b) for every $A \in \mathfrak{A}$, there exists $B \in \mathfrak{A}$ such that $B^{-1} \subset A$,

c) for every $A \in \mathfrak{A}$ there exists $A_1, A_2 \in \mathfrak{A}$ such that $A_1A_2 \subset A$,

d) for every $g \in \tilde{G}$ and $A \in \mathfrak{A}$, there exists $B \in \mathfrak{A}$ such that $gBg^{-1} \subset A$.

For a) we write

$$\tau(g\ell_0,\ell_2,\ell_0) - \tau(g\ell_0,\ell_2',\ell_0) = \tau(g\ell_0,\ell_2,\ell_2') - \tau(\ell_0,\ell_2,\ell_2').$$

So if g is sufficiently near 1, we can deform $g.\ell_0$ to ℓ_0 in the open set $\ell \cap \ell_2 = \ell \cap \ell_2' = 0$. Hence by 1.9.3 we will have $\tau(g\ell_0,\ell_2,\ell_2') = \tau(\ell_0,\ell_2,\ell_2')$, and a) follows.

For b) we have $(g,\mu)^{-1} = (g^{-1},-u)$ so $v(\Omega,\ell_2)^{-1} = \{(g^{-1},\tau(g\ell_0,\ell_2,\ell_0)\}$. We write $\tau(g\ell_0,\ell_2,\ell_0) = -\tau(g^{-1}\ell_0,g^{-1}\ell_2,\ell_0)$. If g is sufficiently near the identity, $g^{-1}\ell_2$ is transverse to ℓ_0, and b) follows.

For c) we compute

$v(\Omega,\ell_2)\cdot v(\Omega',\ell_2)$

$$= \{(\alpha,-\tau(\alpha\ell_0,\ell_2,\ell_0))\cdot(\beta,-\tau(\beta\ell_0,\ell_2,\ell_0));\ \alpha\in\Omega,\ \beta\in\Omega'\}$$

$$= \{(\alpha\beta,-\tau(\alpha\ell_0,\ell_2,\ell_0) - \tau(\beta\ell_0,\ell_2,\ell_0) + \tau(\ell_0,\alpha\ell_0,\alpha\beta\ell_0))\}.$$

We write

$$\tau(\ell_0,\alpha\ell_0,\alpha\beta\ell_0) = \tau(\ell_0,\alpha\ell_0,\ell_2) + \tau(\alpha\ell_0,\alpha\beta\ell_0,\ell_2) + \tau(\alpha\beta\ell_0,\ell_0,\ell_2),$$

and we remark that for every $\beta\in\Omega'$ sufficiently small $\tau(\ell_0,\beta\ell_0,\ell_2) = \tau(\ell_0,\beta\ell_0,\alpha^{-1}\ell_2)$, if α is sufficiently near the identity, as $\alpha^{-1}\ell_2$ can be deformed to ℓ_2 remaining in the set $\ell_0\cap\ell_2 = 0$, $\beta\ell_0\cap\ell_2 = 0$.

d) is verified in a similar way.

Therefore $\widetilde{G}_0$ has the structure of a topological group. It is easy to verify that $\widetilde{G}_0$ acts continuously on $\widetilde{\Lambda}$.

The group $\widetilde{G}_0$ admits a character s (1.7.11) of order 4. We will see that the kernel of this character is the universal covering group of G. Thus $\widetilde{G}_0$ is the union of four connected components, each of them being simply connected. Similarly $\widetilde{\Lambda}$ is the union of two connected components $\widetilde{\Lambda}_+ \cup \widetilde{\Lambda}_-$, where

$$\widetilde{\Lambda}_+ = \{(\ell,\mu);\ \mu\equiv n+\dim(\ell\cap\ell_0),\ \text{mod } 2\}$$

$$\widetilde{\Lambda}_- = \{(\ell,\mu);\ \mu\equiv n+\dim(\ell\cap\ell_0)+1,\ \text{mod } 2\}$$

each of the components $\widetilde{\Lambda}_+$ and $\widetilde{\Lambda}_-$ being simply connected.

1.9.12. Let us consider as a model of a symplectic vector space $V = \mathbb{C}^n = \mathbb{R}^{2n}$ with the bilinear form $B(u,v) = -\mathrm{Im}\langle u,v\rangle$, where

$\langle u,v\rangle$ is the canonical hermitian inner product on $\mathbb{C}^n$. We denote $\mathrm{Re}\langle u,v\rangle$ by $S(u,v)$. Thus S is a positive definite symmetric form. We denote by J the multiplication by i. Let $U(n)$ be the group of complex-linear transformations of $\mathbb{C}^n$ leaving invariant the hermitian scalar product $\langle u,v\rangle$, i.e. $U(n)$ is the group of unitary transformation of $\mathbb{C}^n$. It is clear that $U(n) \subset Sp(B)$.

We consider $\ell_0 = \mathbb{R}^n$ as a fixed Lagrangian space. We denote by $P_1, P_2, \cdots, P_n$ the canonical basis of $\mathbb{R}^n$, and $Q_j = iP_j$. If ℓ is any Lagrangian subspace, we can find a basis $P_1', P_2', \cdots, P_n'$ of ℓ orthonormal for the symmetric form S. As ℓ is Lagrangian, it follows that $P_1', P_2', \cdots, P_n'$ is an orthonormal basis for the hermitian form h; hence there exists a unitary transformation $u \in U(n)$ such that $u(\mathbb{R}^n) = \ell$, i.e. the manifold Λ is homogeneous for the group $U(n)$. Let us denote by $x \to \bar{x}$ the conjugation σ of $\mathbb{C}^n$ with respect to $\mathbb{R}^n$, and $u \to \bar{u} = \sigma u \sigma$ the corresponding conjugation on $GL(n;\mathbb{C})$. It is clear that $u \in U(n)$ leaves stable $\mathbb{R}^n$, if and only if $u = \bar{u}$, i.e. u has real coefficients, so $u \in O(n)$.

We have $u_1(\mathbb{R}^n) = u_2(\mathbb{R}^n)$ if and only if $u_2^{-1}u_1 = \bar{u}_2^{-1}\bar{u}_1$, i.e. $u_1\bar{u}_1^{-1} = u_2\bar{u}_2^{-1}$. We denote for $u \in U(n)$, $v(u) = u\bar{u}^{-1}$ and for $\ell = u(\mathbb{R}^n) \in \Lambda$, $v(\ell) = u\bar{u}^{-1}$. v is a continuous function from Λ into $U(n)$.

1.9.13. <u>Lemma</u>: $\ell \cap \ell' = 0$ if and only if $v(\ell) - v(\ell')$ is invertible.

<u>Proof</u>: If $\ell \cap \ell' \neq \{0\}$, the equations $z = v(\ell)\bar{z}$ and

$z = v(\ell')\bar{z}$ have a common solution, hence $v(\ell) - v(\ell')$ is not invertible. Conversely by applying a suitable element of $U(n)$, we can suppose $\ell = \mathbb{R}^n$ and $\ell' = u\mathbb{R}^n$; if $x = u\bar{u}^{-1}x$, then $x = \bar{u}u^{-1}x$, hence $\bar{x} = u\bar{u}^{-1}\bar{x}$. Hence the 0-eigenspace of $(1 - u\bar{u}^{-1})$ is stable under conjugation and there exist $x \in \mathbb{R}^n$ such that $x = v(u)x$, so $x \in \mathbb{R}^n \cap \ell'$.

We consider the group $\widetilde{U(n)} \subset U(n) \times \mathbb{R}$ defined by $\{(u,\varphi): \det u = e^{i\varphi}\}$. The map $SU(n) \times \mathbb{R} \to \tilde{U}(n)$ defined by $(A,\varphi) \to (Ae^{i\varphi}, n\varphi)$ is an isomorphism. Hence, as $SU(n)$ is simply connected, so is $\widetilde{U(n)}$. If $g \in U(n)$ there exists a complex basis $e_1, e_2, \ldots, e_n$ of $\mathbb{C}^n$ such that $ge_j = e^{i\theta_j}e_j$, with $\theta_1, \theta_2, \ldots, \theta_n \in \mathbb{R}$.

For $(g,\varphi) \in \widetilde{U(n)}$, let $\theta_1, \theta_2, \ldots, \theta_n \in \mathbb{R}$ such that

a) $(e^{i\theta_1}, \ldots, e^{i\theta_n})$ are the eigenvalues of g,

b) $\theta_1 + \theta_2 + \ldots + \theta_n = \varphi$

and let us consider for μ defined by 1.8.7

$$\mu(\theta_1/2) + \mu(\theta_2/2) + \ldots + \mu(\theta_n/2) .$$

This expression does not depend on the choices of $\theta_1, \theta_2, \ldots, \theta_n$ verifying a) and b); as for another choice $(\theta_1', \theta_2', \ldots, \theta_n')$ we will have up to a reordering

$$\theta_i'/2 = \theta_i/2 + k_i\pi, \text{ so } \mu(\theta_i'/2) = \mu(\theta_i/2) + 2k_i$$

but as $\Sigma\, \theta_i' = \Sigma\, \theta_i$, we have $\Sigma\, k_i = 0$, i.e. $\Sigma\, \mu(\theta_i'/2) = \Sigma\, \mu(\theta_i/2)$.

We now can define a function $\alpha(g,\varphi)$ on $\widetilde{U(n)}$ by $\alpha(g,\varphi) = \Sigma\mu(\theta_i/2)$ with θ_i verifying a) and b). If $g \in U(n)$ with $\det g = e^{i\varphi}$, $\det g\,\bar{g}^{-1} = e^{2i\varphi}$, hence $(g,\varphi) \to (g\bar{g}^{-1},2\varphi)$ is a map from $\widetilde{U(n)}$ into $\widetilde{U(n)}$.

We define now the function $\mu\colon \widetilde{U(n)} \to \mathbb{Z}$ by

1.9.14. $$\mu(g,\varphi) = \alpha(g\bar{g}^{-1},2\varphi) .$$

We have $\mu(g,\varphi+2\pi) = \mu(g,\varphi) + 4$. Let us remark that as $\mu(\theta/2) \equiv 1 \bmod 2$, if $\theta \neq 2k\pi$, we have $\mu(g,\varphi) \equiv n + \dim(\ell \cap \ell_0) \bmod 2$

Let $(\ell_1,n_1) \in \widetilde{\Lambda}_+$, i.e. $n_1 \equiv n + \dim(\ell_1 \cap \ell_0) \bmod 2$. There exists $(g_1,\varphi_1) \in \widetilde{U}(n)$ such that $\alpha(v(g_1),2\varphi_1) = n_1$ as follows from the preceeding observation. φ_1 and $v(g_1)$ are uniquely determined by this condition. We denote $v(g_1)$ by $v(\ell_1)$.

We define for a couple $((\ell_1,n_1),(\ell_2,n_2)) \in \widetilde{\Lambda}_+ \times \widetilde{\Lambda}_+$

$$m'((\ell_1,n_1),(\ell_2,n_2)) = \alpha(v(\ell_1)^{-1}v(\ell_2),2(\varphi_2-\varphi_1)).$$

Remark that $m'((\ell_2,n_2),(\ell_1,n_1)) = -m'((\ell_1,n_1),(\ell_2,n_2))$.

1.9.15. <u>Proposition</u>: Let $\widetilde{\ell}_1,\widetilde{\ell}_2,\widetilde{\ell}_3 \in \widetilde{\Lambda}_+ \times \widetilde{\Lambda}_+ \times \widetilde{\Lambda}_+$ then $\tau(\pi(\widetilde{\ell}_1),\pi(\widetilde{\ell}_2),\pi(\widetilde{\ell}_3)) = m'(\widetilde{\ell}_1,\widetilde{\ell}_2) + m'(\widetilde{\ell}_2,\widetilde{\ell}_3) + m'(\widetilde{\ell}_3,\widetilde{\ell}_1)$.

<u>Proof</u>: We remark first that the second member doesn't depend on $\widetilde{\ell}_1,\widetilde{\ell}_2,\widetilde{\ell}_3$ but only of (ℓ_1,ℓ_2,ℓ_3). In fact if $\pi(\widetilde{\ell}'_1) = \pi(\widetilde{\ell}_1)$, $\pi(\widetilde{\ell}'_2) = \pi(\widetilde{\ell}_2)$, $\pi(\widetilde{\ell}'_3) = \pi(\widetilde{\ell}_3)$, then $\varphi'_1 = \varphi_1 + k_1\pi$, $\varphi'_2 = \varphi_2 + k_2\pi$,

$\varphi_3' = \varphi_3 + k_3\pi$. Hence

$$\varphi_2' - \varphi_1' = \varphi_2 - \varphi_1 + (k_2-k_1)\pi$$

$$\varphi_3' - \varphi_2' = \varphi_3 - \varphi_2 + (k_3-k_2)\pi$$

$$\varphi_1' - \varphi_3' = \varphi_1 - \varphi_3 + (k_1-k_3)\pi \ .$$

Therefore the second member does not change.

We remark also that if $u \in U(n)$

$$m'(\widetilde{u\ell_1},\widetilde{u\ell_2}) + m'(\widetilde{u\ell_2},\widetilde{u\ell_3}) + m'(\widetilde{u\ell_3},\widetilde{u\ell_1})$$

$$= m'(\tilde{\ell}_1,\tilde{\ell}_2) + m'(\tilde{\ell}_2,\tilde{\ell}_3) + m'(\tilde{\ell}_3,\tilde{\ell}_1) \ ,$$

where $\widetilde{u\ell_i}$ denotes any element of $\tilde{\Lambda}_+$ whose projection on Λ is $u\ell_i$. This follows easily from the fact that $v(u\ell_i) = uv(\ell_i)\bar{u}^{-1}$, hence $v(u\ell_i)^{-1}v(u\ell_j) = \bar{u}v(\ell_i)^{-1}v(\ell_j)\bar{u}^{-1}$ is conjugate to $v(\ell_i)^{-1}v(\ell_j)$ and so have the same eigenvalues.

Let us now prove the proposition. We assume first that ℓ_1, ℓ_2, ℓ_3 are transverses. Transforming by an appropriate element of $U(n)$, we can suppose that $\ell_1 = \ell_0 = \mathbb{R}^n$. Now as $\ell_1 \cap \ell_2 = 0$, $\ell_1 \cap \ell_3 = 0$, $\ell_2 \cap \ell_3 = 0$, the matrices $v(\ell_2)$, $v(\ell_3)$ as well as $v(\ell_2)^{-1}v(\ell_3)$ does not have the eigenvalue 1. So for $\{\theta_i\}$, $\{\theta_i'\}$, $\{\theta_i''\}$ verifying a) and b) respectively associated to $v(\ell_1)$, $v(\ell_2)$, $v(\ell_2)\ v(\ell_3)^{-1}$, $2k\pi \notin \{\theta_i, \theta_i', \theta_i''\}$. Now consider the connected component U of the open set Ω of $\Lambda \times \Lambda$ defined by

$$\Omega = \{(\ell,\ell');\ \ell \cap \ell_0 = 0,\ \ell' \cap \ell_0 = 0,\ \ell \cap \ell' = 0\}$$

containing ℓ_2, ℓ_3. By the preceeding remark and 1.9.3 both sides of 1.9.15 remain constant on U. Let

$$\ell_k = \sum_{i=1}^{n} \mathbb{R}(P_i + \epsilon_i Q_i)$$

$$\ell_3 = \sum_{i=1}^{n} \mathbb{R}Q_i .$$

By Lemma 1.9.1 it is then sufficient to prove (1.9.15) for (ℓ_1,ℓ_k,ℓ_3). It is enough now to calculate for the 2-dimensional symplectic space. We have $\tau(\mathbb{R}P,\mathbb{R}(P+Q),\mathbb{R}Q) = 1$ and $P+Q = e^{i\pi/4}P$, $Q = e^{i\pi/2}P$. We have $\mu(e^{i\pi/4},\pi/4) = 1$, $\mu(e^{i\pi/2},\pi/2) = 1$, $\mu(e^{i\pi/2},\pi/2) = 1$. Hence the equality is satisfied.

For $\tau(\mathbb{R}P,\mathbb{R}(P-Q),\mathbb{R}Q) = -1$, $\mu(e^{-i\pi/4},-\pi/4) = -1$, $\mu(e^{i\pi/2},\pi/2) = 1$, $\mu(e^{-i\pi/4},\pi/4) = 1$ and the equality is satisfied.

Suppose now that $\ell_1,\ell_2,\ell_3 \in \Lambda \times \Lambda \times \Lambda$ are not transverse. We will prove 1.9.15 by induction: We can find $\ell \in \Lambda$ transverse to $\ell_1,\ell_2,\ell_3, J\ell_1,J\ell_2,J\ell_3$. Applying 1.5.8, it is enough to prove the Proposition 1.9.15 for $\tau(\ell_1,\ell_2,\ell)$ when ℓ is transverse to ℓ_1,ℓ_2 as well as $J\ell$.

Applying an element $u \in U(n)$, we can assume that $\ell_1 = \mathbb{R}^n$. Let $\ell_1 \cap \ell_2 = \rho \neq 0$: let us choose a basis $P_1, P_2, \cdots, P_k$ of ρ orthogonal for S and we complete it in a orthogonal basis $(P_1,P_2,\ldots,P_k,P_{k+1},\ldots,P_n)$ of $\mathbb{R}^n$ (for the form S). Hence $(P_1,\ldots,P_k,P_{k+1},\ldots,P_n),(Q_1,\ldots,Q_n)$ with $Q_i = JP_i$ is a symplectic basis for V.

We consider the orthogonal W_ρ of $\rho + J\rho = \mathbb{C}^k$ with respect to B, then W_ρ has a complex structure isomorphic to $\mathbb{C}^{n-k}$, and W_ρ is isomorphic to $\rho^\perp/\rho$ as a symplectic vector space.

Let us consider first the case where

$$\ell = \ell \cap \mathbb{C}^{n-k} \oplus \ell \cap \mathbb{C}^k .$$

Let
$$\ell_1' = \ell_1 \cap \mathbb{C}^{n-k} = \mathbb{R}^{n-k} = \ell_0' \simeq \ell_1^\rho/\rho$$
$$\ell_2' = \ell_2 \cap \mathbb{C}^{n-k} = \ell_2 \cap W_\rho \simeq \ell_2^\rho/\rho$$
$$\ell' = \ell \cap \mathbb{C}^{n-k} = \ell \cap W_\rho \simeq \ell^\rho/\rho$$

By 1.5.10, $\tau(\ell_1, \ell_2, \ell) = \tau(\ell_1', \ell_2', \ell')$. By the induction hypothesis

$$\tau(\ell_1', \ell_2', \ell') = m(\tilde{\ell}_1', \tilde{\ell}_2') + m(\tilde{\ell}_2', \tilde{\ell}') + m(\tilde{\ell}', \tilde{\ell}_1') .$$

Let $g_1' \in U(n-k)$ such that $g_1'\ell_1' = \ell_2'$

$g' \in U(n-k)$ such that $g'\ell_1' = \ell'$

then $g_1 = \left(\begin{array}{c|c} g_1' & 0 \\ \hline 0 & 1 \end{array}\right)$ is such that $g_1\ell_1 = \ell_2$ and then there exists $g'' \in U(k)$ such that if $g = \left(\begin{array}{c|c} g' & 0 \\ \hline 0 & g'' \end{array}\right)$, $g\ell_1 = \ell$.

Let φ_1' such that $\det g_1' = e^{i\varphi_1'}$, φ', φ'' such that $\det g' = e^{i\varphi'}$, $\det g'' = e^{i\varphi''}$. Then $\det g = e^{i(\varphi'+\varphi'')}$; set $\varphi = \varphi' + \varphi''$.

We have now to verify the equality

$$\alpha(v(g_1');2\varphi_1') + \alpha(v(g_1')^{-1}v(g');2(\varphi' - \varphi_1')) + \alpha(v(g')^{-1};-2\varphi')$$

$$= \alpha(v(g_1),2\varphi_1') + \alpha(v(g_1)^{-1}v(g),2(\varphi - \varphi_1')) + \alpha(v(g)^{-1},-2\varphi) \ .$$

Equality which is clear, as

$$v(g_1)^{-1}v(g) = \left(\begin{array}{c|c} v(g_1')^{-1}v(g') & 0 \\ \hline 0 & v(g'') \end{array}\right)$$

and as $\alpha\left(\left(\begin{array}{c|c} A & 0 \\ \hline 0 & B \end{array}\right); \varphi_1 + \varphi_2\right)$ with $\det A = e^{i\varphi_1}$, $\det B = e^{i\varphi_2}$

$$= \alpha(A,\varphi_1) + \alpha(B,\varphi_2)$$

and $\qquad \alpha(A^{-1},-\varphi) = -\alpha(A,\varphi)$.

We will reduce ourselves to this situation by deformation. By hypothesis ℓ_2 and ℓ are transverse to $J\ell_1 = \Sigma\, \mathbb{R}Q_k$. Let y_2 be the matrix representing ℓ_2 in $\Lambda_{J\ell_1}$, i.e.

$$y_2 \colon \ell_1 \to J\ell_1$$

and:

$$\ell_2 = \{x + y_2 x \colon x \in \ell_1\} \ .$$

Hence, with respect to the basis $P_1,\ldots,P_k,P_{k+1},\ldots,P_n$, $Q_1,Q_2,\ldots,Q_k,Q_{k+1},\ldots,Q_n$, we have

$$y_2 = \left(\begin{array}{c|c} 0 & 0 \\ \hline 0 & y_2' \end{array}\right) \quad \text{where} \quad y_2' = {}^t y_2' \ \text{ is a symmetric matrix.}$$

Let y the matrix representing ℓ in $\Lambda_{J\ell_1}$, then $y = \left(\begin{array}{c|c} u & v \\ \hline {}^t v & w \end{array}\right)$

where u and w are symmetric matrices. The condition $\ell \cap \ell_1 = 0$ $\ell \cap \ell_2 = 0$ are translated in $\det y \neq 0$ $\det(y - y_2) \neq 0$. We can certainly deform u in $u(\epsilon)$ such that $y(\epsilon) = \begin{pmatrix} u(\epsilon) & {}^t v \\ v & w \end{pmatrix}$ still verifies $\det y(\epsilon) \neq 0$ $\det(y(\epsilon)-y_2) \neq 0$ and such that $\det(u(\epsilon_0)) \neq 0$. Hence we can assume u invertible.

Let us consider then $g(\epsilon) = \begin{pmatrix} 1 & 0 \\ \epsilon\lambda & 1 \end{pmatrix}$ with λ an $(n-k) \times k$ matrix. Then $g(\epsilon) y_2 \, {}^t g(\epsilon) = y_2$ and

$$g(\epsilon)\, y \, {}^t g(\epsilon) = \begin{pmatrix} u & v' \\ {}^t v' & w' \end{pmatrix} = y(\epsilon)$$

with $v' = \epsilon u \lambda^t + v$.

Clearly $\det y(\epsilon) = \det(g(\epsilon))^2 \det y = \det y \neq 0$

$$\det(y(\epsilon) - y_2) = \det(y - y_2) \neq 0 .$$

We can choose, as u is invertible, $\lambda = -u^{-1} v$; as $y(1) = \begin{pmatrix} u & 0 \\ 0 & w' \end{pmatrix}$, we have constructed $\ell(\epsilon)$ a deformation of ℓ, such that $\ell(\epsilon) \cap \ell_1 = 0$, $\ell(\epsilon) \cap \ell_2 = 0$, and $\ell(1) = \ell(1) \cap C^{n-k} + (\ell(1)) \cap C^k$.

1.9.16. Corollary. The map $(g,\varphi) \mapsto (g,\mu(g,\varphi))$ is an homomorphism of $\widetilde{U(n)}$ into $\widetilde{G}$.

<u>Proof</u>: We have to verify

$$\tau(\ell_0, g_1\ell_0, g_1g_2\ell_0) = \mu(g_1g_2, \varphi_1+\varphi_2) - \mu(g_1, \varphi_1) + \mu(g_2, \varphi_2)$$

which is clear.

Appendix: A generalization of Maslov index to local fields.

Let k be any local field of characteristic different of 2. Let (V,B) be a 2n-dimensional symplectic space over k. The Heisenberg group is the set $N = V \times k$ with the law:

$$(v,t)\cdot(v',t') = (v + v', t + t' + \tfrac{1}{2} B(v,v')) .$$

The results of Chapter I, corresponding to the case where $k = \mathbb{R}$ have been generalized by Patrice Perrin [11]. Similar results have been obtained independently by R. Rao [12].

Let E be a vector space over k. We fix χ a non-trivial additive character of k. This allows us to identify the Pontriagin dual $\hat{E}$ of $(E,+)$ with the algebraic dual E^* by associating to an element f of E^* the character $x \to \chi(\langle f,x\rangle)$.

Let Λ be the set of Lagrangian subspaces of V, i.e. the set of maximal isotropic linear subspaces of (V,B). For $\ell \in \Lambda$, $L = \ell \times k$ is an abelian subgroup of N and

$$(1 \times \chi)(v,t) = \chi(t) \quad (v \in \ell,\ t \in k)$$

is a character of L. We obtain the Schrödinger representation of N associated to ℓ, by inducing this character of L. We write

$$W(\ell) = \operatorname{Ind}\uparrow_{L}^{N} (1 \times \chi) .$$

The representation $W(\ell)$ acts on the Hilbert space $H(\ell)$ of function φ from N to $\mathbb{C}$ satisfying:

(1) $\varphi(nh) = (1 \times \chi)(h)^{-1} \varphi(n)$, for every $n \in N$, $h \in L$.

(2) $\int_{N/L} |\varphi(\dot{n})|^2 \, d\dot{n} < \infty$, where $d\dot{n}$ is an invariant measure on N/L. The representations $W(\ell)$ is a unitary irreducible representation of N in $H(\ell)$.

If $\ell \oplus \ell' = V$ is a decomposition of (V,B) into a sum of two Lagrangian subspaces, the map $f(y) = f(y,0)$ is an isomorphism of $H(\ell)$ with $L^2(\ell')$.

We now compare the representations $W(\ell_1)$ and $W(\ell_2)$ associated to two different choices ℓ_1, ℓ_2 of Lagrangian subspaces of (V,B).

Let E be a vector space over k and dx a Haar measure on E. We denote by dx^* the Haar measure on E^* which makes the Fourier transform

$$(\mathcal{F}f)(x^*) = \int_E \chi(-\langle x^*,x\rangle) \, f(x) \, dx$$

a unitary operator (dx^* depends of the choice of χ).

If (V,B) is a symplectic vector space, V is identified to V^* by B. We choose on V the unique Haar measure dv such that $dv = dv^*$.

If S is an isomorphism from the vector space (E,dx) to the vector space (F,dy), we denote by $|S|$ the unique positive scalar such that:

$$|S| \int_E f(Sx) \, dx = \int_F f(y) \, dy \,, \quad \text{for } f \in L^1(F) \,.$$

Let (ℓ_1,ℓ_2) be two Lagrangian subspaces of (V,B). The map

$g_{\ell_2,\ell_1}: \ell_1 \to \ell_2^*$ defined by

$$(g_{\ell_2,\ell_1}(x),y) = B(x,y) \quad (x \in \ell_1,\ y \in \ell_2)$$

induces an isomorphism still denoted by g_{ℓ_2,ℓ_1} from $\ell_1/\ell_1 \cap \ell_2$ to $(\ell_2/\ell_1 \cap \ell_2)^*$. Let $dx_1^!$ be a Haar measure on V/ℓ_1, $dx_2^!$ a Haar measure on V/ℓ_2 (i.e. we fix the norm in $H(\ell_1)$ and $H(\ell_2)$). We choose a measure du on $\ell_1 \cap \ell_2$. Then there exists a unique measure $d\dot{x}_1$ on $\ell_1/\ell_1 \cap \ell_2$ (resp. $d\dot{x}_2$ on $\ell_2/\ell_1 \cap \ell_2$) such that:

$$du\, d\dot{x}_1\, dx_1^! = dv\ ; (V \simeq \ell_1 \cap \ell_2 \oplus \ell_1/\ell_1 \cap \ell_2 \oplus V/\ell_1)$$

$$\text{resp.}\quad du\, d\dot{x}_2\, dx_1^! = dv\ ; (V \simeq \ell_1 \cap \ell_2 \oplus \ell_2/\ell_1 \cap \ell_2 \oplus V/\ell_2)\ .$$

Let $|g_{\ell_2,\ell_1}|$ be the module of the isomorphism g_{ℓ_2,ℓ_1} with respect to $d\dot{x}_1, (d\dot{x}_2)^*$. Then $|g_{\ell_2,\ell_1}|^{1/2} d\dot{x}_2$ does not depend on the choice of du. We obtain, as in 1.4, the:

A.1 <u>Theorem</u>: The operator

$$(\mathcal{F}_{\ell_2,\ell_1}\varphi)(n) = \int_{\ell_2/\ell_1 \cap \ell_2} \varphi(n\cdot(x_2,0))\,|g_{\ell_2,\ell_1}|^{1/2}\, d\dot{x}_2$$

defined on a dense subspace of $H(\ell)$ extends to a unitary operator from $H(\ell_1)$ to $H(\ell_2)$, intertwining the irreducible representations $W(\ell_1)$ and $W(\ell_2)$. We have $\mathcal{F}_{\ell_1,\ell_2} = \mathcal{F}^{-1}_{\ell_2,\ell_1}$.

A.2. Let E be a finite dimensional vector space over k. Let $V = E \oplus E^*$, with the alternate bilinear form

$$B(x_1 + f_1,\ x_2 + f_2) = f_2(x_1) - f_1(x_2)\ .$$

Then V is a symplectic vector space and $V = E \oplus E^*$ is a decomposition of V as a sum of two Lagrangian subspaces. Let Q be a symmetric form on E. We denote by s_Q the map between E and E^* defined by Q. The subspace $L_Q = \{x + s_Q x;\ x \in E\}$ is a Lagrangian subspace of V. Let us consider the unitary operators $\mathcal{F}_{E,L_Q} \cdot \mathcal{F}_{L_Q,E^*}$ and $\mathcal{F}_{E,E^*}$. They are both unitary operators intertwining the irreducible representation $W(E^*)$ and $W(E)$ of N, hence they are proportional. We thus define the Weil index $\gamma(Q)$ of Q as being the scalar of modulus one such that:

$$\mathcal{F}_{E,L_Q} \circ \mathcal{F}_{L_Q,E^*} = \gamma(Q)\mathcal{F}_{E,E^*}\ .$$

Let Q be non-degenerate: We calculate $\gamma(Q)$ as follows: let us choose a measure dx on E. We denote by $|Q|$ the module of the transformation s_Q with respect to dx, dx^*. The equality

$$(\mathcal{F}_{E,L_Q} \cdot \mathcal{F}_{L_Q,E^*} \cdot \varpi)(0,0) = \gamma(Q)(\mathcal{F}_{E,E^*}\varpi)(0,0)$$

is then written as follows:

A.3. $$\iint_{E\,E} \varpi(x-y)\ \chi(1/2\ Q(y))\ |Q|^{1/2}\ dy\ dx = \gamma(Q) \int_E \varpi(x)\ dx$$

for ϖ a function in the Schwartz space over E, identified to a function in $H(E^*)$. (Let us remark that $|Q|^{1/2}\ dy$ is independent of dy: it is the unique positive measure d_Q such that the transformation

$$\hat{f}_Q(x) = \int \chi(Q(x,y)) f(y)\, d_Q y \quad \text{is unitary}).$$

The following properties of $\gamma(Q)$ are immediate:

A.4. If $Q = 0$, then $\gamma(Q) = 1$.

A.5. $\gamma(Q)^{-1} = \gamma(-Q)$. (This follows after conjugation of A.3.)

A.6. If $E = E_1 \oplus E_2$, with $Q = Q_1 \oplus Q_2$ is an orthogonal decomposition of E, $\gamma(Q_1 \oplus Q_2) = \gamma(Q_1)\,\gamma(Q_2)$.

Let us consider the Witt group W_k of k defined as follows: We form first the semi group generated by the equivalence class of non-degenerate orthogonal vector space (E,Q), with the additive law $(E_1,Q_1) + (E_2,Q_2) = (E_1 \oplus E_2,\ Q_1 + Q_2)$. We then identify (E,Q) to 0, if $E \simeq (V \oplus V^*, Q_0)$ with $Q_0(x + f) = f(x)$ the duality form. It then follows from the properties A.4, A.5, A.6 that $Q \to \gamma(Q)$ defines a character of W_k.

Remark: If $k = \mathbb{R}$, the map $s(E,Q) = \operatorname{sign} Q$ defines an isomorphism of W_k with $\mathbb{Z}$.

Let (V,B) be a symplectic space. We define as in 1.5 the Kashiwara index $\tau(\ell_1,\ell_2,\ell_3)$ of three Lagrangian subspaces of (V,B), as being the element of the Witt group W_k associated to the 3n-dimensional orthogonal space $\ell_1 \oplus \ell_2 \oplus \ell_3$, with the quadratic form

$$Q_{123}(x_1 + x_2 + x_3) = B(x_1,x_2) + B(x_2,x_3) + B(x_3,x_1)\ .$$

We have (with the same notation and proofs as in 1.5) the following properties of the Kashiwara index:

A.7. Theorem:

a) $\tau(\ell_1,\ell_2,\ell_3)$ is invariant under the action of the symplectic group.

b) $\tau(\ell_1,\ell_2,\ell_3) = -\tau(\ell_2,\ell_1,\ell_3) = -\tau(\ell_1,\ell_3,\ell_2)$.

c) Suppose (ℓ_1,ℓ_3) are transverse Lagrangian spaces. Let Q'_{123} be the quadratic form on ℓ_2 defined by

$$Q'_{123}(x_2) = B(p_{13}x_2, p_{31}x_2) ,$$

then

$$\tau(\ell_1,\ell_2,\ell_3) = (\ell_2, Q'_{123})$$

in W_k.

d) Let $\ell_1,\ell_2,\ell_3,\ell_4$ be 4 Lagrangian subspaces of (V,B) then:

$$\tau(\ell_1,\ell_2,\ell_3) = \tau(\ell_1,\ell_2,\ell_4) + \tau(\ell_2,\ell_3,\ell_4) + \tau(\ell_3,\ell_1,\ell_4).$$

e) Let $\rho \subset \ell_1 \cap \ell_2 + \ell_2 \cap \ell_3 + \ell_3 \cap \ell_1$. Then

$$\tau(\ell_1^\rho,\ell_2^\rho,\ell_3^\rho) = \tau(\ell_1,\ell_2,\ell_3).$$

A.8. Theorem: (Patrice Perrin). Let (ℓ_1,ℓ_2,ℓ_3) be three Lagrangian subspaces of (V,B). We have:

$$\mathcal{F}_{\ell_1,\ell_2} \circ \mathcal{F}_{\ell_2,\ell_3} \circ \mathcal{F}_{\ell_3,\ell_1} = \gamma(\tau(\ell_1,\ell_2,\ell_3)).\mathrm{Id}.$$

Proof: Let (ℓ_1,ℓ_3) be transverse. The form B identifies ℓ_3 to ℓ_1^*. In the notation of A.2, we then have $\ell_2 = L_{Q'_{123}}$. Thus, by definition of $\nu(Q'_{123})$ and A.7 c), the theorem holds. If ℓ_1 and ℓ_3 are not transverse, we prove it by induction on dim V as in 1.6, using A.7 c) and d) and Lemma 1.6.4.

A.9. We define a canonical projective representation $R_\ell(g)$ of the symplectic group G over k, by

$$R_\ell(g) = \mathcal{F}_{\ell,g\ell} \cdot A(g).$$

In coordinates $R_\ell(g)$ are given by the formula 1.6.20. These operators $R_\ell(g)$ satisfy the fundamental relations:

$$R_\ell(g)\ W(\ell)(n)\ R_\ell(g)^{-1} = W(\ell)(g\cdot n)$$

$$R_\ell(g_1,g_2) = c_\ell(g_1,g_2)\ R_\ell(g_1)\ R_\ell(g_2)$$

with $$c_\ell(g_1,g_2) = \gamma(\tau(\ell,g_1,\ell,g_1g_2\ell))^{-1}.$$

<u>Remark</u>: On the open set $\{(g_1,g_2)\} \in G \times G$ with $g_1 = \begin{pmatrix} a_1 & b_1 \\ c_1 & d_1 \end{pmatrix}$, $g_2 = \begin{pmatrix} a_2 & b_2 \\ c_2 & d_2 \end{pmatrix}$, $g_3 = g_1g_2 = \begin{pmatrix} a_3 & b_3 \\ c_3 & d_3 \end{pmatrix}$, with c_1,c_2 invertible, we obtain, as in 1.6.22,

$$c_\ell(g_1,g_2) = \gamma(c_1^{-1}c_3c_2^{-1})^{-1},$$

as in [17].

A.10. It follows, from the Theorem A.7 a) c) that $(g_1,g_2) \to \tau(\ell,g_1\ell,g_1g_2\ell)$ is a cocycle of Sp(B) with values in the abelian group W_k. We then can introduce the extension $\tilde{G}_\ell$ of G by W_k with the exact sequence:

$$1 \longrightarrow W_k \longrightarrow G_\ell \longrightarrow G \longrightarrow 1$$

by defining:

$$\widetilde{G_\ell} = \{(g,q);\ g \in G,\ q \in W_k\}$$

with the multiplicative law:

$$(g_1,q_1)\cdot(g_2,q_2) = (g_1g_2, q_1+q_2+\tau(\ell, g_1\ell, g_1g_2\ell)).$$

A.11. The map $(g,q) \to \nu(q)\ R_\ell(g)$ is a true representation of $\widetilde{G_\ell}$. We will now prove that $R_\ell(g)$ lifts in fact to a true representation of a double covering of G.

Let us define for $a \in k^*$, $\nu(a)$ to be $\nu(Q_a)$, when $Q_a(x) = ax^2$ is the one dimensional quadratic form. $\nu(a)$ is a function on $k^*/(k^*)^2$.

Let us define, for $a,b \in k^*$, the Hilbert symbol (a,b) to be 1, if a is the norm of an element of $k(b^{1/2})$. We need the:

A.12. Proposition: (Weil, [17]). Let a and b be two element of k^*, then

$$(a,b) = \frac{\gamma(ab)\ \nu(1)}{\gamma(a)\ \nu(b)}.$$

Let Q be a non-degenerate quadtratic form on a vector space E of dimension n. We write $Q = \Sigma\, a_i x_i^2$ over a basis. The Hasse invariant $\mathcal{E}(Q)$ is defined to be $\prod_{1\le i<j\le n} (a_i,a_j)$. It is independent of the basis. The discriminant $D(Q)$ of Q is defined in k^*/k^{*2}. By induction on the dimension of V, A.12 implies:

A.13. $$\nu(Q) = (\nu(1))^{r g Q-1}\ \nu(D(Q))\ \mathcal{E}(Q).$$

Let us remark that we have:

$$\gamma(1)^4 = \frac{\gamma(1)\gamma(1)}{\gamma(-1)\gamma(-1)} = \frac{\gamma((-1)^2)\gamma(1)}{\gamma(-1)\gamma(-1)} = (-1,-1). \qquad (A.12)$$

Thus $\gamma(1)$ is a 8th root of unity. We recall that for $k = \mathbb{R}$, and $\chi(x) = e^{2i\pi x}$, we have $\gamma(a) = e^{i\pi/4 \text{ sign } a}$. It is possible to calculate $\gamma(a)$ explicitly for any local field [11],[12].

Let us define oriented vector spaces: Let E be a vector space over k of dimension n. We say that two couples (E,e) and (E,e'), where e and e' are two non-zero elements of $\wedge^n E$, are equivalent, if $e = \lambda e'$ with $\lambda \in k^{*2}$. An oriented vector space is a class of equivalence of such a couple (E,e). If E is oriented, E^* is naturally oriented. Let $\hat{\ell}_1$ and $\hat{\ell}_2$ be two oriented Lagrangian planes. The map g_{ℓ_1,ℓ_2} is an isomorphism from $\ell_1/\ell_1\cap\ell_2$ to $(\ell_2/\ell_1\cap\ell_2)^*$. If one chooses an orientation on $\ell_1 \cap \ell_2$, we obtain, from the orientations of ℓ_1 and ℓ_2, orientations on $\ell_1/\ell_1\cap\ell_2$ and $\ell_2/\ell_1\cap\ell_2$. Then, one can define $\det g_{\tilde{\ell}_1,\tilde{\ell}_2} \bmod (k^*)^2$ and it does not depend on the choice of the orientation on $\ell_1 \cap \ell_2$. As $g_{\ell_1,\ell_2} = -g^t_{\ell_2,\ell_1}$, we obtain from A.13 the relation:

$$\gamma(\det g_{\tilde{\ell}_1,\tilde{\ell}_2})^2 \, \gamma(\det g_{\tilde{\ell}_2,\tilde{\ell}_1})^2 = (\gamma(1))^{4(1-(n-\dim \ell_1 \cap \ell_2))}.$$

A.14. Let $\hat{\ell}_1$ and $\hat{\ell}_2$ be two oriented planes of E. We define

$$m(\tilde{\ell}_1,\tilde{\ell}_2) = (\gamma(1))^{2(1-(n-\dim \ell_1 \cap \ell_2))} \, \gamma(\det g_{\tilde{\ell}_1,\tilde{\ell}_2})^{-2}.$$

We then have:

$$m(\tilde{\ell}_1,\tilde{\ell}_2) = m(\tilde{\ell}_2,\tilde{\ell}_1)^{-1} .$$

The symplectic group G acts on the set $\tilde{\Lambda}$ of oriented Lagrangian planes. Clearly $m(\tilde{\ell}_1,\tilde{\ell}_2)$ is invariant under the action of G.

A.15. Theorem:

$$\gamma(\tau(\ell_1,\ell_2,\ell_3))^2 = m(\tilde{\ell}_1,\tilde{\ell}_2)\, m(\tilde{\ell}_2,\tilde{\ell}_3)\, m(\tilde{\ell}_3,\tilde{\ell}_1) .$$

Proof: If (ℓ_1,ℓ_2,ℓ_3) are mutually transverse, we have $\gamma(\tau(\ell_1,\ell_2,\ell_3)) = \gamma(Q'_{123})$ and Q'_{123} is the symmetric form on ℓ_2 which is equal to $g_{21}g_{31}^{-1}g_{32}$ (1.7.6), thus the Theorem A.15 is a consequence of A.12. If not, we prove it by induction as in 1.7.

A.16. Let $\tilde{\ell}$ be an oriented Lagrangian space above ℓ. Let $s(g) = m(\tilde{\ell},g.\tilde{\ell})$. The function $s(g)$ satisfies:

$$c_\ell(g_1,g_2)^2 = s(g_1)^{-1}\, s(g_2)^{-1}\, s(g_1g_2).$$

Thus the square of the cocyle of the Weil representation is a coboundary. This implies that the representation $R_\ell(g)$ lifts up to a true representation of a double covering of G, the metaplectic group, that we now describe:

Let T = the unit circle in $\mathbb{C}$. We consider as in 1.6.2, the Mackey group $G_c = G \times T$ with the multiplicative law:

$$(g_1,t_1)\cdot(g_2,t_2) = (g_1g_2,t_1t_2c_\ell(g_1,g_2)^{-1}).$$

Then the function $\widehat{R}(g,t) = tR_{\ell}(g)$ is a unitary representation of the group G_c. As in 1.7.9, we define the metaplectic group G_2 to be the subgroup of G_c given by:

$$G_2 = \{(g,t);\ t^2 = s(g)^{-1}\}.$$

The map $(g,t) \to g$ realizes G_2 as a double covering of G.

Let us choose a section $g \to (g,t(g))$ of G in G_2, i.e. let us choose a function $g \to t(g)$ such that $t(g)^2 = s(g)^{-1}$. Then the map $(g,\varepsilon) \to (g, \varepsilon\, t(g))$ is an isomorphism of the set $G \times \mathbb{Z}/2\mathbb{Z}$ with G_2. The multiplicative law on $G \times \mathbb{Z}/2\mathbb{Z}$ becomes:

A.17. $(g_1, \varepsilon_1)(g_2\ \varepsilon_2) = (g_1g_2, \varepsilon_1\ \varepsilon_2 c_t^!(g_1,g_2)^{-1})$,

with

$$c_t^!(g_1,g_2) = \frac{c_{\ell}(g_1,g_2)\ t(g_1g_2)}{t(g_1)\ t(g_2)} = \pm 1, \text{ from A.16.}$$

Choosing $t(u) = \gamma(1)^{(1-(n-\dim(\ell\cap u\cdot\ell))}\ \gamma(\det g_{u.\ell,\ell})^{-1}$ for ex., if $u = \begin{pmatrix} a & b \\ c & d \end{pmatrix}$, with c invertible, $t(u) = \gamma(1)^{1-n}\gamma(\det c)^{-1})$ and using A.12, A.13, it is possible to express $c_t^!$ in function of the Hasse invariant and Hilbert symbols. If

$$g_1 = \begin{pmatrix} a_1 & b_1 \\ c_1 & d_1 \end{pmatrix},\quad g_2 = \begin{pmatrix} a_2 & b_2 \\ c_2 & d_2 \end{pmatrix},\quad g_3 = g_1g_2 = \begin{pmatrix} a_3 & b_3 \\ c_3 & d_3 \end{pmatrix}$$

with $\det c_1 = \delta_1$, $\det c_2 = \delta_2$, $\det c_3 = \delta_3$, all non zero, we have

$$c_t^!(g_1,g_2) = \mathcal{E}(c_1^{-1}c_3c_2^{-1})(\delta_3,-\delta_1\delta_2)(\delta_1,\delta_2)$$

and similar formulas on the full set $G \times G$. In particular, if $G = SL(2,k)$

$$t\begin{pmatrix} a & b \\ c & d \end{pmatrix} = v(c)^{-1} \quad \text{if} \quad c \neq 0$$

$$t\begin{pmatrix} a & b \\ c & d \end{pmatrix} = v(1)\, v(a)^{-1} \quad \text{if} \quad c = 0$$

and

$$c_t'(g_1,g_2) = (v(g_1),v(g_2))(-v(g_1)v(g_2),v(g_1g_2))$$

with

$$v\begin{pmatrix} a & b \\ c & d \end{pmatrix} = \begin{cases} c & \text{if } c \neq 0 \\ d & \text{if } c = 0 \end{cases}$$ which is the formula of Kubota [6].

Bibliographical Notes.

The study of the Heisenberg group and of its commutation relations (Heisenberg's uncertainty principle) was developed with the theory of quantum mechanics. A fundamental result is the theorem of Stone-Von Neumann [1] (Section 1.3).

As discovered by I. Segal [13], the uniqueness of the canonical commutation relations leads to the construction of a projective representation of the symplectic group. This representation was studied by D. Shale [14] for the finite or infinite dimensional real symplectic group and by A. Weil [17] for the symplectic group over a local field.

In connection with asymptotic solutions of systems of differential equations, Maslov [10] introduced an index. The properties of this index were studied by Arnold [1], Hörmander [4] and Leray [7]. We introduced a modified definition due to M. Kashiwara who obtained the results of Section 1.5 and gave the approach of Section 1.9. The link between the Maslov index, half forms and the Shale-Weil representation was observed by J. M. Souriau [15] and R. Blattner-B. Kostant-S. Sternberg [3]. Using the definition of M. Kashiwara, the Theorem 1.6.1 was obtained by the first author. This gives an explicit formula for the cocycle of the Shale-Weil representation ([8] a)). This formula was known on an open subset of the symplectic group [17].

In Section 1.7, the Theorem 1.7.6 is due to the second author (see [7] a)). Results of Section 1.9 are due to Masaki Kashiwara. Results of the appendix are due to Patrice Perrin [11] and Ranga Rao [12]. Of course, underlying implicit references are [14] and [17].

We note that B. Magneron [9] has defined a generalization of the Maslov index for a triple of complex positive Lagrangian planes, which he used to compute the cocycle in the Fock model.

The Weil representation is a privileged example of the deep relation between group representations and symplectic geometry. This relation, known as "the orbit method", was developed by the work of A. A. Kirillov [5] a) (see also b)) on nilpotent groups. The cocycle of the Weil representation is a particular and fundamental case of the Mackey cocycle for extending a representation of a nilpotent Lie group. The corresponding Mackey extension has been determined by M. Duflo [2]. Generalization of the explicit formula 1.6.11 has been given by the first author in [7] a).

References

[1] V. I. Arnold: "Characteristic class entering in quantization conditions," Functional Analysis and its Applications, Vol. 1, 1967, p. 1-13.

[2] M. Duflo: "Sur les extensions des representations irreductibles des groupes de Lie nilpotents," Ann. Scient. Ec. Norm. Sup. 5, 1972, p. 71-120.

[3] V. Guillemin - S. Sternberg: "Geometric asymptotics," Math. Surveys 14, A.M.S., 1977, Providence, R. I.

[4] L. Hörmander: "Fourier integral operators I," Acta Math. 127 (1971), p. 79-183.

[5] A. A. Kirillov: a) "Unitary representations of nilpotent Lie groups," Uspehi Mat. Nauk. 17, 1962, p. 57-110.
b) "Elements of the theory of representations," Springer-Verlag 1976, Berlin-New York.

[6] T. Kubota: "Topological covering of SL(2) over a local field," J. Math. Soc. Japan, 19 (1967), p. 114-121.

[7] J. Leray: "Analyse lagrangienne et mecanique quantique," Publ. Math. I.R.M.A., Vol. 25, Strasbourg, 1978 (English translation by Carolyn Schröeder in M.I.T. Press).

[8] G. Lion: a) "Indices de Maslov et representation de Weil," Publ. Université Paris 7, fasc 2, 1978.
b) "Intégrales d'entrelacement sur des groupes de Lie nilpotents et indices de Maslov," Colloque de Marseille-Luminy. Lecture Notes 587, p. 160-176. Springer-Verlag, Berlin, Heidelberg, New York.

c) "Extension de representations de groupes de Lie nilpotents," C. R. Acad. Sc. Paris, Vol. 288, 26 Mars 1979.

[9] B. Magneron: "Une extension de la notion d'indice de Maslov," Thesis, Université Marie et Pierre Curie, Paris 6, 1979, C. R. Acad. Sc. Paris, Vol.

[10] V. P. Maslov: "Theorie des perturbations et methodes asymptotiques," Dunod 1972, Paris.

[11] P. Perrin: "Representation de Schrödinger et groupe metaplectique sur les corps locaux," Thesis, Université Paris 7, 1979.

[12] R. Rao: "On some explicit formulas in the theory of Weil representation." (preprint).

[13] I. E. Segal: "Foundations of the theory of dynamical systems of infinitely many degrees of freedom," (I), Mat. Fys. Medd. Danske Vid. Selsk. 31, No. 12, 1959, p. 1-39.

[14] D. Shale: "Linear symmetries of free boson fields," Trans. Amer. Math. Soc., 103, p. 149-167, 1962.

[15] J. M. Souriau: "Construction explicite de l'indice de Maslov et applications," Fourth international colloquium on group theoretical methods in physics, University of Nijmegen, 1975.

[16] J. von Neumann: "Die Eindeutigkeit der Schrödingershen Operatoren," Ann. Mat. Pure Appl. 104, 1931, p. 570-578.

[17] A. Weil: Sur certains groupes d'operateurs unitaires, Acta Math. 111, 1964, p. 143-211.

The Shale-Weil representation. Part II

θ-series and correspondences

by

Michèle Vergne

2.0. <u>Introduction:</u> We have constructed in Part I a projective representation R of the symplectic group. As discovered by A. Weil, this representation plays a central role in the transformation properties of the classical Jacobi θ-serie $\theta(z) = \sum_n e^{i\pi n^2 z}$ and higher dimensional θ-series, when interpreted as suitable coefficients of this representation R. We indicate now the nature of the relation between R and theta series:

Let D be the Siegel upper half-plane, i.e.

$$D = \{Z, (n \times n) \text{ complex symmetric matrices, such that } \operatorname{Im} Z >> 0\}.$$

The symplectic group G acts transitively on D, via the fractional linear transformations $Z \to (AZ + B)(CZ + D)^{-1}$.

Let (R,V) be a representation of G (or of a covering group of G) in a topological vector space V (for example, the space of C^∞-vectors of a unitary representation T). We suppose that there exists a covariant map $Z \to v_Z$ of D to V such that

2.0.1. $$R(g)\cdot v_Z = \det(CZ + D)^{-k}\, v_{g\cdot Z}.$$

Let Γ be a discrete subgroup of G. Let $\theta \in V^*$ be a linear functional on V semi-invariant under Γ, i.e. we have:

2.0.2. $$\langle R(\gamma)\cdot\theta, v\rangle = \langle\theta, R(\gamma)^{-1}.v\rangle = \chi(\gamma)\langle\theta, v\rangle, \text{ for } \gamma \in \Gamma, v \in V,$$

with χ a character of Γ.

It is clear from 2.0.1, 2.0.2 that the function $\theta(Z) = \langle \theta, v_Z \rangle$ is a function on the Siegel upper half-plane satisfying the modular relation:

$$\theta((AZ+B)(CZ+D)^{-1}) = \chi(\gamma)^{-1} \det(CZ+D)^k\, \theta(Z), \text{ for } \gamma = \left(\frac{A|B}{C|D}\right) \in \Gamma.$$

We will call $\theta(Z)$ the coefficient of the representation R with respect to v_Z and θ. Thus the transformation properties of such function $\theta(Z)$ are the immediate consequences of the separate transformation properties 2.0.1. for v_Z and 2.0.2. for θ and the search for such theta-functions will be divided in two parts: the construction of covariant functions $Z \to v_Z$ from D to V and the construction of semi-invariant functionals θ under Γ.

Let us consider the first problem: It is clear that a covariant map $Z \to v_Z$ from D to V is completely determined by its value at the point $Z_0 = i\,\mathrm{Id}$ of D. The vector v_{Z_0} has to be an eigenvector for the stabilizer K_0 of Z_0 in G. If we require that the function $Z \to v_Z$ be holomorphic in Z in order for $\theta(Z)$ to be a holomorphic function of Z (the corresponding infinitesimal condition is that v_{Z_0} should be a highest weight vector with respect to a compact Cartan subgroup of G), all such "holomorphic" covariant maps arise in the following way: let us consider the representation $\overline{T}_k$ of G on the space $\overline{\mathcal{F}}(D)$ of antiholomorphic functions on D given by:

2.0.3. $(\overline{T}_k(g^{-1})f)(\overline{U}) = (\det(C\overline{U}+D))^{-k}\, f(AU+B)(CU+D)^{-1})$ for

$$g = \left(\frac{A|B}{C|D}\right).$$

Consider the map $Z \to v_Z$ with

2.0.4. $$v_Z(U) = \det\left(\frac{\bar{U}-Z}{2i}\right)^{-k}$$

then $Z \to v_Z$ is a covariant holomorphic map (with weight k) from D to $(\mathbb{T}_k, \mathcal{O}(D))$.

However, the problem 2.0.2. for $(\mathbb{T}_k, \mathcal{O}(D))$ is tautological to the construction of modular functions. At the contrary, the model of the Weil representation will provide ample examples of θ-distributions semi-invariant under congruence subgroups of $Sp(n,\mathbb{Z})$.

Let us consider (V,B) the canonical real symplectic vector space of dimension $2n$. Let N be the Heisenberg group associated to (V,B). We have constructed in (1.2) a unitary representation $W(\ell)$ of the Heisenberg group by inducing a character of the connected maximal abelian subgroup L associated to a Lagrangian subspace ℓ. As pointed out by P. Cartier [4], there is another model for the representation W especially interesting in the context of θ-series: Let r be a self-dual lattice in (V,B) and χ a quasi-character of r (i.e. $\chi(v+v') = e^{i\pi B(v,v')}\chi(v)\chi(v')$). To (r,χ) is associated the induced representation $W(r,\chi)$ by the subgroup $R = \exp(r + \mathbb{R}E)$ of the Heisenberg group N (section 2.1). This representation $W(r,\chi)$ is irreducible (2.1.13), thus is a model for the representation W.

In some symplectic coordinates, $r = \bigoplus_{i=1}^{n} \mathbb{Z}P_i \oplus \bigoplus_{i=1}^{n} \mathbb{Z}Q_i$ and we choose χ as given by: $\chi(\Sigma\, m_i P_i + \Sigma\, n_i Q_i) = (-1)^{\Sigma\, m_i \cdot n_i}$.

The symplectic group G acts on N. The θ-group $\Gamma(r,\chi)$ is then defined to be the subgroup of the symplectic group leaving stable (r,χ). In the basis precedently chosen, we have

$$\Gamma(r,\chi) = \{ \left(\begin{array}{c|c} A & B \\ \hline C & D \end{array}\right) \in Sp(n,\mathbb{Z}) ;\ A^tB,\ D^tC \text{ have even diagonal coefficients} \} \qquad (2.2.19)$$

As the action of $\Gamma(r,\chi)$ on N leaves stable the inducing datas, $\Gamma(r,\chi)$ acts naturally on the space $H(r,\chi)$ by automorphisms.

Let us choose $\ell = \bigoplus_{i=1}^{n} \mathbb{R}P_i$ as fixed Lagrangian subspace. Thus W (in the Schrödinger model associated to ℓ) is realized in $L^2(\bigoplus_{i=1}^{n} \mathbb{R}Q_i) = L^2(\mathbb{R}^n)$. For $g \in G$, the canonical operator $R(g)$ is the (essentially) unique unitary operator such that $R(g)\, W(n)\, R(g)^{-1} = W(g\cdot n)$.

Let us consider the functional $(\theta,f) = \sum_{\xi\in\mathbb{Z}^n} f(\xi)$. It is immediate to verify that the function:

$$n \to (\theta, W(n)^{-1}.f) = (\theta\cdot f)(n)$$

is naturally an element of $H(r,\chi)$. It follows (2.2.30) from the unicity of R that

2.0.5. $$(\theta\cdot R_\ell(\gamma)\cdot f)(n) = \alpha(\gamma)^{-1}\,(\theta\cdot f)(\gamma^{-1}n)\ .$$

In particular the functional $f \to (\theta,f) = (\theta f)(e)$ where e is the identity of the group N, is semi-invariant under the operators $R(\gamma)$ $(\gamma \in \Gamma(r,\chi))$, and we have solved the question 2.0.2. (This result is essentially equivalent to Poisson summation formula.)

The explicit formula for $\alpha(\gamma)$ is not known. We however express $\alpha(\nu)$ as a Gauss sum (2.2.26), (or 2.2.8, more generally). The formula for the cocycle c_ℓ of the Weil representation determined in Part I, Section 1.6 gives us the relation (2.2.28)

$$\alpha(\gamma_1\gamma_2) = \alpha(\gamma_1)\alpha(\gamma_2)\, e^{\frac{i\pi}{4}\tau(\ell,\gamma_1\ell,\gamma_1\gamma_2\cdot\ell)} .$$

This relation implies reciprocity formula for Gauss sums.

Let us discuss the Problem 2.0.1. Let Z be an element of the Siegel upper half-plane. Then Z defines a complex structure on (V,B), thus a subspace λ_Z of $V^{\mathbb{C}}$.

2.0.6. The function $(v_Z)(\xi) = e^{i\pi(Z\xi,\xi)}$ (which is $L^2(\mathbb{R}^n)$, as $\operatorname{Im} Z >> 0$) is then characterized as being the unique vacuum vector for λ_Z (2.2.23). Thus it follows that the function $Z \to v_Z$ is a covariant map from D to $L^2(\mathbb{R}^n)$, i.e. $R(g)\cdot v_Z$ is proportional to $v_{g.Z}$. It is easy to compute (2.2.35) that:

2.0.7. $$\tilde{R}(g)\cdot v_Z = (\det(CZ + D))^{-1/2}\, v_{g\cdot Z}$$

(where $\tilde{R}$ is the true representation of the metaplectic group).

Thus from properties 2.0.5-2.0.6, we see that $\theta(Z) = (\theta, v_Z) = \sum_{\xi\in\mathbb{Z}^n} e^{i\pi(Z\xi,\xi)}$ is a modular form on D of weight $1/2$, with respect to a double covering group of the θ-group, i.e. we have:

$$\theta((AZ+B)(CZ+D)^{-1}) = \varepsilon(\gamma)\,(\det(CZ+D))^{1/2}\,\theta(Z) \quad \text{for every } \gamma \in \Gamma(r,\chi).$$

The explicit determination (1.7) of the square of the cocycle c_ℓ leads us (2.2.22) to the determination of $\varepsilon(\gamma)^2$, as given by Igusa [13].

In Section 2.3, we establish in more detail the relations between "highest weight" vectors of a representation T of $SL(2,\mathbb{R})$ (or of the universal covering group), coefficients of T and holomorphic modular forms. We also define Poincare series and their basic property with respect to the Petersson inner product.

In Section 2.4, we give as example the construction of modular forms of weight $1/2$ on the upper half-plane by means of θ-series. Let us recall that a theorem of J. P. Serre and H. Stark [28] asserts that all modular forms of weight $1/2$ with respect to some congruence subgroup $\Gamma_0(N)$ arise this way. Section 2.3 and Section 2.4 are not used in the rest of the notes except 2.3.1-2.3.6.

In Section 2.5, we study tensor products of the Weil representation R. The setting is the following: Let (V,B) be a symplectic space and (E,S) an orthogonal space, with a quadratic form S. Then the space $(V \otimes E, B \otimes S)$ is again a symplectic space. It is clear that the direct product $Sp(B) \times O(S)$ is naturally imbedded in $Sp(B \otimes S)$. The pair

$(Sp(B), O(S))$ is a dual pair in the terminology of Roger Howe [11] This imbedding allows us to construct a representation R_S of $Sp(B) \times O(S)$ by considering the restriction of the Weil representation R of $Sp(B \otimes S)$ to $Sp(B) \times O(S)$. The role of the group $O(S)$ in decomposing the representation R_S is analogous to the role of the symmetric group $\mathfrak{S}_k$ in decomposing k-tensors under the action of $GL(n, \mathbb{C})$. Where S is positive definite, each isotypic component of R_S under $O(k)$ parametrize in a one-to-one way an irreducible representation of $Sp(B)$ (Howe, see [33], Kashiwara-Vergne [15]).

If $V = \mathbb{R}P \oplus \mathbb{R}Q$, then $V \otimes E = \mathbb{R}P \otimes E + \mathbb{R}Q \otimes E$ and the space of the representation R_S is $L^2(E)$. Let us suppose that S is positive definite. Let P be an homogeneous harmonic polynomial of degree d, with respect to S. Then the map $z \to v_z^P$, where

2.0.8. $$(v_z^P)(\xi) = P(\xi)\, e^{i\pi z S(\xi,\xi)}$$

is easily seen (2.5.15) to be a covariant map from the upper half-plane to $L^2(E)$ of weight $(k/2 + d)$.

Let L be a lattice in E, integral with respect to S; L^* its dual, then $L^* \underset{\mathbb{Z}}{\otimes} \mathbb{Z}P + L \underset{\mathbb{Z}}{\otimes} \mathbb{Z}Q$ is a self dual lattice r in $V \otimes E$. The image of $\Gamma_0(L) \times O(L)$ in $Sp(V \otimes B)$, for $\Gamma_0(L)$ an appropriate congruence subgroup of $SL(2, \mathbb{Z})$ and $O(L)$ the discrete subgroup of $O(S)$ leaving L stable, is contained in the corresponding θ-group $\Gamma(r,\nu)$. It follows that the distribution

2.0.9. $$(\theta_L, f) = \sum_{\xi \in L} f(\xi)$$

is semi-invariant under $R_S(\Gamma_0(L) \times O(L))$ (here S is of arbitrary signature). We compute explicitly the corresponding multiplicator $\alpha(\nu)$ (2.6.8). Similarly if $\dim V = 2n$, n arbitrary, $\dim E = k$ with k even, the determination of the square of the cocycle of the Weil representation (1.7) give us the multiplicator of the semi-invariant distribution θ_L (2.6.20), as determined by Andrianov [1].

If S is positive definite, the construction of v_z^P and θ_L leads us to the classical theta series associated to positive definite quadratic forms, via:

$$\theta_P(z) = \langle \theta_L, v_z^P \rangle = \sum_{\xi \in L} P(\xi)\, e^{i\pi z S(\xi,\xi)}$$

which are modular forms of weight $(k/2 + d)$ on appropriate congruence subgroups.

Some modular forms considered by Hecke [10] or Zagier [37] are similar to these θ-series, being given by a serie of the form $\sum_{\substack{\xi \in L \\ S(\xi)>0}} p(\xi)\, e^{2i\pi z S(\xi,\xi)}$. However, here $S(\xi,\xi)$ is indefinite and the sum is restricted to the subset of L contained in the cone $S(\xi,\xi) > 0$. (We denote $S(\xi,\xi)$ also by $S(\xi)$.)

S. Rallis and G. Schiffmann [12] were indeed showing that these θ-series arise also by forming coefficients of the representation R_S, where S is indefinite. They explicitly constructed rational function $p(\xi)$ harmonic with respect to S, vanishing on the set $S(\xi,\xi) = 0$ such that, if:

$$(v_z^p)(\xi) = p(\xi)\, e^{i\pi z S(\xi,\xi)} \qquad \text{for } S(\xi,\xi) > 0$$

$$= 0 \qquad \text{if } S(\xi,\xi) \leq 0,$$

the map $z \to v_z^p$ is a covariant map from the upper half-plane to $L^2(E)$. Thus the representation R_S contains discretely some representation of $SL(2,\mathbb{R})$ with highest weight vectors, but in a striking way, this highest weight vectors are supported on the set $S(\xi) > 0$ (Section 2.5).

The corresponding θ-series $(\theta_L, v_z^p) = \sum_{\substack{\xi\in L \\ S(\xi)>0}} p(\xi)\, e^{2i\pi z S(\xi,\xi)}$ reinterpret thus some Hecke-series or the Zagier kernel in the framework of the Weil representation, (2.6.14-2.6.16).

We now indicate some applications of the representation R_S to correspondences between modular forms.

In Section 2.7, we consider the case, where S is of signature (2,1). Then the connected component of group $O(2,1)$ is isomorphic to $PSL_2(\mathbb{R})$. Thus the representation R_S is a representation of $G_2 \times PSL_2(\mathbb{R})$, where G_2 is the double covering of $SL(2,\mathbb{R})$. It follows from the results of Rallis-Schiffmann (Section 2.5) that, for $k \geq 1$, the representation $T_{k+1/2} \otimes T_{2k}$ (2.0.3) is contained discretely in R_S.

If $v(z,\tau)$ is the corresponding covariant map, and L is a lattice in E, the θ series $\theta_L(z,\tau) = \langle \theta_L, v(z,\tau)\rangle$ will be a holomorphic modular form in z of weight $k + 1/2$, and in τ of weight $2k$. Integration of modular forms of half integral weight $k + 1/2$ against $\theta_L(z,\tau)$ for the Petersson inner product produces modular forms of even weight $2k$. For an appropriate choice of L, this correspondence between modular forms of half integral weight and modular forms of even weight is the Shimura-correspondence ([30]). Note that Shintani [31] and Niwa [21] were at the origin of this interpretation of Shimura-correspondence.

In Section 2.8, we consider the case where S is of signature $(2,2)$, in view of the Doi-Naganuma lifting: Let $K = \mathbb{Q}(\sqrt{D})$ be a real quadratic field, let $f(\tau) = \Sigma\, a(n)\, e^{2i\pi n\tau}$ be a modular form of weight k for the full group $SL(2,\mathbb{Z})$. Suppose f is an eigenfunction of all the Hecke operators. Let $L_f(s) = \Sigma\, a(n)\, n^{-s}$ be the associated Dirichlet serie and $L_f(\chi,s) = \Sigma\, a(n)\, (\frac{n}{D})\, n^{-s}$ the twisted serie. Then, Doi-Naganuma [7] proved (under certain restrictions) that the function $L_f(s)\, L_f(\chi,s)$ is the Dirichlet series associated to a Hilbert modular form $F(z_1,z_2)$ of weight (k,k).

Following Zagier [37], we will explicit this correspondence $f \to F$ as given by an explicit kernel $\Omega(\tau;z_1,z_2)$, via the Petersson inner product.

For this, we consider a form S of signature $(2,2)$. The connected component of $O(2,2)$ is a quotient of $SL(2,\mathbb{R}) \times SL(2,\mathbb{R})$. Thus the representation R_S is a representation of $SL(2,\mathbb{R}) \times (SL(2,\mathbb{R}) \times SL(2,\mathbb{R}))$. For $k > 1$, the representation $T_k \otimes (T_k \otimes T_k)$ is contained in R_S with multiplicity one. We explicit the corresponding highest-weight vector as follows:

We choose $E = \{2 \times 2$ matrices $x = \begin{pmatrix} x_1 & x_3 \\ x_2 & x_4 \end{pmatrix}\}$ and $S(x,x) = -2 \det x$. The group $SL(2,\mathbb{R}) \times SL(2,\mathbb{R})$ acts on E via $(g_1,g_2)\cdot x = g_1 x g_2^{-1}$. For $(z_1,z_2) \in P^+ \times P^+$, we consider

$$Q(z_1,z_2) = \begin{pmatrix} -z_1 & z_1 z_2 \\ -1 & z_2 \end{pmatrix} \in E^{\mathbb{C}} .$$

The Rallis-Schiffmann function $v(\tau;z_1,z_2)$ is given by

$$\begin{aligned} v(\tau;z_1,z_2)(\xi) &= S(\xi,Q(z_1,z_2))^k \, S(\xi,\xi)^{k-1} \, e^{i\pi\tau S(\xi,\xi)} \\ &\qquad \text{on } S(\xi,\xi) > 0 \\ &= 0 \quad \text{on } S(\xi,\xi) \leq 0 . \end{aligned}$$

This function satisfies the fundamental relation:

$$\begin{aligned} &R_S(g;(g_1,g_2))\cdot v(\tau,(z_1,z_2)) \\ &\quad = (c\tau+d)^{-k} \, (c_1 z_1+d_1)^{-k} \, (c_2 z_2+d_2)^{-k} \, v(g\cdot\tau;(g_1\cdot z_1,g_2\cdot z_2)) \end{aligned}$$

for $g, g_1, g_2 \in SL(2,\mathbb{R})$, and τ, z_1, z_2 in the upper half-plane.

The lattice

$$L = \{ \begin{pmatrix} \lambda & \sqrt{D}\,a \\ \sqrt{D}\,b & \lambda' \end{pmatrix} ; a,b \in \mathbb{Z} , \lambda \in \mathcal{O}\}$$

is invariant under the action of the Hilbert subgroup $SL(2,\mathcal{O})$ imbedded in $SL(2,\mathbb{R}) \times SL(2,\mathbb{R}) \subset O(2,2)$. This lattice is of level D. Thus we obtain that the distribution θ_L with $(\theta_L,f) = \sum_{\xi \in L} f(\xi)$, is semi-invariant under $\Gamma_0(D) \times SL(2,\mathcal{O})$. In particular, the Zagier kernel $\Omega(\tau,z_1,z_2) = (\theta_L,v(\tau,z_1,z_2))$ is a modular form in τ with respect to $\Gamma_0(D)$, and a Hilbert modular form on (z_1,z_2).

If S is of signature $(1,1)$, the representation R_S is isomorphic to the natural representation of $SL(2,\mathbb{R})$ in $L^2(\mathbb{R}^2)$ via $(g^{-1}.f)(x,y) = f(ax+by,cx+dy)$. This simple fact (2.5.6) allows us to compare θ_L with a distribution related to the orbits of $SL(2,\mathbb{Z})$ in $\mathbb{Z}^2$. This in turn leads to the Zagier identity expressing $\Omega(\tau;z_1,z_2)$ in functions of Poincaré series in τ. This proof of the Zagier identity is based on an idea of Rallis and Schiffmann. As in Zagier, this identity is at the basis of the explicit calculation of the Fourier expansion of f, when $f \to F$ is the Doi-Naganuma map (for $\Gamma_0(D)$).

In the same spirit than Doi-Naganuma, H. Cohen [6] associated to a modular form f with respect to any congruence subgroup $\Gamma_0(N)$ a Hilbert modular form $C_f^K(z_1,z_2)$. He also conjectured the level of the modular forms C_f^K. In Section 2.9 we prove this conjecture of Cohen. Namely, we prove the:

Theorem: Let $K = \mathbb{Q}(\sqrt{D})$. with $D \equiv 1 \bmod 4$. Let k be an integer greater or equal to 3.

Let $f(\tau) = \Sigma\, a(n)\, e^{2i\pi n\tau} \in S_k(\Gamma_0(N),\chi)$ be a cusp form of weight k and character χ on $\Gamma_0(N)$, where χ is a character mod N.

Let us define, for $\mathfrak{A}$ an integral ideal of $\mathcal{O}_K$,

$$c(\mathfrak{A}) = \sum_{\substack{r\in\mathbb{N}^+\\ r|\mathfrak{A}}} r^{k-1}\chi(r)\left(\frac{r}{D}\right) a(N_{K/\mathbb{Q}}(\mathfrak{A}/r)).$$

Then:

$$C_f^K(z_1,z_2) = \sum_{\substack{\nu\in\delta^{-1}\\ \nu>>0}} c(\nu\delta)\, e^{2i\pi(\nu z_1+\nu' z_2)}$$

is a Hilbert modular form of weight k and character $\chi \circ N_{K/\mathbb{Q}}$ on the congruence subgroup

$$\Gamma_0(N,\mathcal{O}_K) = \{\begin{pmatrix}\alpha & \beta\\ \gamma & \delta\end{pmatrix} \in SL(2,\mathcal{O}_K),\ \gamma \in N\mathcal{O}_K\}.$$

Our method is similar to the one of Zagier. We will reinterpret the Cohen map $f \to C_f^K$ as given by the Petersson scalar product with a θ-function $\theta_\chi(\tau,z_1,z_2) = \langle V_\chi, v(\tau,z_1,z_2)\rangle$, where V_χ is a semi-invariant distribution on E carefully chosen. The main difficulty is to prove an identity for $\theta_\chi(\tau,z_1,z_2)$ where the modular properties in the variable τ or (z_1,z_2) are separately evident on each side of the formula--this is done in a similar way as the proof of the Zagier identity.

We have chosen in these notes to minimize the group theoretical background. In particular, at the exception of the study of the Shale-Weil representation R, described in Part I, no a-priori knowledge of unitary representations of $SL(2,\mathbb{R})$ is required.

The results of Section 2.1, 2.2, and 2.6 on transformation properties of the θ-distributions are classical. We have derived here these properties from the explicit calculation of the cocycle of the Weil representation as presented in Part I. We believe this exposition is enlightening. However, a reader mainly interested in the material of Section 2.7, 2.8, 2.9 and familiar with the transformation properties of θ-series can can read only 2.5, 2.7, 2.8, 2.9 (eventually with some glance at earlier paragraphs).

The exposition of the applications of the representation R_S to correspondences is strongly influenced by the work of Rallis and Schiffmann. We have however used also ideas of M. F. Vigneras [35] and of R. Howe [12] to give quite simple-minded proofs. As a result of these simplifications, the Rallis-Schiffmann method for computing the explicit expression of kernels as the Zagier kernel, and their expansion in terms of Poincare series, is an effective and conceptually clear way to give explicit correspondences.

More detailed bibliographical notes, are given at the end

of each section. The list of bibliographical references is at the end of the notes.

I would like to thank Dan Barbasch, Dorian Goldfeld, Victor Kac Dale Peterson, Harold Stark, Audrey Terras, Marie-France Vigneras and Don Zagier for mathematical discussions and references.

As stated before, a part of these notes depends on the discussion of Part I in collaboration with Gerard Lion. Also some roots of this work comes from the collaboration of the author with Masaki Kashiwara on the Weil representation. I would like to thank both of them for sharing their ideas, in these common works.

Also many thanks goes to Sophie Koulouras for the very patient typing of a fluctuating manuscript.

2.1. Lattices and representations of the Heisenberg group.

2.1.1. A lattice r in a real vector space V is a subgroup of V such that r is discrete and V/r is compact. Then, there exists a basis $(e_1, e_2, \cdots, e_n)$ of V over $\mathbb{R}$ such that $r = \mathbb{Z}e_1 \oplus \mathbb{Z}e_2 \oplus \cdots \oplus \mathbb{Z}e_n$.

2.1.2. Let (V,B) be a real symplectic vector space of dimension $2n$. Let r be a lattice in (V,B). We define $r^* = \{\xi \in V,$ such that $B(\xi,r) \subset \mathbb{Z}\}$. If $r = r^*$, r is called a self-dual lattice.

2.1.3. Lemma: Let r be a self-dual lattice in (V,B), m an isotropic subspace of V such that $r \cap m$ generates m as a vector space. Then, there exists a symplectic basis $P_1, P_2, \cdots, P_n, Q_1, Q_2, \cdots, Q_n$ of (V,B) such that:

$$r = \mathbb{Z}P_1 \oplus \mathbb{Z}P_2 \oplus \cdots \oplus \mathbb{Z}P_n \oplus \mathbb{Z}Q_1 \oplus \cdots \oplus \mathbb{Z}Q_n$$

$$m = \mathbb{R}P_1 \oplus \cdots \oplus \mathbb{R}P_k \quad (k \leq n).$$

Proof: We will prove this by induction on $\dim V$. We may assume $m \neq \{0\}$, taking $m = \mathbb{R}\ell$ for some $\ell \in r$. Let us consider x_1 in $r \cap m$. The subgroup $B(x_1,r)$ is a discrete subgroup of $\mathbb{Z}$. Hence there exists an integer N such that $B(x_1,r) = N\mathbb{Z}$. Thus $\frac{x_1}{N} \in r^* = r$ and $B(\frac{x_1}{N},r) = \mathbb{Z}$. We denote by $P_1 = \frac{x_1}{N}$. Let us choose Q_1 an element of r such that $B(P_1,Q_1) = 1$. Consider the orthogonal decomposition

$V = \mathbb{R}P_1 \oplus \mathbb{R}Q_1 \oplus V_0$ of (V,B). It is now easy to prove that $r = \mathbb{Z}P_1 \oplus \mathbb{Z}Q_1 \oplus r \cap V_0$: if $x \in r$ and if $B(x,P) = n_1$, $B(x,Q) = n_2$, $x - n_2P_1 + n_1Q_1 \in r \cap V_0$. Similarly $m = \mathbb{R}P_1 \oplus V_0 \cap m$. Furthermore $r \cap V_0$ is a self-dual lattice in V_0 and $r \cap V_0 \cap m$ generates $V_0 \cap m$ as a vector space. Therefore the lemma follows from our induction hypothesis.

2.1.4. Let (V,B) be a symplectic vector space and $r = r^*$ a self-dual lattice in V. We consider the subgroup $R = \exp(r \oplus \mathbb{R}E)$ of the Heisenberg group attached to (V,B).

We are interested in the representations W of the Heisenberg group satisfying $W(\exp tE) = e^{2i\pi t}\,\mathrm{Id}$. Thus all these representations are trivial on the discrete central subgroup $\exp \mathbb{Z}E$ of N. Hence we may consider them as representations of $\bar{N} = N/\exp \mathbb{Z}E$. We identify $\bar{N}$ with $V \times T$ via $(v,t) \to (v, e^{2i\pi t})$ with the multiplicative law being given by:

$$(u,\tau)\cdot(v,\tau') = (u+v,\ \tau\tau'\, e^{i\pi B(u,v)}).$$

Let $\bar{R}$ be the image of the group R in $\bar{N}$. The fact that the lattice r is self-dual is equivalent to the fact that $\bar{R}$ is a maximal commutative subgroup of $\bar{N}$: the condition $(u,\tau)\cdot(v,\tau') = (v,\tau')\cdot(u,\tau)$ for every $u \in r$ is equivalent to the condition $e^{2i\pi B(u,v)} = 1$ for every $u \in r$, i.e. $v \in r^*$.

The subgroup $T = \{(0,\tau)\}$ is a subgroup of the commutative group $\bar{R}$. Let χ be a character of $\bar{R}$ extending the character $(0,\tau) \to \tau$ of the subgroup T. We consider χ as a character of R such that $\chi(\exp tE) = e^{2i\pi t}$.

2.1.5. Let (r,χ) be a self-dual lattice with a given character χ of R. We consider the representation $W(r,\chi)$ of N induced by the character χ of R, i.e. $W(r,\chi) = \mathrm{Ind}\uparrow_R^N \chi$. By definition $W(r,\chi)$ is realized in the Hilbert space: $H(r,\chi) = \{\varphi$, measurable functions on N with values in $\mathbb{C}$ such that:

2.1.5.a) $\varphi(n\gamma) = \chi(\gamma)^{-1}\,\varphi(n),\ \gamma \in R,\ n \in N$

2.1.5.b) $\int_{N/R} |\varphi|^2\, d\dot{n} < \infty\}$.

Let us remark here that, as $N/R \simeq V/r$ is compact, there is a canonical choice of the measure $d\dot{n}$ on N/R, namely we can choose $d\dot{n}$ such that the volume of the total space N/R is 1.

The map $(I_r f)(x) = f(\exp x)$ identifies $H(r,\chi)$ with the space $L^2(V,r,\chi)$ of measurable functions f on V such that:

2.1.5.a') $f(x+\gamma) = e^{i\pi B(x,\gamma)}\,\chi(\gamma)^{-1} f(x),\ x \in V,\ \gamma \in r$

2.1.5.b') $\int_{V/R} |f(x)|^2\, d\dot{x} < \infty$.

Let r be identified to the canonical lattice $\mathbb{Z}^{2n}$ with respect to some symplectic basis (P_i, Q_j) and let $\Delta = \{(x,y) = (\Sigma\, x_i P_i,\ \Sigma\, y_i Q_i)$, with $0 \le x_i \le 1,\ 0 \le y_j \le 1\}$, then

$$\int_{V/r} |f(x)|^2\, d\dot{x} = \int_{\Delta} |f(x)|^2\, dx .$$

2.1.6. Our aim is to prove that the representation $W(r,\chi)$ is irreducible. Let us first discuss how this model depends of the choice of the character χ. The subgroup R of N is a normal subgroup of N, however the action of N doesn't leave the character χ stable. In fact we have:

2.1.7. Lemma: Let χ and χ' be two characters of R such that $\chi(\exp tE) = \chi'(\exp tE) = e^{2i\pi t}$. There exists an element v of V uniquely determined modulo r, such that $\chi'(\gamma) = \chi((\exp v)\gamma(\exp v)^{-1})$ for every $\gamma \in R$.

Proof: χ'/χ is a character of the discrete subgroup r of the vector space V. So there exists $v \in V$ such that $(\chi'/\chi)(\gamma) = e^{2i\pi B(v,\gamma)}$ and v is uniquely determined mod r by this property. (In the coordinates of Lemma 1.8.3, $\theta = \chi/\chi'$ is a character of $\mathbb{Z}^{2n}$, hence there exists (θ_i, θ_j') uniquely determined mod $\mathbb{Z}$ such that

$$\theta(\Sigma\, m_i P_i + \Sigma\, n_j Q_j) = e^{2i\pi(\Sigma\, m_i\theta_i + \Sigma\, n_j\theta_j')} = e^{2i\pi B(\Sigma\, \theta_j' P_j - \Sigma\, \theta_i Q_i,\, \Sigma\, m_i P_i + \Sigma\, n_j Q_j)}).$$

Then, for this v, we have

$$\chi(\exp v \exp \gamma(\exp v)^{-1}) = \chi(\exp \gamma \exp(B(v,\gamma)E)) = e^{2i\pi B(v,\gamma)}\chi(\gamma) = \chi'(\gamma).$$

2.1.8. Let n_0 be an element of N and let $(n_0\cdot\chi)(\gamma) = \chi(n_0^{-1}\gamma n_0)$. The representation $W(r,n_0\cdot\chi)$ is equivalent to $W(r,\chi)$, as the right translation operator $(\rho(n_0)\varphi)(n) = \varphi(nn_0)$ establishes an isomorphism between $H(r,\chi)$ and $H(r,n_0\cdot\chi)$ commuting with left translations.

2.1.9. Let r be a self-dual lattice in V, $R = \exp(r + \mathbb{R}E)$, and (ℓ,ℓ') two Lagrangian subspaces such that $r = r \cap \ell \oplus r \cap \ell'$. (For a given r, such a couple (ℓ,ℓ') exists by Lemma 2.1.3). Each element γ of R can be written uniquely as $\gamma = \exp u \exp v \exp tE$, with $u \in r \cap \ell$, $v \in r \cap \ell'$, $t \in \mathbb{R}$. We then define $\chi_{\ell,\ell'}(\exp u \exp v \exp tE) = e^{2i\pi t}$, which is a character of R, as it is easy to check. We have $\chi_{\ell,\ell'}(\exp(u + v)) = (-1)^{B(u,v)} = \pm 1$ for $u \in \ell \cap r$, $v \in \ell' \cap r$ (we recall that $B(u,v) \in \mathbb{Z}$).

2.1.10. To prove that $W(r,\chi)$ is irreducible, it is sufficient to prove that the representations $W(\ell)$ (1.2) and $W(r,\chi_{\ell,\ell'})$ are unitarily equivalent. More precisely we will construct a canonical operator between these two representations.

2.1.11. We consider the following data:

- ℓ a Lagrangian subspace of (V,B)
- r a self dual lattice in (V,B) such that $r \cap \ell$ generates ℓ as a vector space
- χ a character of $R = \exp(r \oplus \mathbb{R}E)$ such that $\chi(\exp y) = 1$ if $y \in r \cap \ell$.

We consider the representation $W(\ell)$ of N in $H(\ell)$ (1.2) and the representation $W(r,\chi)$ in $H(r,\chi)$. We want to define $\theta^{\chi}_{r,\ell}$ a canonical isomorphism between $H(\ell)$ and $H(r,\chi)$ intertwing the representations $W(\ell)$ and $W(r,\chi)$. As in 1.4 the formal construction of $\theta^{\chi}_{r,\ell}$ is simple:

A function φ in $H(\ell)$ verifies:

a) $\varphi(n \exp y) = \varphi(n)$ (for $y \in \ell$)

b) $\varphi(n \exp tE) = e^{-2i\pi t}\varphi(n)$.

A function φ' in $H(r,\chi)$ must verify:

a') $\varphi(n \exp y) = \chi(\exp y)^{-1}\, \varphi(n)$ (for $y \in r$)

b') $\varphi(n \exp tE) = e^{-2i\pi t}\varphi(n)$.

We remark that if φ verifies a) then φ verifies a') for $y \in r \cap \ell$ as $\chi(\exp y) = 1$ if $y \in r \cap \ell$. Therefore we will "force" a function φ in $H(\ell)$ to verify a') by forming

2.1.12. $$(\theta^{\chi}_{r,\ell}\varphi)(n) = \sum_{u \in r/r\cap\ell} \chi(\exp u)\, \varphi(n \exp u).$$

Now it is clear that formally $\theta^{\chi}_{r,\ell}$ verifies a') and that $\theta^{\chi}_{r,\ell}$, being a sum of right translations operators, commutes with left translations.

As $r \cap \ell$ is a lattice in ℓ, $r \cap \ell$ defines a canonical element $|e_r|$ of $|\Lambda^n_n \ell|$: for $P_1, P_2, \cdots, P_n$ a basis of ℓ such that $r \cap \ell = \bigoplus_{i=1} \mathbb{Z} P_i$, we define $|e_r| = |P_1 \wedge P_2 \wedge \cdots \wedge P_n|$. This doesn't depend on the choice of the $\mathbb{Z}$ basis of $r \cap \ell$, as the matrix for a change of $\mathbb{Z}$- basis has determinant $\pm$ 1.

We now prove:

2.1.13. <u>Proposition</u>: $\theta^{\chi}_{r,\ell}$ defines a unitary isomorphism between $H(\ell, e_r)$ and $H(r,\chi)$.

<u>Proof</u>: By Lemma 1.8.3, we can choose ℓ' a complementary Lagrangian subspace of ℓ in V such that $r = r \cap \ell \oplus r \cap \ell'$. For the choice of the symplectic basis (P_i, Q_j) as in 2.1.3, the space

$H(\ell, e_r)$ is identified with $L^2(\mathbb{R}^n, dy)$, with $\varphi(y) = \varphi(\exp y\cdot Q)$. For $y = (y_1, y_2, \cdots, y_n)$ and $x = (x_1, x_2, \cdots, x_n)$, we denote $\Sigma\, y_i Q_i$ by $y.Q$, $\Sigma\, x_i P_i$ by $x.P$, and $B(x\cdot P, y\cdot Q)$ by $x\cdot y$. For $u = (u_1, u_2, \cdots, u_n) \in \mathbb{Z}^n$, $u\cdot Q$ describes the lattice $r \cap \ell'$. As $r = r \cap \ell \oplus r \cap \ell'$, our operator $\theta^{\chi}_{r,\ell}$ is then expressed as:

$$(\theta^{\chi}_{r,\ell}\varphi)(\exp x.P+y\cdot Q) = \sum_{u\in\mathbb{Z}^n} \varphi(\exp(x\cdot P+y\cdot Q)\exp u\cdot Q)\ \chi(\exp u\cdot Q).$$

We write

$$\exp(x\cdot P+y\cdot Q)\ \exp u.Q = \exp(y\cdot Q+u\cdot Q)\ \exp(x\cdot P)\ \exp(\tfrac{x\cdot y}{2} + x\cdot u)E.$$

Hence

$$(\theta^{\chi}_{r,\ell}\varphi)(\exp(x\cdot P+y\cdot Q)) = \sum_{u\in\mathbb{Z}^n} \chi(\exp u\cdot Q)e^{-2i\pi x\cdot u}e^{-i\pi x\cdot y}\varphi(y+u).$$

For φ in $\mathcal{S}(\mathbb{R}^n)$, this serie is absolutely convergent. Let us compute the norm of $\theta^{\chi}_{r,\ell}\varphi$ in $L^2(\Delta)$. As the functions $\{e^{-2i\pi x\cdot u}\}$ form an orthonormal basis of $L^2(dx;\ 0 < x_i < 1)$, we have

$$\|\theta^{\chi}_{r,\ell}\varphi\|^2_{L^2(\Delta)} = \int_{0<y_i<1} dy \int_{0<x_i<1} |e^{-i\pi x\cdot y} \sum_{u\in\mathbb{Z}^n} \varphi(y+u)\chi(\exp u\cdot Q)e^{-2i\pi x\cdot u}|^2 dx$$

$$= \int_{0<y_i<1} \sum_{u\in\mathbb{Z}^n} |\varphi(y+u)|^2\ dy = \int_{\mathbb{R}^n} |\varphi(y)|^2\ dy = \|\varphi\|^2 .$$

Hence we see that our operator $\theta^{\chi}_{r,\ell}$ is an isometry. We will now prove that $\theta^{\chi}_{r,\ell}$ is surjective. More precisely, we similarly define a natural operator $\theta^{\chi}_{\ell,r}$ from $H(r,\chi)$ into $H(\ell)$ by "forcing" a function φ in $H(r,\chi)$ to be invariant under right translations by ℓ. This leads us to:

2.1.14. $$(\theta^{\chi}_{\ell,r}f)(n) = \int_{\ell/r\cap\ell} f(n \exp \dot{y})\, d\dot{y}\,.$$

Let us check that:

2.1.15. $$\theta^{\chi}_{r,\ell} \circ \theta^{\chi}_{\ell,r} = \mathrm{Id}.$$

We have to calculate for $\varphi \in H(r,\chi)$

$$(\theta^{\chi}_{r,\ell} \circ \theta^{\chi}_{\ell,r}\varphi)(\exp x\cdot P+y\cdot Q)$$

$$= \sum_{u\in\mathbb{Z}^n} \chi(\exp u\cdot Q)e^{-2i\pi x\cdot u}e^{-i\pi x\cdot y}(\theta^{\chi}_{\ell,r}\varphi)(\exp(y\cdot Q+u\cdot Q))$$

$$= \sum_{u\in\mathbb{Z}^n} \chi(\exp u\cdot Q)e^{-2i\pi x\cdot u}e^{-i\pi x\cdot y} \int_{\ell/r\cap\ell} \varphi(\exp y\cdot Q \exp u\cdot Q \exp \dot{t}\cdot P)d\dot{t}$$

$$= \sum_{u\in\mathbb{Z}^n} \chi(\exp u\cdot Q)e^{-2i\pi x\cdot u}e^{-i\pi x\cdot y} \int_{\ell/r\cap\ell} \varphi(\exp y\cdot Q \exp t\cdot P \exp u\cdot Q)e^{2i\pi u\cdot t}dt$$

$$= e^{-i\pi x\cdot y} \sum_{u\in\mathbb{Z}^n} \int_{\mathbb{R}^n/\mathbb{Z}^n} \varphi(\exp y\cdot Q \exp t\cdot P)e^{2i\pi u\cdot(t-x)}\, dt\,,$$

as $\varphi(n \exp u\cdot Q) = \chi(\exp u\cdot Q)^{-1}\, \varphi(n)$,

$$= e^{-i\pi x\cdot y} \sum_{u\in\mathbb{Z}^n} \int_{\mathbb{R}^n/\mathbb{Z}^n} \varphi(\exp y\cdot Q \exp (t+x)\cdot P)e^{2i\pi u\cdot t}\, dt.$$

The function $\alpha(t) = \varphi(\exp y\cdot Q \exp (t+x)\cdot P)$ is periodic in t ($\varphi(n \exp u\cdot P) = \varphi(n)$ for $u \in \mathbb{Z}^n$, as $\varphi \in H(r,\chi)$).
Hence

$$\sum_{u\in\mathbb{Z}^n} \int_{\mathbb{R}^n/\mathbb{Z}^n} \alpha(t)e^{2i\pi u\cdot t}\, dt = \alpha(0).$$

Now we obtain:

$$(\theta^{\chi}_{r,\ell}\cdot\theta^{\chi}_{\ell,r}\varphi)(\exp x\cdot P+y\cdot Q) = e^{-i\pi x\cdot y}\varphi(\exp y.Q\ \exp x\cdot P)$$

$$= \varphi(\exp x\cdot P+y\cdot Q),\ \text{Q.E.D.}$$

2.1.16. Let us relate the Poisson summation formula to the preceding proposition.

Let U be a real vector space with a lattice Γ.

Let U^* be its dual vector space with the lattice

$$\Gamma^* = \{\xi \in U^*;\ \xi(\Gamma) \subset \mathbb{Z}\} .$$

Let f be a function in the Schwartz space $\mathcal{S}(U)$. We define $\hat{f} \in \mathcal{S}(U^*)$ by:

$$f(\xi) = \int_U e^{-2i\pi(\xi,x)}\ f(x)\ dx ,$$

where dx is the Euclidean measure on U defined by Γ.

2.1.17. <u>Proposition</u> (Poisson Summation Formula): Let $f \in \mathcal{S}(U)$, then:

$$\sum_{\gamma\in\Gamma} f(\gamma) = \sum_{\gamma^*\in\Gamma^*} \hat{f}(\gamma^*) .$$

<u>Proof</u>: The alternate form on $V = U \oplus U^*$ given by $B(x_1 + f_1,\ x_2 + f_2) = f_1(x_2) - f_2(x_1)$ defines on V a structure of symplectic vector space. It is clear that $r = \Gamma \oplus \Gamma^*$ is a self-dual lattice in (V,B). The subspaces $\ell = U$, $\ell' = U^*$ are complementary Lagrangian subspaces of (V,B). We consider $\chi = \chi_{\ell,\ell'}$, $\theta_{r,\ell}$: $H(\ell,e_r) \to H(r,\chi)$ and

$\theta_{\ell,r}: H(r,\chi) \to H(\ell,e_r)$ the unitary operators (2.1.12), (resp. 2.1.14). The operator $\mathcal{F}_{\ell,\ell'}: H(\ell',e_r) \to H(\ell,e_r)$ (1.4) and the operator $\theta_{\ell,r} \circ \theta_{r,\ell'}$ are two unitary operators intertwining the unitary representations $W(\ell')$ and $W(\ell)$, thus they are proportional. In fact it is immediate to check that $\mathcal{F}_{\ell,\ell'} = \theta_{\ell,r} \circ \theta_{r,\ell'}$: we have for φ in $L^2(U) = H(\ell')$

1) $(\mathcal{F}_{\ell,\ell'}\varphi)(e) = \int_U \varphi(u)\, du$

2) $(\theta_{r,\ell'}\varphi)(\exp u) = \sum_{\gamma\in\Gamma} \varphi(u+\gamma),\ u \in U$

3) $(\theta_{\ell,r} \circ \theta_{r,\ell'}\varphi)(e) = \int_{U/\Gamma} \left(\sum_{\gamma\in\Gamma} \varphi(u+\gamma)\right) d\dot{u} = \int_U \varphi(u)\, du.$

Thus we obtain by 2.1.15

$$\theta_{r,\ell}\, \mathcal{F}_{\ell,\ell'} = \theta_{r,\ell'} .$$

In particular, for $\varphi \in \mathcal{S}(U)$

$$(\theta_{r,\ell}\, \mathcal{F}_{\ell,\ell'}\varphi)(e) = (\theta_{r,\ell'}\varphi)(e),$$

which is formula 2.1.17.

2.1.18. Let us remark that if U is a vector space with a given measure dx and if $\hat{f}(\xi) = \int_U e^{-2i\pi(\xi,x)} f(x)dx$ is defined with respect to this measure dx, then as $\frac{dx}{\operatorname{vol}\Gamma} = d_\Gamma x$ when $d_\Gamma(x)$ is the measure on U defined by Γ, we have the formula:

$$\frac{1}{\operatorname{vol}\Gamma} \sum_{\gamma^*\in\Gamma^*} \hat{f}(\gamma^*) = \sum_{\gamma\in\Gamma} f(\gamma) .$$

2.1.19. We will also have to consider the following situation:

Let (r_1,χ_1) and (r_2,χ_2) be two self-dual lattices with their associated characters χ_1, χ_2.

Let us suppose that r_1 and r_2 are commensurable, i.e. the change of basis from r_1 to r_2 has rational coefficients, and let us suppose that χ_1 coincides with χ_2 on $r_1 \cap r_2$. As $r_2/r_1 \cap r_2$ is finite, there exists a natural intertwining operator

$$\theta^{\chi_2,\chi_1}_{r_2,r_1} : H(r_1,\chi_1) \to H(r_2,\chi_2)$$

between the representations $W(r_1,\chi_1)$ and $W(r_2,\chi_2)$ given by:

$$(\theta_{r_2,r_1}\varphi)(n) = \sum_{u \in r_2/r_1 \cap r_2} \chi_2(\exp u)\, \varphi(n \exp u).$$

It would be extremely interesting to analyze as in Chapter I, the composed operator:

$$\theta^{\chi_1\chi_3}_{r_1,r_3} \circ \theta^{\chi_3\chi_2}_{r_3,r_2} \circ \theta^{\chi_2\chi_1}_{r_2,r_1} = \alpha(r_1,r_2,r_3;\chi_1,\chi_2,\chi_3)\ \mathrm{Id}.$$

We will determine the scalar α in special cases.

2.2. The multiplier of θ.

2.2.1. Let us consider a self-dual lattice r in the symplectic vector space (V,B). Let χ be a character of the associated group $R = \exp(r \oplus \mathbb{R})$. Equivalently χ is a function on r satisfying $\chi(v + v') = \chi(v)\chi(v')e^{-i\pi B(v,v')}$. The couple (r,χ) will be fixed in this paragraph.

2.2.2. Let ℓ be a Lagrangian subspace such that

a) $r \cap \ell$ generates ℓ as a vector space,

b) $\chi(\exp y) = 1$ if $y \in r \cap \ell$.

We have constructed in (2.1) canonical unitary operators $\theta^{\chi}_{\ell,r}$ and $\theta^{\chi}_{r,\ell}$ intertwining the irreducible representations $W(\ell)$ and $W(r,\chi)$.

Let ℓ_1,ℓ_2 be two Lagrangian subspaces satisfying the conditions 2.2.2 a) and b). There exists a scalar $b(\ell_1,\ell_2;(r,\chi)) = b(\ell_1,\ell_2)$ of modulus one such that $\theta^{\chi}_{\ell_2,r}\cdot\theta^{\chi}_{r,\ell_1} = b(\ell_1,\ell_2)\mathcal{F}_{\ell_1,\ell_2}$, as both members of the equality are unitary operators intertwining the irreducible representations $W(\ell_1)$ and $W(\ell_2)$. (The operator $\mathcal{F}_{\ell_1,\ell_2}$ has been defined in 1.4.)

We will express $b(\ell_1,\ell_2)$ as a Gauss sum.

2.2.3. Let us first consider the case where $\ell_1 \cap \ell_2 = 0$. Then $r \cap \ell_1 + r \cap \ell_2$ is a sublattice of r. We denote by $F = F(\ell_1,\ell_2;r)$ the quotient $r/r\cap\ell_1+r\cap\ell_2$. This is a finite abelian group.

Let e_1 and e_2 be the canonical elements of $\Lambda^n \ell_1$ and $\Lambda^n \ell_2$ associated with $r \cap \ell_1$ and $r \cap \ell_2$ (2.1.12). We have $e_1 \wedge e_2 = cw$, where w is the canonical form of (V,B). By 2.1.3, $|w|$ is the volume form associated to a $\mathbb{Z}$-basis of r. As $e_1 \wedge e_2$ is associated to a $\mathbb{Z}$ basis of $r \cap \ell_1 + r \cap \ell_2$, we have $|e_1 \wedge e_2| = f|w|$ when f is the number of elements of F. Hence the element $\delta(e_1, e_2)$ defined in 1.4.13 is equal to $f^{-1/2}|e_1|$ where $e_1 \in \Lambda^n \ell_1 \simeq \Lambda^n \ell_2^*$.

We denote by (p_1, p_2) the projections of V on (ℓ_1, ℓ_2) according to the decomposition $V = \ell_1 \oplus \ell_2$.

2.2.4. __Lemma__: The function $q_{1,2,\chi}(z) = \chi(\exp z) e^{-i\pi B(p_1(z), p_2(z))}$ is a function on $F = r/r\cap\ell_1 + r\cap\ell_2$.

__Proof__: Let us consider $u_1 \in r \cap \ell_1$, $u_2 \in r \cap \ell_2$ and $z = z_1 + z_2 \in r$ (with $z_1 = p_1(z)$, $z_2 = p_2(z)$). Then

$$\chi(\exp z+u_1+u_2) e^{-i\pi B(z_1+u_1, z_2+u_2)}$$

$$= \chi(\exp z) e^{-i\pi B(z, u_1+u_2)} e^{-i\pi B(u_1,u_2)} e^{-i\pi B(z_1+u_1, z_2+u_2)}$$

as $\exp(z+u_1+u_2) = \exp z \exp u_1 \exp u_2 \exp - \dfrac{B(z,u_1+u_2) + B(u_1,u_2)}{2} E$. But $e^{-2i\pi B(u_1,u_2)} = 1$, as $B(u_1,u_2) \in B(r,r) \subset \mathbb{Z}$. Similarly

$$e^{-i\pi B(z,u_2)} e^{-i\pi B(z_1,u_2)} = e^{-i\pi B(z,u_2)} e^{-i\pi B(z,u_2)} = e^{-2i\pi B(z,u_2)} = 1$$

and the lemma follows.

2.2.5. __Proposition__: Let ℓ_1, ℓ_2 be two transverse Lagrangian subspaces:

a) Let $G(\ell_1,\ell_2;r,\chi) = \sum_{z\in F} \chi(\exp z)e^{-i\pi B(p_1(z),p_2(z))}$

then $|G(\ell_1,\ell_2;r,\chi)| = f^{1/2}$.

b) $b(\ell_1,\ell_2) = f^{-1/2}G(\ell_1,\ell_2;r,\chi)$.

<u>Proof</u>: Let $\varphi \in H(\ell_1)$, we have

$$(\mathcal{F}^{\delta}_{\ell_2,\ell_1}\varphi)(e) = f^{1/2} \int_{\ell_2} \varphi(\exp v)\, d_2v.$$

We compute:

$$(\theta^{\chi}_{\ell_2,r} \circ \theta^{\chi}_{r,\ell_1}\varphi)(e) = \int_{\ell_2/\ell_2\cap r} (\theta^{\chi}_{r,\ell_1}\varphi)(\exp v)\, dv$$

$$= \int_{v\in\ell_2/\ell_2\cap r} \Big(\sum_{u\in r/\ell_1\cap r} \varphi(\exp v \exp u)\chi(\exp u)\Big)\, d\dot{v}.$$

Consider the inclusions $\ell_1 \cap r \subset \ell_1 \cap r + \ell_2 \cap r \subset r$. We can write any element of $r/\ell_1\cap r$ as $u = z + \delta$, where z varies over a system of representatives of $F = r/\ell_1\cap r+\ell_2\cap r$ in r and δ varies in $\ell_2 \cap r \simeq (\ell_1 \cap r + \ell_2 \cap r)/\ell_1\cap r$. We write our integral as:

$$\int_{v\in\ell_2/\ell_2\cap r} \sum_{z\in F} \sum_{\delta\in\ell_2\cap r} \varphi(\exp v \exp(\delta+z))\, \chi(\exp(\delta+z))\, d\dot{v}\,.$$

Now $\varphi(\exp v \exp(\delta+z))\, \chi(\exp(\delta+z)) = \varphi(\exp v \exp \delta \exp z)\, \chi(\exp z)$

$$= \varphi(\exp v \exp \delta \exp(z_2+z_1))\, \chi(\exp z)$$

$$= \varphi(\exp v \exp \delta \exp z_2)\, e^{-i\pi B(z_1,z_2)}\, \chi(\exp z)$$

as $\exp(z_2+z_1) = \exp z_2 \exp z_1 \exp -\dfrac{B(z_2,z_1)}{2} E$

and $\varphi \in H(\ell_1)$,

$$= \varphi(\exp(v+\delta+z_2))\, \chi(\exp z)\, e^{-i\pi B(z_1,z_2)}, \text{ as } v,\delta,z_2 \in \ell_2.$$

Then:

$$\int_{v\in\ell_2/\ell_2\cap r}\sum_{z\in F}\sum_{\delta\in\ell_2\cap r}\varphi(\exp v+\delta+z_2)\chi(\exp z)e^{-i\pi B(z_1,z_2)}d\dot{v}$$

$$=\sum_{z\in F}\chi(\exp z)e^{-i\pi B(z_1,z_2)}\left(\int_{v\in\ell_2/\ell_2\cap r}\sum_{\delta\in\ell_2\cap r}\varphi(\exp(v+\delta+z_2))d\dot{v}\right.$$

Clearly

$$\int_{v\in\ell_2/\ell_2\cap r}\sum_{\delta\in\ell_2\cap r}\varphi(\exp(v+\delta+z_2))d\dot{v}=\int_{v\in\ell_2}\varphi(\exp(v+z_2))dv$$

$$=\int_{v\in\ell_2}\varphi(\exp v)dv.$$

Therefore we obtain:

$$(\theta^{\chi}_{\ell_2,r}\cdot\theta^{\chi}_{r,\ell_1}\varphi)(e)=G(\ell_1,\ell_2;r,\chi)\int_{v\in\ell_2}\varphi(\exp v)dv$$

$$=G(\ell_1,\ell_2;r,\chi)(f)^{-1/2}(\mathcal{F}^{\delta}_{\ell_2,\ell_1}\varphi)(e).$$

As we know that both operators $\mathcal{F}^{\delta}_{\ell_2,\ell_1}$ and $\theta^{\chi}_{\ell_2,r}\circ\theta^{\chi}_{r,\ell_1}$ are unitary, we obtain the proposition.

Let us now consider the case where $\ell_1\cap\ell_2=\rho\neq 0$. We form similarly to 1.5.9, $r^{\rho}=(r\cap\rho^{\perp})+\rho=(r+\rho)\cap\rho^{\perp}$.

2.2.6. <u>Lemma</u>: a) $r\cap\rho$ is a lattice in ρ.

b) r^{ρ}/ρ is a self-dual lattice in the symplectic vector space $\rho^{\perp}/\rho$.

<u>Proof</u>: a) The condition $x\in\rho$ is equivalent to the equations $B(x,r\cap\ell_1)=B(x,r\cap\ell_2)=0$, as $r\cap\ell_1$ and $r\cap\ell_2$ generate

ℓ_1 and ℓ_2 as a vector space. Let us write $x = \sum_{i=1}^{2n} x_i e_i$ where $e_i \in r$ (r is a lattice in V); then $x \in \rho$ if and only if (x_i) is the solution of a linear system of equations in x_i with integral coefficients. Clearly the space of solutions is generated by the ones with integral coefficients, i.e. ρ is generated by $\rho \cap r$.

b) We can choose a symplectic basis of (V,B) such that

$$r \cap \rho = \bigoplus_{i=1}^{k} \mathbb{Z} P_i \quad \text{and} \quad r = \bigoplus_{i=1}^{n} \mathbb{Z} P_i + \bigoplus_{i=1}^{n} \mathbb{Z} Q_i .$$

Thus r^ρ/ρ is given by the lattice $\bigoplus_{i=k+1}^{n} \mathbb{Z} P_i \oplus \bigoplus_{i=k+1}^{n} \mathbb{Z} Q_i$ in $\rho^\perp/\rho \simeq \bigoplus_{i=k+1}^{n} \mathbb{R} P_i \oplus \bigoplus_{i=k+1}^{n} \mathbb{R} Q_i$.

2.2.7. Let $(\ell_1,\ell_2,(r,\chi))$ satisfy 2.2.2 a),b). We consider $\rho = \ell_1 \cap \ell_2$, $V' = \rho^\perp/\rho$, $N' = \exp(\rho^\perp/\rho \oplus \mathbb{R}E)$, $\ell_1' = \ell_1/\rho$, $\ell_2' = \ell_2/\rho$, $r' = r^\rho/\rho$ and $R' = \exp(r' \oplus \mathbb{R}E)$. As the character χ is trivial on $\exp(r \cap \rho)$, we can define χ' on R' by $\chi'(\exp \dot{u}_1 + tE) = \chi(\exp(u_1 + tE))$ for $u_1 \in r \cap \rho^\perp \bmod r \cap \rho$ without ambiguity.

2.2.8. <u>Proposition:</u>

$$b(\ell_1,\ell_2;(r,\chi)) = b(\ell_1',\ell_2';(r',\chi'))$$

$$= (f')^{-1/2} \sum_{z \in r\cap(\ell_1+\ell_2)/r\cap\ell_1+r\cap\ell_2} \chi(\exp z)e^{-i\pi B(z_1,z_2)}$$

where $z = z_1 + z_2$, with $z_1 \in \ell_1, z_2 \in \ell_2$, and f' = cardinal of $F_{12}' = r\cap(\ell_1+\ell_2)/r\cap\ell_1+r\cap\ell_2$.

Proof: Let us use the coordinates of Lemma 2.2.6. We write: $V = V_0 \oplus V'$, with

$$V_0 = \bigoplus_{i=1}^{k} \mathbb{R}P_i \oplus \bigoplus_{i=1}^{k} \mathbb{R}Q_i$$

$$V' = \bigoplus_{i=k+1}^{n} \mathbb{R}P_i \oplus \bigoplus_{i=k+1}^{n} \mathbb{R}Q_i$$

$$\rho = \bigoplus_{i=1}^{k} \mathbb{R}P_i \quad .$$

$$r = \bigoplus_{i=1}^{n} \mathbb{Z}P_i \oplus \bigoplus_{i=1}^{n} \mathbb{Z}Q_i \ .$$

As $\ell_1, \ell_2 \supset \rho$, we have $\ell_1 = \rho \oplus \ell_1'$ and $\ell_2 = \rho \oplus \ell_2'$ where ℓ_1' and ℓ_2' are Lagrangian subspaces of V'. The space ρ is Lagrangian in V_0. We have $r = r \cap V_0 \oplus r \cap V' = r_0 \oplus r'$.

According to the decomposition $V = V_0 \oplus V'$, we can write:

$$H(\ell_1) \simeq H_0(\rho) \otimes H'(\ell_1')$$

$$H(r,\chi) \simeq H_0(r_0,\chi_0) \otimes H'(r',\chi')$$

$$H(\ell_2) = H_0(\rho) \otimes H'(\ell_2') .$$

Clearly

$$\theta^{\chi}_{\ell_2,r} \simeq \theta^{\chi_0}_{\rho,r_0} \otimes \theta^{\chi'}_{\ell_2',r'}$$

$$\theta^{\chi}_{r,\ell_1} \simeq \theta^{\chi_0}_{r_0,\rho} \otimes \theta^{\chi'}_{r',\ell_1'} \ .$$

As $\theta^{\chi_0}_{\rho,r_0} \circ \theta^{\chi_0}_{r_0,\rho} = Id_{H_0}$, we obtain our Proposition.

2.2.9. Corollary:

$$\Big| \sum_{z \in r \cap (\ell_1+\ell_2)/r\cap\ell_1 + r\cap\ell_2} \chi(\exp z) e^{-i\pi B(z_1,z_2)} \Big| = f'^{1/2} .$$

We have hence expressed $b(\ell_1,\ell_2)$ in function of a Gauss sum G over F' and determined the absolute value of G.

Now let ℓ_1,ℓ_2 and ℓ_3 be three Lagrangian planes such that $(\ell_i;(r,\chi))$ satisfies the conditions 2.2.2 a) and b). The Theorem 1.6.1 implies:

2.2.10. Theorem:

$$b(\ell_1,\ell_2)b(\ell_2,\ell_3)b(\ell_3,\ell_1) = e^{\frac{i\pi}{4}\tau(\ell_1,\ell_2,\ell_3)}.$$

2.2.11. We will now calculate $b(\ell_1,\ell_2)$ when $\dim V = 2$.

Let $V = \mathbb{R}P \oplus \mathbb{R}Q$, $r = \mathbb{Z}P \oplus \mathbb{Z}Q$, and $\chi(\exp(mP+nQ)) = (-1)^{mn}$. Let (c,d) be two relatively prime integers with cd even. We consider $\ell_1 = \mathbb{R}P$ and $\ell_2 = \mathbb{R}(cP+dQ)$; ℓ_1,ℓ_2 satisfy the relation 2.2.2 a) b), relative to (r,χ). We have $r \cap \ell_2 = \mathbb{Z}(cP+dQ)$. Now $r \cap \ell_1 + r \cap \ell_2 = \mathbb{Z}P + \mathbb{Z}dQ$. Therefore

$$F_{12} = r/r\cap\ell_1+r\cap\ell_2 \simeq \mathbb{Z}Q/\mathbb{Z}dQ \simeq \mathbb{Z}/d\mathbb{Z}.$$

We write $nQ = -\frac{cn}{d}P + \frac{n}{d}(cP+dQ) = z_1 + z_2$. Hence

$$b(\ell_1,\ell_2) = b(c,d) = |d|^{-1/2} \sum_{n\in\mathbb{Z}/d\mathbb{Z}} e^{\frac{i\pi cn^2}{d}}.$$

Let $\ell_3 = \mathbb{R}Q$, then $F_{23} = r/r\cap\ell_2+r\cap\ell_3 \simeq \mathbb{Z}P/\mathbb{Z}cQ \simeq \mathbb{Z}/c\mathbb{Z}$. We write $nP = \frac{n}{c}(cP+dQ) - \frac{nd}{c}Q$, therefore

$$b(\ell_2,\ell_3) = |c|^{-1/2} \sum_{n\in\mathbb{Z}/c\mathbb{Z}} e^{\frac{i\pi dn^2}{c}} = b(d,c).$$

As $F_{31} = 0$, $b(\ell_3,\ell_1) = 1$. Thus by 2.2.10, $b(c,d)b(d,c) = e^{i\pi/4 \operatorname{sign} cd}$, as $\tau(\ell_1,\ell_2,\ell_3) = \operatorname{sign}(cd)$.

We define for any integers (c,d)

$$g(2c,d) = \sum_{k\in\mathbb{Z}/d\mathbb{Z}} e^{\frac{2i\pi k^2 c}{d}},$$

i.e. if $(2c,d)$ are relatively prime, $g(2c,d) = |d|^{1/2}\, b(2c,d)$.

2.2.12. __Proposition__: For $d > 0$

$$g(2,d) = \sum_{k\in\mathbb{Z}/d\mathbb{Z}} e^{\frac{2i\pi k^2}{d}}$$

is given by:

1) $g(2,d) = (1+i)\sqrt{d}$ if $d \equiv 0 \text{ Mod } 4$
2) $g(2,d) = \sqrt{d}$ if $d \equiv 1 \text{ Mod } 4$
3) $g(2,d) = 0$ if $d \equiv 2 \text{ Mod } 4$
4) $g(2,d) = i\sqrt{d}$ if $d \equiv 3 \text{ Mod } 4$.

__Proof__: Let d be an odd positive number. We have from 2.2.11

$$g(2,d) = |d|^{1/2}\, b(2,d) = |d|^{1/2}\, e^{\frac{i\pi}{4}}\, b(d,2)^{-1}.$$

But

$$b(d,2) = \frac{1}{\sqrt{2}}\,(1 + e^{\frac{i\pi d}{2}}).$$

Therefore

$$b(d,2) = e^{\frac{i\pi}{4}} \quad \text{if } d \equiv 1 \text{ Mod } 4$$

$$= e^{-\frac{i\pi}{4}} \quad \text{if } d \equiv 3 \text{ Mod } 4,$$

and we obtain the equalities 2) and 4).

Now let d be even. We consider the translation

$k \to k + \frac{d}{2}$ on $\mathbb{Z}/d\mathbb{Z}$. This implies that

$$g(2,d) = \sum_{k \in \mathbb{Z}/d\mathbb{Z}} e^{2i\pi(k+d/2)^2/d} = e^{\frac{i\pi d}{2}} g(2,d).$$

Thus, if $d \equiv 2 \text{ Mod } 4$, then $g(2,d) = -g(2,d)$ implying 3.
If $d \equiv 0 \text{ Mod } 4$, then

$$g(2,d) = 2 \sum_{h \in \mathbb{Z}/\frac{d}{2}\mathbb{Z}} e^{2i\pi k^2/d} = 2\sqrt{d/2}\, b(1,\tfrac{d}{2}) ,$$

$(1,\frac{d}{2})$ are relatively prime and $\frac{d}{2}$ is even)

$$= 2\,(\sqrt{d/2})\, e^{\frac{i\pi}{4}}\, b(\tfrac{d}{2},1)^{-1}$$

$$= \sqrt{2}\, e^{\frac{i\pi}{4}} \sqrt{d} = (1+i)\sqrt{d} \text{ , proving 1)}.$$

2.2.13. We recall now some definitions (see [18]).

Let m be an integer. We consider the group $(\mathbb{Z}/m\mathbb{Z})^*$ of invertible elements of $\mathbb{Z}/m\mathbb{Z}$. An element of $(\mathbb{Z}/m\mathbb{Z})^*$ is the image of an integer a in $\mathbb{Z}$ such that $(a,m) = 1$. Let χ be a character of the group $(\mathbb{Z}/m\mathbb{Z})^*$. We may consider χ as a function defined on the set of integers prime to m such that

$$\chi(ab) = \chi(a)\chi(b)$$
$$\chi(a+mN) = \chi(a) \quad .$$

We extend $\chi(a)$ as a function on $\mathbb{Z}$, by defining $\chi(a) = 0$ if a is not prime to m. Such a χ is called a character Mod m.

Let p be a prime. We define for an integer x prime to p, $(\frac{x}{p}) = 1$ if x is a square in $\mathbb{F}_p = \mathbb{Z}/p\mathbb{Z}$ and $(\frac{x}{p}) = -1$ if not. We have $(\frac{x}{p}) = x^{p-1/2}$ Mod p. The map $x \to (\frac{x}{p})$ is a character mod p, called the quadratic residue mod p.

We define now, for d an odd integer, the number ε_d to be

$$\varepsilon_d = 1 \text{ if } d \equiv 1 \text{ Mod } 4$$
$$\varepsilon_d = i \text{ if } d \equiv -1 \text{ Mod } 4 .$$

We have $\varepsilon_d \varepsilon_{-d} = i$

$$\varepsilon_d^2 = (-1)^{(d-1)/2} = (\frac{-1}{d})$$

$$\varepsilon_{pq} = \varepsilon_p \varepsilon_q (-1)^{(\frac{p-1}{2})(\frac{q-1}{2})} .$$

2.2.14. <u>Lemma</u>: Let p be an odd positive prime:

1) $b(2x,p) = |p|^{-1/2} \sum_{h\in\mathbb{Z}/p\mathbb{Z}} e^{2i\pi xk^2/p} = (\frac{x}{p}) \varepsilon_p$

2) $b(x,2p) = |p|^{-1/2} \frac{1}{\sqrt{2}} \sum_{h\in\mathbb{Z}/2p\mathbb{Z}} e^{i\pi xk^2/2p} = (\frac{1+e^{i\pi xp/2}}{\sqrt{2}}) (\frac{x}{p}) \varepsilon_p .$

<u>Proof</u>: a) The expression

$$\sum_{k\in\mathbb{Z}/p\mathbb{Z}} e^{\frac{2i\pi k^2 x}{p}}$$

depends only of the class of x modulo the squares in $\mathbb{F}_p$: if $x = a^2$ mod p, we can replace k by ak in the sum. Therefore if $x = a^2$, we obtain $b(2x,p) = b(2,p) = |p|^{-1/2} \varepsilon_p$ from 2.2.12.

Let us now suppose that x is not a square mod p. Using

the fact that the sum of all the p-roots of unity is zero, we get

$$2 \sum_{u \in \mathbb{Z}/p\mathbb{Z}} e^{\frac{2i\pi u}{p}} = 0 = (1 + 2 \sum_{\substack{k \text{ square} \\ \text{in } \mathbb{F}_p}} e^{\frac{2i\pi k}{p}}) + (1 + 2 \sum_{\substack{k' \text{ not} \\ \text{a square}}} e^{\frac{2i\pi k'}{p}})$$

$$= \sum_{k \in \mathbb{Z}/p\mathbb{Z}} e^{\frac{2i\pi k^2}{p}} + \sum e^{\frac{2i\pi k^2 x}{p}} .$$

Hence $b(2x,p) = -b(2,p) = (\frac{x}{p}) |p|^{-1/2} \varepsilon_p$.

b) We write an element $k \in \mathbb{Z}/2p\mathbb{Z}$ as $k = 2k_1 + k_2 p$ where $k_1 \in \mathbb{Z}/p\mathbb{Z}$ and $k_2 = 0$ or 1. Now

$$e^{i\pi x \frac{(2k_1 + k_2 p)^2}{2p}} = e^{\frac{2i\pi x k_1^2}{p}} e^{2i\pi k_1 k_2 x} e^{i\pi \frac{x k_2^2 p}{2}}$$

$$= e^{\frac{2i\pi\, x k_1^2}{p}} e^{\frac{i\pi x p k_2^2}{2}} .$$

So

$$\sum e^{\frac{i\pi x k^2}{2p}} = \sum_{k_1 \in \mathbb{Z}/p\mathbb{Z}} e^{\frac{2i\pi x k_1^2}{p}} (1 + e^{\frac{i\pi x p}{2}})$$

which is formula 2).

2.2.15. <u>Corollary</u>: (Quadratic Reciprocity Law).

a) Let (p,q) be two positive odd primes, then

$$(\frac{p}{q})(\frac{q}{p}) = (-1)^{(\frac{p-1}{2})(\frac{q-1}{2})} .$$

b) $(\frac{2}{p}) = 1$ if $p \equiv 1$ or $7 \bmod 8$

$(\frac{2}{p}) = -1$ if $p \equiv 3$ or $5 \bmod 8$.

Proof: a) We have

$$b(2p,q) = \left(\frac{p}{q}\right) \varepsilon_q$$

$$b(q,2p) = \left(\frac{1 + e^{i\pi pq}}{\sqrt{2}}\right)\left(\frac{q}{p}\right) \varepsilon_p .$$

Then, as $b(2p,q)b(q,2p) = e^{\frac{i\pi}{4}}$ by 2.2.11, we have

$$\left(\frac{p}{q}\right)\left(\frac{q}{p}\right) \varepsilon_q \varepsilon_p \left(\frac{1 + e^{\frac{i\pi pq}{2}}}{\sqrt{2}}\right) = e^{\frac{i\pi}{4}} .$$

The formula then follows by direct computation, as:

$$e^{\frac{i\pi}{4}} \left(\frac{1 - e^{-\frac{i\pi pq}{2}}}{\sqrt{2}}\right) = \varepsilon_{pq}$$

b) We have $b(4,p) = \left(\frac{2}{p}\right) \varepsilon_p$ from 2.2.14. Using $b(4,p)b(p,4) = e^{\frac{i\pi}{4}}$ and the explicit computation of $b(p,4)$ we obtain b).

2.2.16. We define now the generalized quadratic residue symbol $\left(\frac{a}{b}\right)$ for (a,b) integers, b odd as in ([30]) by the formula:

1) $\left(\frac{a}{b}\right) = 0$ if $(a,b) \neq 1$

2) $\left(\frac{a}{-1}\right) = \text{sign } a$

3) if $b > 0$, $b = \pi b_i$, b_i primes not necessarily distinct

$$\left(\frac{a}{b}\right) = \pi_i \left(\frac{a}{b_i}\right)$$

4) $\left(\frac{a}{-b}\right) = \left(\frac{a}{-1}\right)\left(\frac{a}{b}\right)$

5) $\left(\frac{0}{\pm 1}\right) = 1$.

It is clear that $(\frac{a}{b})$ is bimultiplicative in (a,b), i.e.

$$(\frac{a_1a_2}{b}) = (\frac{a_1}{b})(\frac{a_2}{b})$$

$$(\frac{a}{b_1b_2}) = (\frac{a}{b_1})(\frac{a}{b_2}) .$$

2.2.17. We have the following properties:

1) For $b > 0$, $(\frac{a}{b})$ depends only of a Mod b.

2) $(\frac{-1}{b}) = (-1)^{\frac{b-1}{2}}$.

3) $(\frac{a}{b}) = (-1)^{(\frac{b-1}{2})(\frac{a-1}{2})}(\frac{b}{a})$. (For $a > 0$ and $b > 0$, this follows from developing a and b in product of primes, and apply 2.2.15. For a and b arbitrary, we use 2.2.16 2) and 4).)

4) $b \to (\frac{a}{b})$ is a character Mod $4a$. (This follows from the property 3.)

We now express the Gauss sum

$$b_{12} = b(c,d) = |d|^{-1/2} \sum_{k\in\mathbb{Z}/d\mathbb{Z}} e^{\frac{i\pi c k^2}{d}}$$

in function of the quadratic residue symbol $(\frac{c}{d})$. We recall that $b(c,d)$ is defined for $cd \equiv 0$ Mod 2, $(c,d) = 1$.

2.2.18. Proposition: Let (c,d) be two integers, with $(c,d) = 1$ and d an odd positive integer. Then $b(2c,d) = (\frac{c}{d})\,\varepsilon_d$.

Proof: If d is prime, the proposition has already been established. Let $d = p^r$, with $r > 1$. We prove it by induction on r. We have to compute

$$\sum_{k\in\mathbb{Z}/p^r\mathbb{Z}} e^{\frac{2i\pi k^2 c}{d}}.$$

We write every element k in $\mathbb{Z}/p^r\mathbb{Z}$ as $k = y + p^{r-1}z$ where $z \in \mathbb{Z}/p\mathbb{Z}$ and y varies over a system of representatives of $\mathbb{Z}/p^{r-1}\mathbb{Z}$. Then

$$e^{2i\pi c(y+p^{r-1}z)^2/p^r} = e^{2i\pi y^2 c/p^r}\, e^{2i\pi(2ycz)/p}.$$

If y is not divisible by p, then $z \to 2ycz$ is a permutation of the classes of $\mathbb{Z}/p\mathbb{Z}$. Thus

$$\sum_{z\in\mathbb{Z}/p\mathbb{Z}} e^{2i\pi(\frac{2ycz}{p})} = 0.$$

If y is divisible by p, then

$$\sum_{z\in\mathbb{Z}/p\mathbb{Z}} e^{2i\pi(2ycz)/p} = p.$$

Hence we see that we have only to consider the classes y of the form py' where $y' \in \mathbb{Z}/p^{r-2}\mathbb{Z}$. We obtain:

$$\sum_{k\in\mathbb{Z}/p^r\mathbb{Z}} e^{2i\pi k^2c/p^r} = p \sum_{y'\in\mathbb{Z}/p^{r-1}\mathbb{Z}} e^{2i\pi y'^2c/p^{r-2}} .$$

Thus we see that

$$b(2c,p^r) = b(2c,p^{r-2}) = \left(\frac{c}{p^{r-2}}\right) \varepsilon_{p^{r-2}}$$

by induction hypothesis. But

$$\left(\frac{c}{p^{r-2}}\right) = \left(\frac{c}{p^r}\right),\ \varepsilon_{p^{r-2}} = \varepsilon_{p^r}\ ,\ (\text{as } p^2 \equiv 1 \text{ Mod } 4).$$

Therefore we obtain the Proposition 2.2.18 in this case.

Now let $d = d_1d_2$ with d_1 and d_2 mutually prime. We have to compute

$$\sum_{k\in\mathbb{Z}/d_1d_2\mathbb{Z}} e^{\frac{2i\pi ck^2}{d_1d_2}} .$$

The map $(k_1,k_2) \to k_1d_2 + k_2d_1$ is an isomorphism of $\mathbb{Z}/d_1\mathbb{Z} \times \mathbb{Z}/d_2\mathbb{Z}$ onto $\mathbb{Z}/d_1d_2\mathbb{Z}$. As

$$e^{\frac{2i\pi c(k_1d_2 + k_2d_1)^2}{d_1d_2}} = e^{\frac{2i\pi cd_2k_1^2}{d_1}}\, e^{2i\pi\frac{cd_1k_2^2}{d_2}} ,$$

it follows that:

$$\begin{aligned} b(2c,d_1d_2) &= b(2cd_2,d_1)\cdot b(2cd_1,d_2) \\ &= \left(\frac{cd_2}{d_1}\right) \varepsilon_{d_1} \left(\frac{cd_1}{d_2}\right) \varepsilon_{d_2}, \text{ by induction,} \\ &= \left(\frac{c}{d_1d_2}\right)\left(\frac{d_2}{d_1}\right)\left(\frac{d_1}{d_2}\right) \varepsilon_{d_1} \varepsilon_{d_2}, \\ &= \left(\frac{c}{d}\right) \varepsilon_d, \text{ as } \varepsilon_{d_1d_2} = \varepsilon_{d_1} \varepsilon_{d_2} (-1)^{\left(\frac{d_1-1}{2}\right)\left(\frac{d_2-1}{2}\right)} \\ &\text{and } \left(\frac{d_2}{d_1}\right)\left(\frac{d_1}{d_2}\right) = (-1)^{\left(\frac{d_1-1}{2}\right)\left(\frac{d_2-1}{2}\right)}. \end{aligned}$$

This ends the proof of the proposition.

2.2.19. Let (V,B) be a 2n-dimensional symplectic vector space. Let $r = \mathbb{Z}P_1 \oplus \mathbb{Z}P_2 \oplus \cdots \oplus \mathbb{Z}P_n \oplus \mathbb{Z}Q_1 \oplus \cdots \oplus \mathbb{Z}Q_n$, $\chi(\exp(m\cdot P + n\cdot Q)) = (-1)^{m\cdot n}$, $\ell = \mathbb{R}P_1 \oplus \cdots \oplus \mathbb{R}P_n$ and $\ell' = \mathbb{R}Q_1 \oplus \cdots \oplus \mathbb{R}Q_n$.

We consider the group $G = Sp(B)$. The subgroup Γ of G leaving the lattice r stable is the subgroup $Sp(n,\mathbb{Z})$ of matrices $g \in G$ with integral coefficients in the basis P_i, Q_j. Let us write $g = \left(\begin{array}{c|c}\alpha & \beta \\ \hline \gamma & \delta\end{array}\right)$ in $Sp(n;\mathbb{Z})$. Then $g^{-1} = \left(\begin{array}{c|c}{}^t\delta & -{}^t\beta \\ \hline -{}^t\gamma & {}^t\alpha\end{array}\right)$.

We have
$$\begin{aligned} (g\cdot\chi)(\exp u) &= \chi(\exp({}^t\delta u - {}^t\gamma u), \text{ for } u \in r \cap \ell \\ &= (-1)^{B({}^t\delta u, {}^t\gamma u)}. \end{aligned}$$

$$\begin{aligned} (g\cdot\chi)(\exp v) &= \chi(\exp {}^t\beta v - {}^t\alpha v) \quad \text{for } v \in r \cap \ell' \\ &= (-1)^{B({}^t\beta v, {}^t\alpha v)} \quad \text{for } v \in r \cap \ell'. \end{aligned}$$

Hence g leaves the character χ stable if and only if the symmetric matrices $\alpha^t\beta$ and $\delta^t\gamma$ have even diagonal coefficients in the basis P_i, Q_j.

2.2.20. For every $g \in \Gamma(r,\chi)$ the subgroup of G leaving the pair (r,χ) stable, the Lagrangian subspace $g\cdot\ell$ clearly satisfies the conditions 2.2.2 a) and b). Hence we define $b(g) = b(\ell, g\cdot\ell; (r,\chi))$ as a function on $\Gamma(r,\chi)$. The Theorem 2.2.10 implies

$$b(g_1)b(g_2)b(g_1g_2)^{-1} = e^{\frac{i\pi}{4}\tau(\ell, g_1\ell, g_1g_2\ell)}.$$

Therefore the function $g \to b(g)^8$ is a character of the group $\Gamma(r,\chi)$. Let (t_i, s_i, σ_i) be the elements of $\Gamma(r,\chi)$ acting on

$$\sum_{j\neq i} \mathbb{R}P_j \oplus \mathbb{R}Q_j$$

by the identity and on $\mathbb{R}P_i \oplus \mathbb{R}Q_i$ by the transformations

$$\left(\begin{pmatrix}1 & 2\\ 0 & 1\end{pmatrix}, \begin{pmatrix}-1 & 0\\ 0 & -1\end{pmatrix}, \begin{pmatrix}0 & 1\\ -1 & 0\end{pmatrix}\right), \text{ respectively.}$$

It is known [8] that $\Gamma(r,\chi)$ is generated by the elements $\{t_i, s_i, \sigma_i\}$. As for each of these elements γ, $b(\gamma) = 1$, we obtain:

2.2.21. <u>Proposition</u>: $b(g)$ is an 8th-root of unity on $\Gamma(r,\chi)$.

We will precise the value of $b(g)$ up to sign.

Let us consider an orientation ℓ^+ on ℓ. We recall that we have defined a function $s_\ell(g) = s(\ell^+, g.\ell^+)$ (1.7.7) satisfying

$$s_\ell(g_1g_2) = s_\ell(g_1)s_\ell(g_2)e^{-\frac{i\pi}{2}\tau(\ell, g_1\ell, g_1g_2\ell)} .$$

Hence we obtain:

2.2.22. <u>Proposition</u>: The function $k(g) = b(g)^2 s_\ell(g)^{-1}$ is a character of the group $\Gamma(r,\chi)$, with values in the 4th-roots of unity.

We will identify this character on a special subgroup of $\Gamma(r,\chi)$. We define:

$$\Gamma_0^0(2) = \{M = \left(\begin{array}{c|c}\alpha & \beta\\ \hline \gamma & \delta\end{array}\right) \in \Gamma \; ; \; \beta \equiv \gamma \equiv 0 \text{ Mod } 2\} .$$

Clearly $\Gamma_0^0(2) \subset \Gamma(r,\chi)$. The element $\sigma = \left(\begin{array}{c|c}0 & 1\\ \hline -1 & 0\end{array}\right)$ belongs also to $\Gamma(r,\chi)$ and acts by automorphisms on $\Gamma_0^0(2)$:

$$\sigma\cdot\begin{pmatrix}\alpha & \beta\\ \gamma & \delta\end{pmatrix}\sigma^{-1} = \left(\begin{array}{c|c}\delta & -\gamma\\ \hline -\beta & \alpha\end{array}\right) .$$

We define $\hat{\Gamma}_0^0(2)$ to be the semi-direct product of $\Gamma_0^0(2)$ by σ. As a set

$$\hat{\Gamma}_0^0(2) = \left\{ g = \left(\begin{array}{c|c}\alpha & \beta\\ \hline \gamma & \delta\end{array}\right) \in \Gamma \text{ such that either } \begin{array}{l}\alpha \equiv \delta \equiv 0 \text{ Mod } 2\\ \text{either } \beta \equiv \gamma \equiv 0 \text{ Mod } 2\end{array}\right\} .$$

Let us consider the following special elements of $\hat{\Gamma}_0^0(2)$ given by:

$u(S) = \left(\begin{array}{c|c} 1 & S \\ \hline 0 & 1 \end{array}\right)$, when S is a symmetric matrix with even coefficients.

$g(A) = \begin{pmatrix} A & 0 \\ 0 & {}^tA^{-1} \end{pmatrix}$, with $A \in GL(n;\mathbb{Z})$.

2.2.23. It is known that the group $\hat{\Gamma}_0^0(2)$ is generated by $\{u(S), g(A), \sigma\}$ [8].

2.2.24. <u>Proposition</u> [13]: On $\hat{\Gamma}_0^0(2)$

$$k(g) = \left(\frac{-1}{\det \delta}\right) \quad \text{if} \quad g \in \Gamma_0^0(2),\ g = \left(\begin{array}{c|c} \alpha & \beta \\ \hline \gamma & \delta \end{array}\right)$$

$$k(\sigma) = i^n .$$

<u>Proof</u>: Let $g = \left(\begin{array}{c|c} \alpha & \beta \\ \hline \gamma & \delta \end{array}\right)$ in $\Gamma_0^0(2)$. As ${\alpha}^t\delta - {\beta}^t\gamma = 1$ we have ${\alpha}^t\delta \equiv 1 \,(\mathrm{Mod}\ 4)$. In particular $(\det \delta)$ in an odd number. It is easy to see that the function

$$g \to \left(\frac{-1}{\det \delta}\right) = (-1)^{\frac{(\det \delta)-1}{2}}$$

is a character of $\Gamma_0^0(2)$, as for

$$g_1 = \left(\begin{array}{c|c} \alpha_1 & 2\beta_1 \\ \hline 2\gamma_1 & \delta_1 \end{array}\right), \quad g_2 = \left(\begin{array}{c|c} \alpha_2 & 2\beta_2 \\ \hline 2\gamma_2 & \delta_2 \end{array}\right), \quad g_1g_2 = \left(\begin{array}{c|c} * & * \\ \hline * & \delta_1\delta_2+4\gamma_1\beta_2 \end{array}\right) .$$

If $(\det \alpha)(\det \delta) \equiv 1$, then $\left(\frac{-1}{\det \alpha}\right) = \left(\frac{-1}{\det \delta}\right)$; hence this character is stable by the action of σ on $\Gamma_0^0(2)$. Thus the formula

$$k'(g) = \left(\frac{-1}{\det \delta}\right), \quad k'(\sigma) = i^n$$

defines a character of $\hat{\Gamma}_0^0(2)$. It is then sufficient to prove that $k = k'$ on the given set of generators of $\hat{\Gamma}_0^0(2)$.

Let $g = u(S)$, then clearly $b(g) = 1 = s(g)$; hence $k(g) = 1 = k'(g)$. Let $g = g(A)$ with $A \in GL(n;\mathbb{Z})$. Clearly $b(g) = 1$ (as $g\cdot\ell = \ell$)

$$s(g) = \text{sign}(\det A) = \det A, \text{ as } \det A = \pm 1 .$$

Thus $k(g) = s(g)^{-1} = (\det A) = (\frac{-1}{\det A})$. Let $g = \sigma$ then $b(g) = 1$ as $F_{\ell,g\cdot\ell} = 1$, $s(g) = i^n(-1)^n$, $s(g)^{-1} = i^n$, q.e.d.

2.2.25. Let $g = \left(\frac{\alpha}{\gamma}\middle|\frac{\beta}{\delta}\right) \in \Gamma(r,\chi)$. We now express $b(g) = b(\ell,g\cdot\ell)$ as a Gauss sum in function of (α,γ).

By 2.2.8

$$b(g) = (f')^{-1/2} \sum_{z\in r\cap(\ell+g\cdot\ell)/r\cap\ell+r\cap g\cdot\ell} \chi(\exp z)\, e^{-i\pi B(z_1,z_2)}$$

where

$$f' = \text{cardinal of } r\cap(\ell+g\cdot\ell)/r\cap\ell+r\cap g\cdot\ell ,$$

$z \in r\cap(\ell+g\ell)$, $z = z_1 + z_2$, with $z_1 \in \ell$, $z_2 \in g\cdot\ell$.

We have:

$$\ell + g\cdot\ell = \ell + \gamma\ell$$

$$r\cap(\ell+g\ell) = r\cap\ell + \gamma\ell\cap r$$

$$r\cap\ell + r\cap g\ell = r\cap\ell + \gamma(\ell\cap r) \quad .$$

Therefore

$$F_g = r\cap(\ell+g\cdot\ell)/r\cap\ell+r\cap g\cdot\ell \simeq (\gamma\ell)\cap r/\gamma(\ell\cap r) = \gamma(\mathbb{R}^n) \cap \mathbb{Z}^n/\gamma(\mathbb{Z}^n)$$

Let $z \in \gamma\ell \cap r$, we have $z = \gamma x$ for $x \in \ell$. We write symbolically $x = \gamma^{-1}z$ (i.e. x is __any__ element satisfying $\gamma x = z$). Then $z = -\alpha x + (\alpha x + \gamma x)$ with $\alpha x \in \ell$, $\alpha x + \gamma x \in g(\ell)$. So

$$B(z_1,z_2) = -B(\alpha x,\gamma x) = -"B(\alpha\gamma^{-1}z,z)" .$$

(It is easy to see that this formula is indeed well defined for $z \in \gamma\ell$, as if $\gamma x' = 0$,

$$B(\alpha x',\gamma\cdot u) = -B({}^t\gamma\alpha x',u) = -B({}^t\alpha\gamma x',u) = 0.)$$

Hence we obtain the formula:

2.2.26. $$b(g) = f^{-1/2} \sum_{z\in\gamma\ell\cap r/\gamma(\ell\cap r)} e^{i\pi B(\alpha\gamma^{-1}z,z)} = b(\alpha,\gamma)$$

where $f = \#(\gamma\ell\cap r/\gamma(\ell\cap r))$.

We recall that we have

2.2.27. $$|b(\alpha,\gamma)| = 1$$

2.2.28. $$b(g_1g_2) = b(g_1)b(g_2)e^{-\frac{i\pi}{4}\tau(\ell,g_1\ell,g_1g_2\ell)} .$$

In particular let $\sigma = \left(\begin{array}{c|c} 0 & 1 \\ \hline -1 & 0\end{array}\right)$. Then $b(\sigma) = 1$. Now if

$$g = \left(\begin{array}{c|c} \alpha & \beta \\ \hline \gamma & \delta\end{array}\right), \qquad \sigma g = \left(\begin{array}{c|c} \gamma & \delta \\ \hline -\alpha & -\beta\end{array}\right) .$$

Thus

$$b(\sigma g) = b(g)\, e^{-\frac{i\pi}{4}\tau(\ell,\, \sigma\ell, {}^{\sigma}g\ell)}$$

$$= b(g)\, e^{-\frac{i\pi}{4}\tau(\ell, g\ell, \sigma^{-1}\ell)} .$$

As $b(\gamma,-\alpha) = \overline{b(\gamma,\alpha)} = b(\gamma,\alpha)^{-1}$, we obtain the reciprocity formula:

2.2.29. <u>Proposition:</u> $$b(\alpha,\gamma)b(\gamma,\alpha) = e^{\frac{i\pi}{4}\operatorname{sign}({}^{t}\gamma\alpha)} .$$

2.2.30. Let us consider the projective representation R_ℓ of the group $G = Sp(B)$. As $\Gamma(r,\chi)$ leaves the pair (r,χ) stable, it is clear that, if we define $(A_r(\gamma)\varphi)(n) = \varphi(\gamma^{-1}\cdot n)$, $A_r(\gamma)$ for $\gamma \in \Gamma(r,\chi)$ is a unitary operator satisfying

$$A_r(\gamma)\; W(r,\chi)(n)\; A_r(\gamma)^{-1} = (W(r,\chi))(\gamma\cdot n) .$$

Hence $A_r(\gamma)$ is proportional to the operator $R_\ell(\gamma)$ of the projective Weil representation. More precisely there exists a scalar $\alpha(\gamma)$ such that the following diagram is commutative:

$$\begin{array}{ccc}
H(\ell) & \xrightarrow{\;\alpha(\gamma)R_\ell(\gamma)\;} & H(\ell) \\
\Big\downarrow {\scriptstyle \theta^{\chi}_{r,\ell}} & & \Big\downarrow {\scriptstyle \theta^{\chi}_{r,\ell}} \\
H(r,\chi) & \xrightarrow{\;A_r(\gamma)\;} & H(r,\chi)
\end{array} .$$

Let us compute $\alpha(\gamma)$. We recall that R_ℓ is canonically definedby $R_\ell(\gamma) = \mathcal{F}_{\ell,\gamma\cdot\ell} \circ A(\gamma)$, where $(A(\gamma)\cdot\varphi)(n) = \varphi(\gamma^{-1}\cdot n)$ is a unitary operator from $H(\ell,e)$ to $H(\gamma\cdot\ell,\gamma\cdot e)$. Now it is clear that the diagram:

$$\begin{array}{ccc} H(\ell) & \xrightarrow{A(\gamma)} & H(\gamma\cdot\ell) \\ \downarrow{\scriptstyle \theta^{\chi}_{r,\ell}} & & \downarrow{\scriptstyle \theta^{\chi}_{r,\gamma\ell}} \\ H(r,\chi) & \xrightarrow{A(\gamma)} & H(r,\chi) \end{array}$$

is commutative (γ leaves the pair (r,χ) stable). Hence to calculate $\alpha(\gamma)$, we have to calculate the scalar α such that the following diagram is commutative:

$$\begin{array}{ccc} H(\gamma\ell) & \xrightarrow{\alpha\mathcal{F}_{\ell,\gamma\ell}} & H(\ell) \\ & {\scriptstyle \theta^{\chi}_{r,\ell}} \searrow \quad \nearrow {\scriptstyle \theta^{\chi}_{\ell,r}} & \\ & H(r,\chi) & \end{array}$$

i.e. $\alpha(\gamma) = b(\gamma\ell,\ell) = b(\ell,\gamma^{-1}\ell) = b(\gamma^{-1}) = b(\gamma)^{-1}$.

2.2.31. We now relate the model $H(r,\chi)$ of the representation W and θ-series. We have fixed (V,B) with its symplectic basis (P_i,Q_j), $r = \oplus\, \mathbb{Z} P_i \oplus \mathbb{Z} Q_j$, $\ell = \bigoplus_{i=1}^{n} \mathbb{R}P_i$, $\ell' = \bigoplus_{i=1}^{n} \mathbb{R}Q_j$.

Let D be the Siegel upper half-plane. By definition $D = \{\lambda;$ n-dimensional complex subspaces of $V^{\mathbb{C}}$, such that

1) $B(x,y) = 0$ for $x,y \in \lambda$

2) $iB(x,\bar{x}) > 0$ for $x \in \lambda - \{0\}$.$\}$

Let $\lambda \in D$, then the condition 2) implies that $\lambda \cap (\ell^{\mathbb{C}}) = 0$. Thus there exists a map $Z\colon (\ell')^{\mathbb{C}} \to (\ell^{\mathbb{C}})$ such that

$\lambda = \{(Zx+x);\ x \in (\ell')^{\mathbb{C}}\}$. Identifying ℓ' with ℓ^* as in 1.1.8, the condition 1) is translated by the fact that Z is a symmetric form, i.e. $Z = {}^tZ$, the condition 2) by the condition that $(\mathrm{Im}\ Z)$ is a positive definite symmetric form. Thus, considering the basis $(P_i),(Q_j)$ of ℓ and ℓ', we can parametrize D as:

$$D = \{Z,\ n \times n \text{ complex symmetric matrices, such that } \mathrm{Im}\ Z \gg 0\}.$$

Clearly $G = Sp(B)$ acts on D via $\lambda \to g\cdot\lambda$. If $\lambda = \{Zx + x,\ x \in (\ell')^{\mathbb{C}}\}$ and $g = \left(\begin{array}{c|c} A & B \\ \hline C & D \end{array}\right)$ then $g\cdot\lambda = \{(AZ+B)x + (CZ+D)x;\ x \in (\ell')^{\mathbb{C}}\}$. So we see that if $Z \in D$ the matrix $(CZ+D)$ is invertible and, as

$$(g\cdot\lambda) = \{(AZ+B)(CZ+D)^{-1}x + x;\ x \in (\ell')^{\mathbb{C}}\},$$

the action of G on D is given in the Z-coordinates by $g\cdot Z = (AZ+B)(CZ+D)^{-1}$.

2.2.32. Let us define for $g = \left(\begin{array}{c|c} A & B \\ \hline C & D \end{array}\right) \in Sp(B)$ and $Z \in D$, $j(g,Z) = \det(CZ+D)$. It is immediate to verify that:

$$j(g_1g_2,Z) = j(g_1,g_2\cdot Z)\ j(g_2,Z).$$

Let us consider the representation $W(\ell)$ of N in $H(\ell)$. We identify $H(\ell)$ with $L^2(\ell') = L^2(dy_1\ dy_2\ \cdots\ dy_n)$. Let us consider the infinitesimal representation dW of η in $\mathcal{S}(\mathbb{R}^n)$ ($\mathcal{S}(\mathbb{R}^n)$ is the space of C^∞-vectors of the representation W).

Let $\lambda \subset V^{\mathbb{C}}$, we say that v is a vacuum vector for λ if v is a C^∞-vector such that $dW(\lambda)\cdot v = 0$.

2.2.33. <u>Proposition</u>: Let $\lambda \in D$, then the space of vacuum vectors for λ is one-dimensional and spanned by

$$v_\lambda(y) = e^{i\pi(Zy,y)} \qquad (y \in \ell').$$

(Z corresponds to λ, under the above parametrization.)

<u>Proof</u>: λ has the basis $Q_i + ZQ_i$. As

$$dW(Q_i)\varphi = -\frac{\partial}{\partial y_i}\varphi$$

$$dW(ZQ_i)\varphi = 2i\pi B(ZQ_i,y)\varphi ,$$

it is immediate to verify that v_λ is the unique solution of the equations $(\frac{\partial}{\partial y_i}\varphi)(y) = 2i\pi B(ZQ_i,y)\varphi(y)$. Clearly, as $\operatorname{Im} Z >> 0$, $v_\lambda(y)$ is in the Schwartz space $\mathcal{S}(\mathbb{R}^n)$.

For $\lambda \in D$, we also denote v_λ by v_Z. We have defined the canonical projective representation $R_\ell(g)$ of $Sp(B)$ on $H(\ell)$. From the fundamental property $R_\ell(g)W_\ell(n)R_\ell(g)^{-1} = W_\ell(g\cdot n)$, it follows that if v_Z is a vacuum vector for $Z = \lambda$, $R_\ell(g)\cdot v_Z$ is a vacuum vector for $g\cdot\lambda$. Hence there exists a scalar $m(g,Z)$ such that

2.2.34. $R_\ell(g)\cdot v_Z = m(g,Z)\cdot v_{g\cdot Z}$. The relation $R_\ell(g_1g_2) = c_\ell(g_1,g_2)\, R_\ell(g_1)\cdot R_\ell(g_2)$ (1.6.11) implies:

$$m(g_1g_2, Z) = c_\ell(g_1,g_2)m(g_1,g_2\cdot Z)m(g_2,Z).$$

In particular, using 1.7.8:

$$m(g_1g_2,Z)^2\cdot s(g_1g_2)^{-1} = (s(g_1)m(g_1,g_2\cdot Z)^2)(s(g_2)m(g_2,Z)^2).$$

2.2.35. Proposition: $m(g,Z)^2 = s(g)\, j(g,Z)^{-1}$.

Proof: Both functions $m(g,Z)^2 s(g)^{-1} = u_1(g,Z)$ and $j(g,Z)^{-1}$ verify the cocycle relation: $c(g_1g_2,Z) = c(g_1,g_2\cdot Z)\; c(g_2,Z)$. It is then sufficient to prove the required equality on a set of generators of $Sp(B)$.

a) Let

$$g(a) = \left(\begin{array}{c|c} a & 0 \\ \hline 0 & {}^t a^{-1} \end{array}\right),$$

with $a \in GL(\mathbb{R}^n)$, then as

$$(R_\ell(g(a))\cdot\varphi)(y) = |\det a|^{1/2}\, \varphi({}^t a y) \qquad (1.6.21)$$

$$m(g(a),Z)^2 = |\det a|$$

$$s(g(a)) = \operatorname{sign}(\det a)$$

$$j(g(a),Z) = (\det a)^{-1}$$

the equality is satisfied.

b) Let

$$u(x) = \left(\begin{array}{c|c} 1 & x \\ \hline 0 & 1 \end{array}\right),$$

with $x = {}^t x$, then

$$m(u(x),Z) = 1$$

$$s(u(x)) = 1$$

$$j(u(x),Z) = 1$$

and the equality is satisfied.

c) Let

$$\sigma = \left(\begin{array}{c|c} 0 & -1 \\ \hline +1 & 0 \end{array}\right),$$

then $(R_\ell(\sigma)\varphi)(y) = \int \varphi(y') \, e^{-2i\pi(y,y')} \, dy'$. Thus from a calculation similar to 1.6.2 it follows that

$$m(\sigma,Z) = (\det(\tfrac{Z}{I}))^{-1/2}.$$

We have (1.7.4)

$$s(\sigma) = i^n$$

$$j(\sigma,Z) = (\det Z)$$

and the formula is satisfied.

As $\{g(a),u(x),\sigma\}$ form a system of generators of $Sp(B)$, our proposition is proven.

Remark: As "formally" $\tilde{R}(g) = s(g)^{-1/2} R_\ell(g)$, we have $\tilde{R}(g)\cdot v_Z = (\det(CZ+D))^{-1/2} \, v_{g\cdot Z}$, where the determination of $(\det(CZ+D))^{-1/2}$ is well defined for g belonging to the metaplectic group.

2.2.36. Let (r,χ) be our lattice with its given character χ. The operator $\theta^{\chi}_{r,\ell}: H(\ell) \to H(r,\chi)$ is such that

$$(\theta^{\chi}_{r,\ell}\cdot\varphi)(0) = \sum_{\xi\in\mathbb{Z}^n} \varphi(\xi) .$$

We define the function:

$$\theta(Z) = (\theta^{\chi}_{r,\ell}\cdot v_Z)(0) = \sum_{\xi\in\mathbb{Z}^n} e^{i\pi(Z\xi,\xi)} .$$

It is clear that $\theta(Z)$ is a holomorphic function on D.

2.2.37. <u>Theorem</u>: Let $\gamma \in \Gamma(r,\chi)$, then (for $\gamma = \left(\frac{A|B}{C|D}\right)$),

$$\theta(\gamma \cdot Z) = \mathcal{E}(\gamma)\ \det(CZ+D)^{1/2}\ \theta(Z)$$

when $\mathcal{E}(\gamma)^2 = k(\gamma)$ is the character of $\Gamma(r,\chi)$ defined in 2.2.22. In particular if $\gamma \in \Gamma_0^0(2)$, $\mathcal{E}(\gamma)^2 = \left(\frac{-1}{\det D}\right)$.

<u>Proof</u>: As $\det(CZ+D)^{1/2}$ is defined up to ± 1, $\mathcal{E}(\gamma)^2$ is well defined. Now we have:

$$\theta(\gamma \cdot Z) = (\theta^{\chi}_{r,\ell} \cdot v_{\gamma \cdot Z})(0)$$

$$v_{\gamma \cdot Z} = m(\gamma, Z)^{-1}\ R_{\ell}(\gamma) \cdot v_Z .$$

If $\gamma \in \Gamma(r,\chi)$,

$$\theta^{\chi}_{r,\ell}\ R_{\ell}(\gamma) = b(\gamma)\ A_r(\gamma)\ \theta^{\chi}_{r,\ell} . \qquad (2.2.30)$$

Thus:

$$\begin{aligned}(\theta^{\chi}_{r,\ell}\ R_{\ell}(\gamma) \cdot v_Z)(0) &= b(\gamma)(A_r(\gamma)\ \theta^{\chi}_{r,\ell} \cdot v_Z)(0) \\ &= b(\gamma)(\theta^{\chi}_{r,\ell} \cdot v_Z)(0) \\ &= b(\gamma)\theta(Z) .\end{aligned}$$

Hence $\theta(\gamma \cdot Z) = m(\gamma, Z)^{-1}\ b(\gamma)\ \theta(Z)$ and our proposition follows from 2.2.35, 2.2.22.

2.3. Modular forms on the upper half-plane.

2.3.1. Let us consider the action of $SL(2,\mathbb{R})$ on the upper half-plane. Let Γ be a discrete subgroup of $SL(2,\mathbb{R})$ and χ a character of Γ. We wish to construct holomorphic functions f on P^+ such that:

$$f\left(\frac{az+b}{cz+d}\right) = \chi(\gamma)(cz+d)^k f(z)$$

for every $\gamma = \begin{pmatrix} a & b \\ c & d \end{pmatrix}$ in Γ. (If f satisfies also the additional condition to be holomorphic at the cusps, f is called a modular form of type (k,χ) for Γ.)

We wish also to consider forms of half-integral weight $\frac{k}{2}$:

Let $\widetilde{G}$ be the universal covering group of $SL(2,\mathbb{R})$ and $\Gamma \subset \widetilde{G}$ a discrete subgroup of $\widetilde{G}$. We consider for any real number k (in fact k will be a half-integer)

$$M(\Gamma,k,\chi) = \{f \text{ holomorphic on } P^+ \text{ such that for every } \tilde{\gamma} = (\gamma,\varphi) \text{ in } \Gamma \subset \widetilde{G}, \text{ we have } f(\gamma\cdot z) = \chi(\tilde{\gamma})e^{k\varphi(z)}f(z)\}.$$

We indicate first, in a sketchy form, how to construct certain theta-functions satisfying these conditions by taking appropriate coefficients of representations of $\widetilde{SL}(2,\mathbb{R})$:

2.3.2. Let $G = SL(2,\mathbb{R})$ and $K = \{\begin{pmatrix} \cos\theta & -\sin\theta \\ \sin\theta & \cos\theta \end{pmatrix}\}$.

Let (R,H) be a representation of $SL(2,\mathbb{R})$ in a (topological) vector space H (we will avoid carefully here all delicate questions of continuity). Let $v \in H$ be an eigenvector for K of weight k, i.e. satisfying $R(u(\theta))\cdot v = e^{-ik\theta}v$. It is immediate to check that, for $g = \begin{pmatrix} a & b \\ c & d \end{pmatrix} \in SL(2,\mathbb{R})$, the vector $v_g = (ci + d)^k R(g)\cdot v$ depends only of $z = g\cdot i$. We denote it by v_z. We have in particular:

2.3.3. $v_z = y^{-k/2} R(b(z)\cdot v)$, where $z = x + iy$ and $b(z) = \begin{pmatrix} 1 & x \\ 0 & 1 \end{pmatrix}\begin{pmatrix} y^{1/2} & 0 \\ 0 & y^{-1/2} \end{pmatrix}$, as $b(z)\cdot i = z$.

2.3.4. The function $z \to v_z$ satisfies the fundamental property:

$$R(g)\cdot v_z = (cz+d)^{-k}\, v_{g\cdot z}\,.$$

Let $J^- = \frac{1}{2}\begin{pmatrix} i & 1 \\ 1 & -i \end{pmatrix}$ in $\mathfrak{sl}(2,\mathbb{C})$ and let $v \in H$ be a eigenvector of weight k for K. If v is annihilated by J^- under the infinitesimal action dR, v is called a lowest weight vector of weight k. In this case the corresponding H-valued function $z \mapsto v_z$ is holomorphic in z. (The Cauchy-Riemann equation for $z \to (f,v_z)$, $f \in H'$ dual vector space of H, corresponds to the equation $J^-\cdot v = 0$, as we will explain.)

The typical example for the construction of such a function v_z is as follows: We consider the representation T_k of $SL(2,\mathbb{R})$ on $\mathcal{O}(P^+)$, space of holomorphic functions on P^+, given by:

$$(T_k(g^{-1})f)(z) = (cz+d)^{-k} f(g\cdot z), \text{ for } g = \begin{pmatrix} a & b \\ c & d \end{pmatrix}.$$

Similarly, we denote by $\overline{T}_k$ the representation of $SL(2,\mathbb{R})$ on the space $\overline{\mathcal{O}}(P^+)$ of antiholomorphic functions on P^+ given by:

$$(T_k(g^{-1})f)(z) = (c\bar{z}+d)^{-k} f(g\cdot z).$$

Let us consider the function $\psi_w(z) = (\bar{z}-w)^{-k}$. From the relation

2.3.5. $$(g\cdot z-\overline{g\cdot w}) = (cz+d)^{-1} (z-\bar{w})(c\bar{w}+d)^{-1},$$

it is obvious to verify that:

$$(T_k(g)\cdot \psi_w) = (cw+d)^{-k} \psi_{g\cdot w}$$

In fact each representation (R,H) with a lowest weight vector v of weight k is isomorphic to a subrepresentation of $(T_k,\mathcal{O}(P^+))$, the isomorphism being obtained by sending v_w to ψ_w.

2.3.6. Let Γ be a discrete subgroup of G. Let $\theta \in H'$ be a semi-invariant functional under Γ, i.e.: $R(\gamma)\cdot\theta = \chi(\gamma)^{-1}\theta$, for all $\gamma \in \Gamma$, where χ is a character of the group Γ.

For v lowest weight vector of R of weight k, we can then form the "coefficient": $\theta(z) = (\theta, v_z)$. Then θ is a holomorphic function of z. Properties 2.3.4, 2.3.6 assure that:

$$\theta(\gamma\cdot z) = \chi(\gamma)(cz+d)^k \theta(z), \text{ for every } \gamma = \begin{pmatrix} a & b \\ c & d \end{pmatrix} \in \Gamma.$$

Therefore a method to construct modular functions in $M(\Gamma,k,\chi)$ is to construct representations (R,H) of $SL(2,\mathbb{R})$ such that there exists

1) a vector $v \neq 0$ in H satisfying

$$R(u(\theta))\cdot v = e^{-ik\theta}v$$

$$dR(J^-)\cdot v = 0 .$$

2) a functional $\theta \in H'$ such that $R(\gamma)\cdot\theta = \chi(\gamma)^{-1}\theta$, for $\gamma \in \Gamma$

We will consider both of these questions separately. In the model $(T_k, \mathcal{O}(P^+))$ where the choice of v_z is apparent, the construction of θ is equivalent to the initial problem: a semi-invariant functional on $\overline{\mathcal{O}}(P^+)$ is produced by a modular form g, via $(g,f) = \iint_{P^+} g(z)f(\overline{z})y^{k-2}\,dxdy$. However in the model of the Weil representation, we have already seen in 2.2 that non-tautological answers appear, via the construction of θ distributions θ_L associated to self dual lattices. In the next chapters, we will make a detailed study of the Weil representation associated to a quadratic form and study in this model questions 1) and 2).

We explicit now the isomorphism between the space $M(\Gamma,k,\chi)$ and a space of functions on G and relate the infinitesimal action of J^- to the Cauchy-Riemann equations on P^+.

Let us consider the identification $g \to g\cdot i$ of G/K with P^+.

When Φ is a function on P^+, the function $g \to \Phi(g\cdot i) = (I\Phi)(g)$ is a function on G. The group G acts on functions on P^+ by $(g\cdot\varphi)(z) = \varphi(g^{-1}\cdot z)$, preserving the space of holomorphic functions. We consider the left regular action of G on functions on G given by $(g_0\cdot\varphi)(g) = \varphi(g_0^{-1}g)$. We identify functions on P^+ with functions on G invariant by right translations by K, by $(I\varphi)(g) = \varphi(g\cdot i)$. We have $g_0\cdot I\varphi = I(g_0\cdot\varphi)$.

2.3.7. Let us consider the Lie algebra $\mathfrak{g}$ of G. $\mathfrak{g}$ consists of the 2 x 2 matrices with zero trace. A basis of $\mathfrak{g}$ is $H = \begin{pmatrix}1 & 0\\ 0 & -1\end{pmatrix}$, $X = \begin{pmatrix}0 & 1\\ 0 & 0\end{pmatrix}$, $Y = \begin{pmatrix}0 & 0\\ 1 & 0\end{pmatrix}$, with relations $[H,X] = 2X$, $[H,Y] = -2Y$ and $[X,Y] = H$.

The corresponding one-parameter subgroups of $SL(2;\mathbb{R})$ are

$$\exp tH = \begin{pmatrix}e^t & 0\\ 0 & e^{-t}\end{pmatrix},\ \exp tX = \begin{pmatrix}1 & t\\ 0 & 1\end{pmatrix},\ \exp tY = \begin{pmatrix}1 & 0\\ t & 1\end{pmatrix}.$$

The generator of the compact one-parameter subgroup $u(\theta) = \begin{pmatrix}\cos\theta & -\sin\theta\\ \sin\theta & \cos\theta\end{pmatrix}$ is $J_0 = \begin{pmatrix}0 & -1\\ 1 & 0\end{pmatrix} = Y - X$.

Let us consider the complexification $\mathfrak{g}^{\mathbb{C}}$ of $\mathfrak{g}$, and

$$c = \frac{1}{\sqrt{2}}\begin{pmatrix}1 & i\\ i & 1\end{pmatrix}.$$

Then $cHc^{-1} = iJ_0 = Z$. In particular $cHc^{-1} = Z$, $cXc^{-1} = J^+$, $cYc^{-1} = J^-$ form a basis of $\mathfrak{g}^{\mathbb{C}}$ with relations:

$$[Z,J^+] = 2J^+,\ [Z,J^-] = -2J^- \text{ and } [J^+,J^-] = Z.$$

We have:

$$J^+ = \frac{1}{2}\begin{pmatrix}-i & 1\\ 1 & i\end{pmatrix},\quad J^- = \frac{1}{2}\begin{pmatrix}i & 1\\ 1 & -i\end{pmatrix},\ \text{i.e.}\ J^- = \overline{J^+}.$$

We define

$$b^+ = \mathbb{C}Z + \mathbb{C}J^+$$

$$b^- = \mathbb{C}Z + \mathbb{C}J^-,\ \text{i.e.}\ b^- = \overline{b^+}.$$

For $X \in \mathfrak{g}$, we define the left invariant vector field $r(X)$ on G by $(r(X)\varphi)(g) = \frac{d}{d\varepsilon}\varphi(g(\exp\varepsilon X))_{|\varepsilon=0}$. The map $X \to r(X)$ is a homomorphism of $\mathfrak{g}$ into the vector fields on G. If $X \in \mathfrak{g}^{\mathbb{C}}$, $X = U + iV$ with U and V in $\mathfrak{g}$, we define $r(X)$ by $r(X) = r(U) + ir(V)$. Similarly, we define the right invariant vector field $\ell(X)$ on G by $(\ell(X)\varphi)(g) = \frac{d}{d\varepsilon}\varphi((\exp -\varepsilon X)g)_{|\varepsilon=0}$. We extend ℓ to $\mathfrak{g}^{\mathbb{C}}$ by linearity.

2.3.8. <u>Lemma</u>: A function Φ on P^+ is holomorphic if and only if $r(X)(I\varphi) = 0$ for every $X \in b^-$.

<u>Proof</u>: The condition $(r(J_0)\cdot f) = 0$, for a function f on G, is equivalent to the fact that f is right invariant under the group K, i.e. f is a function on $G/K = P^+$. Let us now analyze the condition $r(J^-)\cdot I\varphi = 0$:

Let φ be a function $\varphi(z,\overline{z})$ on P^+. We compute $(r(J^-)I\varphi)(e)$ where e is the identity component of G. By definition:

$$(r(J^-)I\varphi)(e) = \frac{1}{2}\frac{d}{d\varepsilon}\varphi(\exp\varepsilon\begin{pmatrix}0 & 1\\ 1 & 0\end{pmatrix}\cdot i)$$

$$+ \frac{i}{2}\frac{d}{d\varepsilon}\varphi(\exp\varepsilon\begin{pmatrix}1 & 0\\ 0 & -1\end{pmatrix}\cdot i).$$

We have modulo ε^2

$$\exp \varepsilon\begin{pmatrix}0 & 1\\ 1 & 0\end{pmatrix}\cdot i = \begin{pmatrix}1 & \varepsilon\\ \varepsilon & 1\end{pmatrix}\cdot i = \frac{i+\varepsilon}{\varepsilon i+1} = i + 2\varepsilon$$

$$\exp \varepsilon\begin{pmatrix}1 & 0\\ 0 & -1\end{pmatrix}\cdot i = \begin{pmatrix}1+\varepsilon & 0\\ 0 & 1-\varepsilon\end{pmatrix}\cdot i = (1+\varepsilon)(1-\varepsilon)^{-1}\cdot i = i + 2\varepsilon i$$

Hence

$$(r(J^-)I\varphi)(e) = \left(\left(\frac{\partial}{\partial x} + i\,\frac{\partial}{\partial y}\right)\cdot\varphi\right)(i)$$

$$= \left(\frac{\partial}{\partial \bar z}\cdot\varphi\right)(i)$$

If φ is holomorphic on P^+, $(r(J^-)I\varphi)(e) = 0$. As $r(J^-)$ is a left invariant vector field, $r(J^-)$ commutes with the left action of G, hence:

$$(r(J^-)I\varphi)(g_0) = (g_0^{-1}\cdot r(J^-)I\varphi)(e)$$

$$= r(J^-)I(g_0^{-1}\cdot\varphi)(e) = 0$$

as $g_0^{-1}\cdot\varphi$ is again a holomorphic function on P^+.

Reciprocally, if $(r(J^-)I\varphi)(g) = 0$, we have that, for every $g_0 \in G$, $g_0 = \begin{pmatrix}a & b\\ c & d\end{pmatrix}$

$$\frac{\partial}{\partial \bar z}\left(z \to \varphi((az+b)(cz+d)^{-1})\right)_{z=i} = 0$$

$$= (c\bar z+d)^{-2}\left(\frac{\partial}{\partial \bar z}\,\varphi\right)((ai+b)(ci+d)^{-1}) = 0.$$

Hence $\left(\frac{\partial}{\partial \bar z}\,\varphi\right) = 0$, i.e. φ is holomorphic on P^+.

2.3.9. Let us denote by $\mathcal{O}\!\ell(P^+)$ the space of real analytic functions on P^+ and by $\mathcal{O}(P^+)$ the space of holomorphic functions on P^+.

For α a real number, we define the representation $T_{\alpha,0}$ of $\mathcal{G}$ in the space of functions on P^+ by:

$$(T_{\alpha,0}((g,\varphi)^{-1})F)(z) = e^{-\alpha\varphi(z)} F(g\cdot z).$$

It is immediate to verify that $T_{\alpha,0}(g_1)\cdot T_{\alpha,0}(g_2) = T_{\alpha,0}(g_1 g_2)$ and that $T_{\alpha,0}$ leaves the space $\mathcal{O}(P^+)$ of holomorphic functions on P^+ stable. (If $\alpha = k$ is an integer, $T_{\alpha,0}$ is indeed the representation of the group $SL(2,\mathbb{R})$ given by:

$$(T_{k,0}(g^{-1})F)(z) = (cz+d)^{-k} F((az+b)(cz+d)^{-1})$$

for $g = \begin{pmatrix} a & b \\ c & d \end{pmatrix}$.).

2.3.10. Let $\mathcal{G}$ be the universal covering group of $SL(2,\mathbb{R})$. We now consider the one parameter subgroup of $\mathcal{G}$ with generator J_0. We have $\exp \theta J_0 = \delta(\theta)$, where $\delta(\theta) = (u(\theta),\varphi_\theta)$, with $\varphi_\theta(i) = i\theta$ (see 1.8.20), as $\delta(\theta)$ is a one parameter subgroup above $\exp \theta J_0$ in $SL(2;\mathbb{R})$.

For any real number α, we define:

2.3.11. $M(\mathcal{G},\alpha) = \{f,$ analytic on $\mathcal{G}$; $f(\sigma\delta(\theta)) = e^{-i\alpha\theta} f(\sigma)\}$.

The group $\mathcal{G}$ acts by left translations on $M(\mathcal{G},\alpha)$. We denote this representation by ℓ_α.

We consider the function $a_\alpha(g,\varphi) = e^{\alpha\varphi(i)}$ on $\mathcal{G}$ where

(g,φ) is an element of $\tilde{G}$. (If $\alpha = k$, $a_\alpha\begin{pmatrix} a & b \\ c & d \end{pmatrix} = (ci+d)^k$.)

We have: $$a_\alpha(\sigma\delta(\theta)) = e^{i\alpha\theta}a_\alpha(\sigma)$$

as for $\sigma = (g,\varphi)$, $\sigma\delta(\theta) = (gu(\theta),\varphi')$, with $\varphi'(i) = \varphi(i) + i\theta$. Hence if $f \in M(\tilde{G},\alpha)$, the function $(I_\alpha f)(\sigma) = a_\alpha(\sigma)f(\sigma)$ is invariant under right translations by $\delta(\theta)$. Let us denote by $\pi: \tilde{G} \to G$ the covering map. We can thus find a function on P^+ still denoted by $I_\alpha f$ such that $(I_\alpha f)(\sigma) = (I_\alpha f)(\pi(\sigma)\cdot i)$.

Let, for z in P^+, $b(z) = \begin{pmatrix} 1 & x \\ 0 & 1 \end{pmatrix}\begin{pmatrix} y^{1/2} & 0 \\ 0 & y^{-1/2} \end{pmatrix}$ be the unique element of B_0 such that $b(z)\cdot i = z$. We still denote by $b(z) = \tilde{b}(z) = (b(z), \operatorname{Log} y^{-1/2})$ the corresponding element of $\tilde{G}$ under the isomorphism $b \to \delta(b)$ of B_0 with its image in $\tilde{G}$ (1.8.20). We have:

$$(I_\alpha f)(z) = (I_\alpha f)(b(z)\cdot i) = (I_\alpha f)(b(z)) = a_\alpha(b(z))\, f(b(z))$$

i.e.

2.3.12. $$(I_\alpha f)(z) = y^{-\alpha/2}\, f(b(z))$$

with $z = x + iy$, and $b(z) = \begin{pmatrix} 1 & x \\ 0 & 1 \end{pmatrix}\begin{pmatrix} y^{1/2} & 0 \\ 0 & y^{-1/2} \end{pmatrix}$.

2.3.13. <u>Lemma</u>: I_α intertwines the representations ℓ_α and $T_{\alpha,0}$.

<u>Proof</u>: We have for $\sigma \in \tilde{G}$, $\sigma = (g,\varphi)$,

$$I_\alpha(\ell(g_0,\varphi_0)^{-1}\cdot f)(g) = a_\alpha(\sigma)\, f((g_0,\varphi_0)\cdot\sigma) = e^{\alpha\varphi(i)}\, f(g_0 g,\varphi'),$$

$$\text{with } \varphi'(i) = \varphi_0(g\cdot i) + \varphi(i).$$

$$(T_{\alpha,0}(g_0,\varphi_0)^{-1}\cdot I_\alpha f)(g) = e^{-\alpha\varphi_0(g\cdot i)}(I_\alpha f)(g_0 g)$$
$$= e^{-\alpha\varphi_0(g\cdot i)}\, e^{\alpha\varphi'(i)}\, f((g_0,\varphi_0)\cdot\sigma),\ \text{q.e.d.}$$

For $X \in \mathfrak{g}^{\mathbb{C}}$, we now consider $r(X)$ to be a left invariant vector field on $\mathcal{G}$.

2.3.14. <u>Lemma</u>: The function a_α satisfies

$$r(J^-)\cdot a_\alpha = 0.$$

<u>Proof</u>: Let us compute $(r(J^-)\cdot a_\alpha)(e)$. We write

$$2J^- = iH + X + Y = iH + J_0 + 2X.$$

The one parameter subgroup of $\mathcal{G}$ corresponding to X in $\mathcal{G}$ is $((\begin{smallmatrix}1 & t\\ 0 & 1\end{smallmatrix}),0)$.

The one parameter subgroup of $\mathcal{G}$ corresponding to H is $((\begin{smallmatrix}e^t & 0\\ 0 & e^{-t}\end{smallmatrix}),-t)$.

The one parameter subgroup corresponding to J_0 in $\mathcal{G}$ is $\delta(t) = (u(t),\varphi_t)$, with $\varphi_t(i) = it$. We have

$$a_\alpha((\begin{smallmatrix}1 & t\\ 0 & 1\end{smallmatrix}),0) = 1$$

$$a_\alpha((\begin{smallmatrix}e^t & 0\\ 0 & e^{-t}\end{smallmatrix}),-t) = e^{-\alpha t}$$

$$a_\alpha(\delta(t)) = e^{i\alpha t}.$$

Hence

$$2(r(J^-)\cdot a_\alpha)(e) = (ir(H) + r(J_0) + 2r(X)\cdot a_\alpha)(e) = -i\alpha + i\alpha = 0.$$

Now for $b \in B_0$, $b = (\begin{pmatrix} a & 0 \\ 0 & a^{-1}\end{pmatrix}, \text{Log } a^{-1})$, $a_\alpha(b\sigma) = e^{-\alpha \text{Log} a}\, a_\alpha(\sigma)$, hence $(r(J^-)a_\alpha)(b\sigma) = e^{-\alpha \text{Log} a}\, a_\alpha(\sigma)$. Therefore we only have to check that $(r(J^-)a_\alpha(\sigma) = 0$ for $\sigma = \delta(t)$. We recall that J^- satisfies $[J_0, J^-] = 2iJ^-$. It follows that:

$$(r(J^-)a_\alpha)(\delta(t)) = -e^{2it}(\ell(J^-)a_\alpha)(\delta(t)) = e^{2it}e^{-i\alpha t}(r(J^-)a_\alpha)(e),$$

(as we have the equality:

$$(r(X)\cdot\varphi)(g) = \frac{d}{d\varepsilon}\varphi(g \exp \varepsilon X) = \frac{d}{d\varepsilon}\varphi(g \exp \varepsilon X g^{-1} g)$$

$$= \frac{d}{d\varepsilon}\varphi(\exp \varepsilon(g\cdot X)\cdot g) = -(\ell(g\cdot X)\varphi)(g).$$

Thus $(r(J^-)a_\alpha)(\delta(t)) = 0$.

2.3.15. <u>Corollary</u>: $I_\alpha f$ is a holomorphic function on P^+ if and only $r(J^-)f = 0$.

<u>Proof</u>: We have $(I_\alpha f)(\sigma) = a_\alpha(\sigma)f(\sigma)$. $I_\alpha f$ is invariant by right multiplication by K, hence by 2.3.8, $I_\alpha f$ is holomorphic on P^+ if and only if $r(J^-)\cdot(I_\alpha f) = 0$. As $r(J^-)$ is a vector field annihilating the function a_α, $r(J^-)\cdot(I_\alpha f) = a_\alpha(\sigma)(r(J^-)\cdot f)$, and we obtain our corollary.

2.3.16. Let Γ be a discrete subgroup of $\mathcal{G}$ and χ be a character of Γ. We denote by $M_\sigma(\Gamma,\chi,\alpha,J^-)$ the space of analytic functions on $\mathcal{G}$ satisfying

a) $$f(\sigma\delta(t)) = e^{-i\alpha t} f(\sigma),$$

b) $\quad r(J^-)\cdot f = 0$

c) $\quad f(\gamma\sigma) = \chi(\gamma)f(\sigma)$ for $\gamma \in \Gamma$.

The condition c) can be restated as $\ell_\alpha(\gamma^{-1})\cdot f = \chi(\gamma)f$ for $\gamma \in \Gamma$. Therefore, the correspondence $f \to I_\alpha f$ is a isomorphism between $M_{\mathcal{G}}(\Gamma,\chi,\alpha,J^-)$ and the space of holomorphic functions on P^+ satisfying

$$f\left(\frac{az+b}{cz+d}\right) = \chi(\gamma)\ \text{"}cz+d\text{"}^{\alpha}\ f(z),$$

for every $\gamma = \left(\begin{pmatrix} a & b \\ c & d \end{pmatrix},\varphi\right) \in \Gamma$, where $\text{"}cz+d\text{"}^{\alpha} = e^{\alpha\varphi(z)}$. Hence we have identified our space $M(\Gamma,k,\chi)$ (2.3.I) with the subspace $M_{\mathcal{G}}(\Gamma,\chi,k,J^-)$ of functions on $\mathcal{G}$ verifying the invariance conditions a), b) under the right action of b^-, and c) under the left action of Γ.

Let (R,V) be a representation of $\mathcal{G}$ in a topological vector space V (V will be for example the space of C^∞ vectors of some unitary representation (R,H)). Then for $v \in V$, $\xi \in V^*$ we denote by $c_{\xi,v}(\sigma) = (\xi, R(\sigma)\cdot v)$ the associated coefficient. We consider $c_{\xi,v}$ as a functions on $\mathcal{G}$. We have:

$$\begin{aligned} c_{\xi,v}(g_1^{-1}gg_2) &= (\xi,\ R(g_1)^{-1}R(g)R(g_2)\cdot v) \\ &= (g_1\cdot\xi,\ R(g)g_2\cdot v) \\ &= c_{g_1\cdot\xi,\,g_2\cdot v}(\sigma)\,. \end{aligned}$$

If v satisfies

1) $R(\delta(t))\cdot v = e^{-i\alpha t}v$

(for $t \in \mathbb{R}$), $c_{\xi,v}$ verifies $c_{\xi,v}(\sigma\delta(t)) = e^{-i\alpha t}\, c_{\xi,v}(\sigma)$.

Let $x \in \mathfrak{g}$, $u \in V$. We denote by $dT(x)\cdot u$ the limit (if it exists) of $\left(\frac{T(\exp \varepsilon x) - 1}{\varepsilon}\right)\cdot u$ when $\varepsilon \to 0$. For $x \in \mathfrak{g}^{\mathbb{C}}$, we define $dT(x)\cdot u$ by linearity. We have

$$r(J^-)\cdot c_{\xi,v} = c_{\xi,dR(J^-)\cdot v} .$$

Hence, if v satisfies

2) $dR(J^-)\cdot v = 0$

$c_{\xi,v}$ satisfies $r(J^-)\cdot c_{\xi,v} = 0$.

As in 2.3.3, we consider, for v satisfying 1), the vector $v_z = a_\alpha(\sigma)\ (R(\sigma)\cdot v)$ which doesn't depend of the choice of $\sigma = (g,\varphi)$ such that $g\cdot i = z$. We keep in mind the formula:

2.3.17. If $v_z = y^{-\alpha/2}\, R(b(z))\cdot v$, for $b(z) = \begin{pmatrix} 1 & x \\ 0 & 1\end{pmatrix}\begin{pmatrix} y^{1/2} & 0 \\ 0 & y^{-1/2}\end{pmatrix}$,

$$R(g)\cdot v_z = \text{"}cz + d\text{"}^{\alpha}\, v_{g\cdot z} .$$

Let ξ be any element of V^*, we have $I_\alpha(c_{\xi,v})(z) = (\xi, v_z)$. Thus the function (ξ, v_z) is holomorphic in z, if v satisfies the relation 2), i.e. if v is a lowest weight vector of the representation R.

Let $\theta \in V^*$ a functional satisfying $R(\gamma)\cdot\theta = \chi(\gamma)^{-1}\cdot\theta$ and v a lowest weight vector of the representation R of weight α. We finally see that $(\theta, v_z) = \theta(z)$ will be a holomorphic function of z satisfying

$$\theta(\frac{az+b}{cz+d}) = \chi(\gamma) "cz + d"^{\alpha} \theta(z) \quad \text{for} \quad \begin{pmatrix} a & b \\ c & d \end{pmatrix} \in \Gamma.$$

We now indicate another important way to construct modular forms, i.e. the notion of Poincare series.

2.3.18. Let $\Gamma \subset SL(2,\mathbb{Z})$ a subgroup of finite index, containing $\begin{pmatrix} -1 & 0 \\ 0 & -1 \end{pmatrix}$.

We define the notion of "cusps": Let us consider the action of $SL(2,\mathbb{Z})$ on $\mathbb{R} \cup \{\infty\} \subset \mathbb{P}^+$ given by $x \to \frac{ax + b}{cx + d}$. By definition, a cusp P of Γ is an equivalence class of $\mathbb{Q} \cup \{\infty\}$ under this action of Γ. If $\Gamma = SL(2,\mathbb{Z})$, each element of $\mathbb{Q}$ can be transformed to (∞) by Γ, hence there is just one cusp: the point of infinity.

For each cusp P, we can fix a matrix $A_P \in SL(2,\mathbb{Z})$ transforming P to ∞, i.e. $A_P(P) = \infty$. If $P = \infty$ we naturally choose $A_\infty = \mathrm{Id}$. We denote by $S_P = A_P^{-1}\begin{pmatrix} 1 & \mathbb{Z} \\ 0 & 1 \end{pmatrix} A_P \cap \Gamma$. Clearly if $\gamma \in S_P$, then $\gamma \cdot P = P$, as the subgroup $\begin{pmatrix} 1 & \mathbb{Z} \\ 0 & 1 \end{pmatrix}$ leaves stable the point at ∞. The subgroup $\Gamma_P = A_P S_P A_P^{-1} \subset \begin{pmatrix} 1 & \mathbb{Z} \\ 0 & 1 \end{pmatrix}$ is of the form $\Gamma_P = \{\begin{pmatrix} 1 & nw_P \\ 0 & 1 \end{pmatrix};\ n \in \mathbb{Z}\}$. The integer w_P is called the width of P.

Let Γ be a discrete subgroup of finite index in $SL(2,\mathbb{Z})$ and $\tilde{\Gamma}$ its reciproc image in $\tilde{G}$. Let χ be a character of $\tilde{\Gamma}$ and f a function in $M(\tilde{\Gamma},\chi,\alpha)$. For $\hat{A}_P \in \tilde{G}$ above A_P, the function $T_\alpha(\hat{A}_P)\cdot f$ is then semi-invariant under the action T_α of $\tilde{\Gamma}_P$ on $\mathcal{H}(P^+)$, with character $A_P\cdot\chi$. We say that f is holomorphic at the cusp P, if $(T_\alpha(\tilde{A}_P)\cdot f)(z)$ admits a development of the form

$$\sum_{n\geq 0} a_P^n e^{\frac{2i\pi nz}{w_P}} .$$

f is said to be a cusp form at P, if $a_P^0 = 0$, i.e., if

$$(T_\alpha(\hat{A}_P)f)(z) = \sum_{n>0} a_P^n e^{\frac{2i\pi nz}{w_P}} .$$

2.3.19. In fact, as each function $e_P^n(z) = e^{\frac{2i\pi nz}{w_P}}$ is an invariant function under the action $f(z) \to f(z + w_P)$ of the generator $\begin{pmatrix} 1 & w_P \\ 0 & 1 \end{pmatrix}$ of Γ_P, there is another important "natural" way to construct functions in $M(\tilde{\Gamma},\chi,\alpha)$: We suppose for simplicity that α is an integer (the modifications for arbitrary α being obvious) and consider Γ a subgroup of $SL(2,\mathbb{Z})$. The function e_P^n is invariant under $T_\alpha(\Gamma_P)$, thus the function $T_\alpha(A_P^{-1})\cdot e_P^n$ is invariant under $T_\alpha(S_P)$. Let us suppose $\chi(S_P) \equiv 1$. We then can form formally the series:

2.3.20. $$G_{P,\alpha}^n = \frac{1}{2} \sum_{\gamma\in\Gamma/S_P} \chi(\gamma)\; T_\alpha(\gamma)\cdot T_\alpha(A_P^{-1})\cdot e_P^n .$$

It is clear that formally G_P^n verifies:

$$T_\alpha(\gamma)\cdot G_P^n = \chi(\gamma)^{-1}\cdot G_P^n \text{, i.e.:}$$

$$G_P^n\left(\frac{az+b}{cz+d}\right) = \chi(\gamma)\; (cz+d)^\alpha\; G_P^n(z), \text{ for } \gamma \in \Gamma.$$

2.3.21. Lemma: The series $G_{P,\alpha}^n$ (2.3.20) is convergent for $\alpha > 2$.

Proof: After conjugation by A_P, it is sufficient to prove that if Γ' is a subgroup of $SL(2,\mathbb{Z})$, the series

$$\sum_{\gamma\in\Gamma'/\Gamma'\cap\begin{pmatrix}1 & \mathbb{Z}\\ 0 & 1\end{pmatrix}} |(T_\alpha(\gamma)\cdot e_P^n)(z)|$$

is absolutely convergent. We have

$$(T_\alpha(\gamma)\cdot e_P^n)(z) = |a - cz|^{-\alpha}\ |e_P^n(\gamma^{-1}.z)|\ .$$

As $\gamma^{-1}z \in P^+$, $|e_P^n(\gamma^{-1}z)| < 1$. Thus we have to check that

$$\sum_{\gamma\in\Gamma'/\Gamma'\cap\begin{pmatrix}1 & \mathbb{Z}\\ 0 & 1\end{pmatrix}} |a - cz|^{-\alpha} < \infty\ .$$

The map $\begin{pmatrix}a & b\\ c & d\end{pmatrix} \to \begin{pmatrix}a\\ c\end{pmatrix}$ is an injection of $\Gamma'/\Gamma'\cap\begin{pmatrix}1 & \mathbb{Z}\\ 0 & 1\end{pmatrix}$ on $\mathbb{Z}^2 - \{0,0\}$. Thus this series is majorized by the serie

$$\sum_{(m,n)\in\mathbb{Z}^2-\{0,0\}} \left|\frac{1}{m + nz}\right|^\alpha\ .$$

Now, for $z \in P^+$, the elements $m + nz$ describe the lattice in $\mathbb{R}^2$ of basis $(1,z)$. As $\frac{1}{|x|^\alpha}$ is integrable at ∞ for $\alpha > 2$, we obtain our lemma. Similar estimates shows that

$$\sum_{\substack{m,n\\ n\neq 0}} \left|\frac{1}{m + nz}\right|^\alpha \to 0$$

when $z = iy$, $y \to \infty$. Thus it is easy to see that $G_{P,\alpha}^n$ is a cusp form when $n \neq 0$. $G_{P,\alpha}^n$ is called a Poincare series.

2.3.22. Let us now define the Petersson scalar product. The measure $y^{-2}\,dxdy$ on P^+ is invariant under the action of $SL(2,\mathbb{R})$ on P^+, i.e. we have:

$$\int_A f(z)y^{-2}\,dxdy = \int_{g(A)} (g\cdot f)(z)y^{-2}\,dxdy$$

for $(g\cdot f)(z) = f(g^{-1}z)$, as follows by change of variables. Let Γ be a discrete subgroup of $SL(2,\mathbb{R})$ containing $\begin{pmatrix}-1 & 0\\ 0 & -1\end{pmatrix}$ and let $\mathcal{F}$ be a fundamental domain for the action of Γ on P^+. Thus $\gamma\mathcal{F}\cap\mathcal{F}$ is of measure 0 (except if $\gamma = -1$, as then γ acts by the identity on P^+). If f is a function on P^+ invariant by Γ $(\gamma\cdot f = f)$, the integral (if it exists) $\int f(z)y^{-2}\,dxdy$ is then independent of the choice of $\mathcal{F}$.

Let us consider Γ a discrete subgroup of $SL(2,\mathbb{R})$, $\tilde{\Gamma}$ its reciproc image in $\mathcal{G}$ and the space $M_{\mathcal{G}}(\tilde{\Gamma},\chi,\alpha,J^-)$; if f_1 and f_2 are two functions in $M_{\mathcal{G}}(\tilde{\Gamma},\chi,\alpha,J^-)$, the function $f(\sigma)\overline{g(\sigma)}$ is a function on $\tilde{\Gamma}\backslash\mathcal{G}/K = \Gamma\backslash P^+ = \mathcal{F}$. Thus we can form (if it exists) the integral $\int_{\mathcal{F}} f(\sigma)\overline{g(\sigma)}\,y^{-2}\,dxdy$, over a fundamental domain $\mathcal{F}$ for Γ. Let us consider the isomorphism I_α of $M_{\mathcal{G}}(\Gamma,\chi,\alpha,J^-)$ with $M(\Gamma,\chi,\alpha)$ given by $(I_\alpha f)(z) = y^{-\alpha/2} f(b(z))$. We have:

$$\int f(\sigma)\overline{g(\sigma)}\,y^{-2}\,dxdy = \int (I_\alpha f)(z)\,\overline{(I_\alpha g)(z)}\,y^{\alpha-2}\,dxdy\,.$$

Thus, we define the Petersson scalar product of two functions in $M(\tilde{\Gamma},\chi,\alpha)$ by

$$\langle f,g\rangle = \int f(z)\overline{g(z)}\,y^{\alpha-2}\,dxdy\,,$$

where $\mathcal{F}$ is a fundamental domain for the action of Γ on P^+. (We recall that this formula is independent of the choice of $\mathcal{F}$.)

Let Γ be a subgroup of finite index of $SL(2,\mathbb{Z})$ containing

$\begin{pmatrix}-1 & 0\\ 0 & -1\end{pmatrix}$. A fundamental domain $\mathcal{F}$ for the action of $SL(2,\mathbb{Z})$ on Γ is $\{z,\ |z| > 1,\ |\mathrm{Im}\, z| \le \frac{1}{2}\}$.

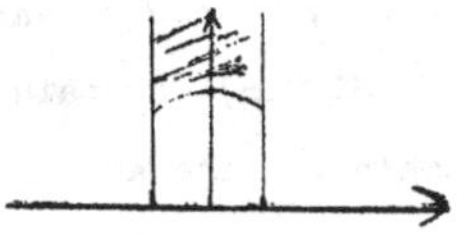

It follows that $\langle f,g\rangle$ exists, if $y^{\alpha-2}\, f(z)\overline{g(z)}$ is integrable in y, when $y \to \infty$; thus, if f or g are cusp forms at ∞ (or if $\alpha = 1$), the Petersson scalar product $\langle f,g\rangle$ is well defined. Let us denote by $S_k(\Gamma,\chi,\alpha)$ the space of cusp forms in $M_k(\Gamma,\chi,\alpha)$ (i.e. f vanishes at the cusps). We have:

2.3.23. <u>Proposition</u>. Let $\alpha > 2$. Let $f \in S_k(\Gamma,\chi,\alpha)$ and

$$(T_\alpha(A_P)\cdot f)(z) = \sum_{n>0} a_P^n(f)\, e^{\frac{2i\pi nz}{w_P}}$$

Then:

$$\langle f, G_{P,\alpha}^n\rangle = w_P\left(\frac{w_P}{4\pi n}\right)^{\alpha-1} \Gamma(\alpha-1)\, a_P^n(f)\ .$$

<u>Proof</u>: We have:

$$\langle f, G_{P,\alpha}^n\rangle = \frac{1}{2}\int_{\mathcal{F}} f(z)\ \overline{\left(\sum_{\gamma\in\Gamma/S_P} \chi(\gamma)T_\alpha(\gamma)\cdot T_\alpha(A_P^{-1})\cdot e_P^n\right)(z)}\,y^{\alpha-2}dxdy\ .$$

The integral is absolutely convergent and we can reverse the order of summation. Now:

$$\int_{\mathcal{F}} f(z)\ \overline{\chi(\gamma)(T_\alpha(\gamma)T_\alpha(A_P^{-1})\cdot e_P^n)(z)}\ y^{\alpha-2}dxdy$$
$$= \int_{\gamma^{-1}(\mathcal{F})} (T_\alpha(\gamma)^{-1}.f)(z)\ \overline{\chi(\gamma)(T_\alpha(A_P^{-1})\cdot e_P^n)(z)}\ y^{\alpha-2}dxdy$$

as follows from the invariance property of $y^{\alpha-2}\,dxdy$,

$$= \int_{\gamma^{-1}(\mathcal{F})} f(z)\,\overline{(T_\alpha(A_P^{-1})\cdot e_P^n)(z)}\; y^{\alpha-2}\,dxdy$$

as follows from the relation $T_\alpha(\gamma)^{-1}\cdot f = \chi(\gamma)\cdot f$, for $\gamma \in \Gamma$.

Thus we have to compute:

$$\frac{1}{2} \sum_{\gamma\in\Gamma/S_P} \int_{\gamma^{-1}(\mathcal{F})} f(z)\,\overline{(T_\alpha(A_P^{-1})\cdot e_P^n)(z)}\; y^{\alpha-2}\,dxdy .$$

Writing $\Gamma = \bigcup_{\gamma_i\in\Gamma/S_P} S_P\gamma_i^{-1}$, we see that $\Delta_P = \bigcup_{\gamma\in\Gamma/S_P} \gamma^{-1}(\mathcal{F})$ is a fundamental domain for the action of S_P on P^+.

As $\gamma^{-1}\mathcal{F} = \gamma'^{-1}\mathcal{F}$ if and only if $\gamma = \pm\gamma'$, our integral is exactly:

$$\int_{\Delta_P} f(z)\,\overline{(T_\alpha(A_P^{-1})\cdot e_P^n)(z)}\; y^{\alpha-2}\,dxdy$$

$$= \int_{A_P(\Delta_P)} (T_\alpha(A_P)\cdot f)(z)\,\overline{e_P^n(z)}\; y^{\alpha-2}\,dxdy .$$

Now $A_P(\Delta_P)$ is a fundamental domain for the action of $\Gamma_P = A_P(S_P) = \{\begin{pmatrix} 1 & nw_P \\ 0 & 1\end{pmatrix};\ n \in \mathbb{Z}\}$ on P^+. As both functions $T_\alpha(A_P)\cdot f$ and e_P^n are invariant under Γ_P, the integral doesn't depend of the choice of the fundamental domain. We then can choose as fundamental domain Δ' for Γ_P the set $\{z,\ 0 \leq |\mathrm{Re}\ z| \leq w_P\}$. Our integral becomes:

$$\int_{\substack{0\leq x\leq w_P \\ 0<y<\infty}} (T_\alpha(A_P)\cdot f)(z)\cdot e^{-\frac{2i\pi n\bar{z}}{w_P}}\; y^{\alpha-2}\,dxdy .$$

Using the development $(T_\alpha(A_P)\cdot f)(z) = \sum_{n>0} a_P^n(f)\, e^{\frac{2i\pi nz}{w_P}}$

we obtain finally:

$$\langle f, G^{n}_{P,\alpha} \rangle = \int_{0<y<\infty} a^{n}_{P}(f)\ w_{P}\ e^{-\frac{4\pi n y}{w_{P}}}\ y^{\alpha-2}\ dy$$

$$= w_{P}\left(\frac{w_{P}}{4\pi n}\right)^{(\alpha-1)}\ \Gamma(\alpha-1)\ a^{n}_{P}(f) \qquad \text{Q.E.D.}$$

2.4. Modular forms of weight 1/2.

We will construct θ-series of weight 1/2 and 3/2 for some congruence subgroups of $SL(2,\mathbb{Z})$ by taking appropriate coefficients of the representation $\widehat{R}$ of $SL(2;\mathbb{R})$, illustrating the method sketched in 2.3. Our method will be the following:

1) Analyze the possible lowest weight vectors in the Shale-Weil representation $\tilde{R}$ of $\tilde{G} = \widetilde{SL}(2,\mathbb{R})$

2) For a given congruence subgroup Γ of $SL(2,\mathbb{Z})$, produce some semi-invariant linear functionals under Γ, using the model $H = H(r,\chi)$ of the representation W.

We consider $V = \mathbb{R}P \oplus \mathbb{R}Q$ our canonical symplectic space.

2.4.1. Let us consider the following elements:

$$\sigma = \begin{pmatrix} 0 & 1 \\ -1 & 0 \end{pmatrix},\ u(x) = \begin{pmatrix} 1 & x \\ 0 & 1 \end{pmatrix},\ g(a) = \begin{pmatrix} a & 0 \\ 0 & a^{-1} \end{pmatrix},\ a \in \mathbb{R} - \{0\},$$

of $SL(2;\mathbb{R})$.

We denote by B the subgroup: $\{\begin{pmatrix} a & u \\ 0 & a^{-1} \end{pmatrix}, a \neq 0\}$. Each element of B is written uniquely as $g(a)u(x)$.

2.4.2. <u>Lemma</u>: $G = B \cup B\sigma B$.

<u>Proof</u>: Let $\ell = \mathbb{R}P$ be our Lagrangian plane of reference. Then B is the subgroup of G leaving ℓ stable. We have $\sigma\ell = \ell' = \mathbb{R}Q$. It is clear that $B\cdot\ell'$ consists of all the lines other than $\mathbb{R}P$. Let $g \in G$: if $g\cdot\ell = \ell$, g belongs

to B; otherwise there exists $b \in B$ with $g\cdot\ell = b\cdot\sigma\cdot\ell$, i.e. $g \in B\sigma B$.

2.4.3. We recall the formulas 1.6.21 for the projective representation $R = R_\ell$ of G on the set of generators $(\sigma, g(a), u(x))$ of $SL(2;\mathbb{R})$. We identify $H(\ell)$ with $L^2(\mathbb{R})$ by $f(y) = f(\exp yQ)$, then:

$$(R(g(a))f)(y) = |a|^{1/2} f(a\cdot y) \quad \text{for} \quad g(a) = \begin{pmatrix} a & 0 \\ 0 & a^{-1} \end{pmatrix},$$

$$(R(u(t))f)(y) = e^{i\pi t y^2} f(y), \quad \text{for} \quad u(t) = \begin{pmatrix} 1 & t \\ 0 & 1 \end{pmatrix},$$

$$(R\begin{pmatrix} 0 & 1 \\ -1 & 0 \end{pmatrix}f)(y) = \int e^{2i\pi xy} f(x)dx.$$

We see that the space $L(0)$ of even functions in $L^2(\mathbb{R})$ as well as the space $L(1)$ of odd functions is stable under R.

Let us consider $\mathcal{S}(\mathbb{R})$, the Schwartz space of rapidly decreasing functions on $\mathbb{R}$. If $f \in \mathcal{S}(\mathbb{R})$, f is a C^∞ vector for the representation $\widehat{R}$.

Let

$$H = \begin{pmatrix} 1 & 0 \\ 0 & -1 \end{pmatrix}, \quad X = \begin{pmatrix} 0 & 1 \\ 0 & 0 \end{pmatrix}, \quad Y = \begin{pmatrix} 0 & 0 \\ 1 & 0 \end{pmatrix}$$

be the standard basis of $\mathfrak{sl}(2,\mathbb{R})$. Then on $\mathcal{S}(\mathbb{R})$:

$$d\tilde{R}\,(X) = i\pi y^2$$

(where $i\pi y^2$ denotes the multiplication operator)

$$d\tilde{R}\,(H) = y\,\frac{\partial}{\partial y} + \frac{1}{2}$$

$$d\tilde{R}\,(Y) = \frac{i}{4\pi}\left(\frac{\partial}{\partial y^2}\right)$$

(For $\sigma = \begin{pmatrix} 0 & 1 \\ -1 & 0 \end{pmatrix}$, σ acts by the Fourier transform and $\sigma X \sigma^{-1} = Y$.)

Let us consider the basis J_0, J^-, J^+ of $\mathfrak{g}^{\mathbb{C}}$ given in 2.3.3. We are interested in finding the lowest weight vectors of the representation $\widetilde{R}$, i.e. the solution of the equations $d\widetilde{R}\,(J^-)\cdot f_0 = 0$.

2.4.4. __Lemma:__

1) The function $f_0(y) = e^{-\pi y^2}$ satisfies
   a) $d\widetilde{R}\,(J_0)\cdot f_0 = -\frac{i}{2}\, f_0$,
   b) $d\widetilde{R}\,(J^-)\cdot f_0 = 0$.
2) The function $f_1(y) = ye^{-\pi y^2}$ satisfies
   a) $d\widetilde{R}\,(J_0)\cdot f_1 = -\frac{3i}{2}\, f_1$,
   b) $d\widetilde{R}\,(J^-)\cdot f_1 = 0$.

__Proof:__ We have $d\widetilde{R}\,(J_0) = d\widetilde{R}\,(Y) - d\widetilde{R}\,(X)$.

Since $\quad \dfrac{\partial^2}{\partial y^2}\,(e^{-\pi y^2}) = (4\pi y^2 - 2\pi)e^{-\pi y^2}$,

$$(d\widetilde{R}\,(Y) - d\widetilde{R}\,(X)\cdot f_0)(y) = (4\,\frac{i\pi y^2}{4} - \frac{i}{2} - i\pi y^2)e^{-\pi y^2}$$

$$= -\frac{i}{2}\, f_0(y).$$

The other calculations are similar.

__Remark:__ According to the decomposition $L^2(\mathbb{R}) = L(0) \oplus L(1)$ in even and odd functions, we have $\widetilde{R} = \widetilde{R}_0 \oplus \widetilde{R}_1$ and both representations $\widetilde{R}_0$ and $\widetilde{R}_1$ are unitary irreducible representations of $\widetilde{SL}(2,\mathbb{R})$ (thus isomorphic respectively to $\overline{T}_{1/2}, \overline{T}_{3/2}$ acting on some Hilbert spaces of functions on P^+). The vector f_0 (resp. f_1) is "the" lowest weight vector of $\widetilde{R}_0$ (resp. $\widetilde{R}_1$). We hence have accomplished the first part of our program.

Let us now analyze the second part.

2.4.5. We consider the lattice $r = \mathbb{Z}P \oplus \mathbb{Z}Q$ with the character $\chi(\exp(mP + nQ + tE)) = (-1)^{nm} e^{2i\pi t}$ of the corresponding subgroup $R = \exp(r \oplus \mathbb{R}E)$ of $N = \exp(\mathbb{R}P \oplus \mathbb{R}Q \oplus \mathbb{R}E)$. To these data is associated the model $H(r,\chi)$ for the irreducible representation W of the Heisenberg group.

We have constructed the operator $\theta^{\chi}_{r,\ell}: H(\ell) \to H(r,\chi)$. $\theta^{\chi}_{r,\ell}$ is given by the formula $(\theta^{\chi}_{r,\ell}\varphi)(g) = \sum_{n\in\mathbb{Z}} \varphi(g \exp nQ)$ for $\varphi \in H(\ell)$.

2.4.6. Let

$$\Gamma = SL(2,\mathbb{Z}), \text{ i.e. } \Gamma = \left\{\begin{pmatrix} a & b \\ c & d\end{pmatrix}, a,b,c,d \in \mathbb{Z}, ad - bc = 1\right\}$$

be the subgroup of G leaving the lattice r stable. Let $\Gamma(r,\chi)$ be the subgroup of Γ leaving χ stable. We have:

$$\Gamma(r,\chi) = \{\begin{pmatrix} a & b \\ c & d\end{pmatrix} \in \Gamma, ac \equiv 0 \text{ Mod } 2, bd \equiv 0 \text{ Mod } 2\}.$$

Let us consider the model $W = W(r,\chi)$ of the representation W acting on $H(r,\chi)$. We also denote $\Gamma(r,\chi)$ by $\Gamma(\chi)$

2.4.7. Let us define for $\gamma \in \Gamma(\chi)$, $(A_r(\gamma)\varphi)(n) = \varphi(\gamma^{-1}\cdot n)$ as a unitary operator on $H(r,\chi)$. As calculated in 2.2.30, we have $\theta^{\chi}_{r,\ell} \circ R_\ell(\gamma) \circ (\theta^{\chi}_{r,\ell})^{-1} = \alpha(\gamma)^{-1} A_r(\gamma)$, with

$$\alpha(\gamma) = b(\gamma\ell,\ell) = b(\ell,\gamma^{-1}\ell) = b(\mathbb{R}P,\mathbb{R}(dP-cQ)) \text{ (as } \gamma^{-1} = \begin{pmatrix} d & -b \\ -c & a\end{pmatrix}).$$

Thus $\alpha(\gamma) = b(d,-c) = |c|^{-1/2} \sum_{n\in\mathbb{Z}/c\mathbb{Z}} e^{-\frac{i\pi cn^2}{d}}$ (2.2.11)

2.4.8. Let us consider the universal covering group $\mathcal{G}^{\sim}$ of $SL(2,\mathbb{R})$ and the true representation $\widehat{R}(\sigma)$ of $\mathcal{G}^{\sim}$ in $H(\ell)$. We denote by $\widehat{\Gamma}(\chi)$ the reciprocal image of $\Gamma(\chi)$ in $\mathcal{G}^{\sim}$. For any $\sigma \in \widehat{\Gamma}(\chi)$ the Weil representation $\widehat{R}(\sigma)$ acting on $H(\ell)$ and the representation $A_r(\sigma)$ acting on $H(r,\chi)$ are proportional. Thus there exists a character λ of $\widehat{\Gamma}(\chi)$ such that:

$$\theta^{\chi}_{r,\ell} \circ \tilde{R}_{\ell}(\gamma,\varphi) \circ \theta^{\chi}_{\ell,r} = \lambda(\gamma,\varphi)\, A_r(\gamma)$$

for every $\sigma = (\gamma,\varphi)$ of $\widetilde{\Gamma}(\chi)$.

We still denote by $\widehat{R}$ the true representation of $\mathcal{G}^{\sim}$ given by $\theta^{\chi}_{r,\ell} \circ \widehat{R} \circ \theta^{\chi}_{\ell,r}$. We will determine $\lambda(\gamma,\varphi)$. We recall that we have defined a particular section $\gamma \to \tilde{\gamma} = (\gamma, \mathrm{Log}(cz+d))$ of $G = SL(2,\mathbb{R})$ in $\mathcal{G}^{\sim}$, where $\mathrm{Log}(cz+d)$ is the principal determination of $\log(cz+d)$.

2.4.9. Theorem: For $\gamma = \begin{pmatrix} a & b \\ c & d \end{pmatrix} \in \Gamma(\chi)$, $\widehat{R}(\tilde{\gamma}) = \lambda(\tilde{\gamma})\, A_r(\gamma)$ with:

1) $\lambda(\tilde{\gamma}) = \varepsilon_d^{-1} \left(\frac{2c}{d}\right)$, if d is odd (c even),

2) $\lambda(\tilde{\gamma}) = e^{i\pi/4} \left(\frac{2d}{c}\right) \varepsilon_c$, if c is odd, $c > 0$,

3) $\lambda(\tilde{\gamma}) = e^{-i\pi/4} \left(\frac{2d}{c}\right) \varepsilon_c \operatorname{sign} d$, if c is odd, $c < 0$.

Proof: We recall (1.8.24): $\widehat{R}(\tilde{\gamma}) = e^{\frac{i\pi}{4}m} R(\gamma)$, where,

$$\text{for } \gamma = \begin{pmatrix} a & b \\ c & d \end{pmatrix} \quad \left|\begin{array}{lll} m = -\operatorname{sign} c & \text{if} & c \neq 0 \\ m = 0 & \text{if} & c = 0 \quad a > 0 \\ m = -2 & \text{if} & c = 0 \quad a < 0 \end{array}\right. .$$

Thus $\tilde{R}(\tilde{\gamma}) = e^{\frac{i\pi}{4}m} b(d,-c)^{-1} A_r(\gamma)$.

If c is even, $c \neq 0$, then

$$\tilde{R}(\tilde{\gamma}) = e^{-\frac{i\pi}{4}\operatorname{sign} c}(b(d,-c))^{-1} A_r(\gamma)$$

$$= e^{-\frac{i\pi}{4}\operatorname{sign} c}\, e^{\frac{i\pi}{4}\operatorname{sign} cd}\, b(-c,d)\, A_r(\gamma)$$

$$\text{as } b(d,-c)b(-c,d) = e^{-\frac{i\pi}{4}\operatorname{sign} cd} \qquad (2.2.11)$$

$$= e^{\frac{i\pi}{4}\operatorname{sign} cd}\, e^{-\frac{i\pi}{4}\operatorname{sign} c}\, b(c,d)^{-1} A_r(\gamma)$$

$$\text{as } b(-c,d) = \overline{b(c,d)} = b(c,d)^{-1}$$

$$= b(2(\tfrac{c}{2}),d)^{-1} A_r(\gamma), \text{ if } d > 0$$

$$= \left(\frac{\frac{c}{2}}{d}\right) \varepsilon_d^{-1} A_r(\gamma), \text{ if } d > 0 . \qquad (2.2.18)$$

The other formulas are proven similarly.

<u>Remark</u>: The group $\Gamma(r,\chi)$ contains the congruence subgroup

$$\Gamma_0(2,2) = \{\begin{pmatrix} a & b \\ c & d \end{pmatrix}, \begin{matrix} c \equiv 0 \ \text{Mod } 2 \\ d \equiv 0 \ \text{Mod } 2 \end{matrix}\} .$$

For this group the formula 2.4.9 is then written as

$$\tilde{R}(\gamma) = \varepsilon_d^{-1} \left(\frac{2c}{d}\right) A_r(\gamma) .$$

2.4.10. Now we are ready to consider question 2). Let us first consider the group $\Gamma(r,\chi)$. It is clear that the distribution $\delta_0(\varphi) = \varphi(e)$ on $H(r,\chi)$ (where e is the identity element of N) is an invariant functional under the representation A_r of $\Gamma(r,\chi)$ in $H(r,\chi)$. Hence δ_0 is semi-invariant with character λ with respect to the representation $\widetilde{R}$ of $\widetilde{\Gamma}(r,\chi)$ in $H(r,\chi)$. We consider the isomorphism $\theta^{\chi}_{r,\ell}$ of $H(\ell)$ with $H(r,\chi)$. Then $\theta^{\chi}_{r,\ell}\, f_0$ is a lowest weight vector in $H(r,\chi)$. Thus, if we form the coefficient

$$\theta(\sigma) = (\delta_0, \widetilde{R}(\sigma)\cdot\theta^{\chi}_{r,\ell} f_0) = (\delta_0, \theta^{\chi}_{r,\ell}\widetilde{R}_{\ell}(\sigma)\cdot f_0),$$

we obtain a function in the space $M_{\sigma}(\Gamma(\chi),\lambda,\frac{1}{2},J^-)$. Using the isomorphism $I_{1/2}$ we may consider θ as a holomorphic function $\theta(z)$ on the upper half-plane. We have

$$\widetilde{R}(b(z))\cdot f = \widetilde{R}_{\ell}(\begin{pmatrix}1 & x\\ 0 & 1\end{pmatrix}\begin{pmatrix}y^{1/2} & 0\\ 0 & y^{-1/2}\end{pmatrix})\cdot f, \quad \text{i.e.:}$$

2.4.11. $$(\widetilde{R}(b(z))\cdot f)(\xi) = y^{1/4} e^{i\pi x\xi^2} f(y^{1/2}\xi),$$

hence

2.4.12. $$\widetilde{R}(b(z)\cdot f_0)(\xi) = y^{1/4} e^{i\pi z\xi^2}.$$

We compute

$$\begin{aligned}(I_{1/2}\theta)(z) &= y^{-1/4}(\delta^0, \theta^{\chi}_{r,\ell}(\xi \to y^{1/4} e^{i\pi z\xi^2}))\\ &= \sum_{n\in\mathbb{Z}} e^{i\pi z n^2}.\end{aligned}$$

So we obtain:

$$\theta(z) = \sum_{n\in\mathbb{Z}} e^{i\pi z n^2}$$

and the.

2.4.13. <u>Theorem</u>: Let $\theta(z) = \sum_{n\in\mathbb{Z}} e^{i\pi z n^2}$, then, for every $\gamma = \begin{pmatrix} a & b \\ c & d \end{pmatrix}$, with $a,b,c,d \in \mathbb{Z}$ $ac \equiv 0$ Mod 2, $bd \equiv 0$ Mod 2,

$$\theta\left(\frac{az+b}{cz+d}\right) = \lambda(\gamma)(cz+d)^{1/2}\,\theta(z)$$

where $(cz+d)^{1/2}$ is the principal determination of $(cz+d)^{1/2}$ and λ is given by 2.4.9. In particular, if c is even, $\lambda(\gamma) = \varepsilon_d^{-1}\left(\frac{2c}{d}\right)$.

2.4.14. We now define the congruence subgroup

$$\Gamma_0(N) = \{\begin{pmatrix} a & b \\ c & d \end{pmatrix} \in \Gamma,\ c \equiv 0 \text{ Mod } N\}$$

and more generally,

$$\Gamma_0(N_1,N_2) = \{\begin{pmatrix} a & b \\ c & d \end{pmatrix} \in \Gamma,\ c \equiv 0 \text{ mod } N_1,\ b \equiv 0 \text{ Mod } N_2\}.$$

2.4.15. Let $g(\sqrt{t}) = \begin{pmatrix} \sqrt{t} & 0 \\ 0 & (\sqrt{t})^{-1} \end{pmatrix}$. Then

$$\begin{pmatrix} \sqrt{t} & 0 \\ 0 & (\sqrt{t})^{-1} \end{pmatrix} \begin{pmatrix} a & b \\ c & d \end{pmatrix} \begin{pmatrix} (\sqrt{t})^{-1} & 0 \\ 0 & \sqrt{t} \end{pmatrix} = \begin{pmatrix} a & tb \\ c/t & d \end{pmatrix}.$$

Hence

$$g\left(\frac{1}{\sqrt{N_2}}\right) \Gamma_0(N_1,N_2)\, g\left(\frac{1}{\sqrt{N_2}}\right)^{-1} = \Gamma_0(N_1N_2).$$

So the group $\Gamma_0(N_1,N_2)$ is conjugated to $\Gamma_0(N_1N_2)$.

Let $\gamma = \begin{pmatrix} a & b \\ c & d \end{pmatrix}$ in $\Gamma_0(N)$. As $ad \equiv 1$ Mod N, d Mod N is an ivertible element of the ring $\mathbb{Z}/N\mathbb{Z}$. If ψ is a character mod N, the function $\psi(\gamma) = \psi(d)$ is a character of $\Gamma_0(N)$. We say that ψ is even if $\psi(-1) = 1$, odd if $\psi(-1) = -1$.

2.4.16. Let N be an integer. We consider

$$\Gamma_0(2,2N^2) \subset \Gamma_0(2,2) \subset \Gamma(r,\chi)$$

Hence we may consider the representation A_r of $\Gamma_0(2,2N^2)$ in $H(r,\chi)$. We will now construct linear functional on $H(r,\chi)$ invariant under this subgroup. Let k be an integer and consider the functional $(\delta_k,\varphi) = \varphi(\exp\frac{k}{N}Q)$ on $H(r,\chi)$. We have $\delta_{k+N} = \delta_k$.

2.4.17. __Lemma:__ $A_r(\gamma)\delta_k = \delta_{dk}$ for every $\gamma = \begin{pmatrix} a & b \\ c & d\end{pmatrix} \in \Gamma_0(2,2N^2)$.

__Proof:__ Let $\varphi \in H(r,\chi)$, we have:

$$(A_r(\gamma)\cdot\delta_k,\varphi) = (\delta_k, A_r(\gamma)^{-1}\cdot\varphi) = (A_r(\gamma)^{-1}\cdot\varphi)(\exp\frac{k}{N}Q)$$

$$= \varphi(\exp\frac{k}{N}\gamma Q) = \varphi(\exp\frac{bk}{N}P + \frac{dk}{N}Q).$$

We write:

$$\exp(\frac{bk}{N}P + \frac{dk}{N}Q) = \exp\frac{dk}{N}Q \exp\frac{bk}{N}P \exp\frac{bdk^2}{2N^2}E.$$

As $\varphi \in H(r,\chi)$, $\frac{bk}{N} \in \mathbb{Z}$ ($b \equiv 0$ Mod $2N^2$) and $\frac{bdk^2}{2N^2} \in \mathbb{Z}$ ($b \equiv 0$ Mod $2N^2$), the lemma follows.

2.4.18. __Corollary:__ Let ψ be a character mod N, we define

$$\delta_\psi = \sum_{k\in(\mathbb{Z}/N\mathbb{Z})^*} \psi(k)\,\delta_k .$$

Then $A_r(\gamma)\cdot\delta_\psi = \psi(d)^{-1}\delta_\psi$ for $\gamma = \begin{pmatrix} a & b \\ c & d \end{pmatrix} \in \Gamma_0(2,2N^2)$.

<u>Proof</u>: We have $A_r(\gamma)\cdot\delta_\psi = \sum_{k\in(\mathbb{Z}/N\mathbb{Z})^*} \psi(k)\delta_{dk}$. As d is invertible mod $4N^2$, d is invertible mod N. Changing k to $d^{-1}k$ the corollary follows.

As in 2.4.10 we now form the coefficients:

$$\theta_\psi^{1/2}(\sigma) = (\delta_\psi, \theta^\chi_{r,\ell}\widetilde{R}_\ell(\sigma)\cdot f_0) \quad \text{for } \psi \text{ even}$$

$$\theta_\psi^{3/2}(\sigma) = (\delta_\psi, \theta^\chi_{r,\ell}\widetilde{R}_\ell(\sigma)\cdot f_1) \quad \text{for } \psi \text{ odd}$$

We have:

$$(\widetilde{R}_\ell(b(z))\cdot f_0)(\xi) = y^{1/4}e^{i\pi z\xi^2}$$

$$(\widetilde{R}_\ell(b(z))\cdot f_1)(\xi) = y^{3/4}\xi e^{i\pi z\xi^2}.$$

We identify, via the isomorphisms $I_{1/2}$ and $I_{3/2}$, $\theta_\psi^{1/2}$ and $\theta_\psi^{3/2}$ with holomorphic functions $\theta_\psi^{1/2}(z), \theta_\psi^{3/2}(z)$ on the upper half-plane, i.e.

$$(\theta^\psi_{1/2})(z) = \sum_{n\in\mathbb{Z}} \psi(n)e^{i\pi z\frac{n^2}{N^2}}$$

$$(\theta^\psi_{3/2})(z) = \sum_{n\in\mathbb{Z}} \psi(n)ne^{i\pi z\frac{n^2}{N^2}}$$

(it is clear that we could define $\theta^\psi_{1/2}$ for any character ψ mod N, but, if ψ is odd, $\theta^\psi_{1/2} = 0$. Similarly $\theta^\psi_{3/2}$ is 0 if ψ is even). It is more standard to consider the modular forms

$$\theta_{\psi}^{+}(z) = \sum_{n\in\mathbb{Z}} \psi(n)e^{2i\pi z n^2} = \theta_{1/2}^{\psi}(2N^2 z),$$

or more generally, for an integer t, the modular forms

$$\theta_{\psi,t}^{+}(z) = \sum_{n\in\mathbb{Z}} \psi(n)e^{2i\pi z n^2 t} = \theta_{1/2}^{\psi}(2N^2 tz).$$

When $\begin{pmatrix} a & b \\ c & d \end{pmatrix} \in \Gamma_0(2,2N^2t)$, $\begin{pmatrix} a & \frac{b}{2N^2t} \\ 2N^2tc & d \end{pmatrix} \in \Gamma_0(4N^2t)$ and thus we obtain:

2.4.19. <u>Theorem</u>: a) Let ψ be an even character mod N, then

$$\theta_{\psi,t}^{+}(z) = \sum_{n\in\mathbb{Z}} \psi(n)e^{2i\pi t n^2 z}$$

satisfies:

$$\theta_{\psi,t}^{+}\left(\frac{az+b}{cz+d}\right) = \psi(d)\left(\frac{t}{d}\right) \varepsilon_d^{-1}\left(\frac{c}{d}\right)(cz+d)^{1/2}\ \theta_{\psi,t}(z)$$

for any $\begin{pmatrix} a & b \\ c & d \end{pmatrix} \in \Gamma_0(4N^2t)$.

b) Let ψ be an odd character mod N, then

$$\theta_{\psi,t}^{-}(z) = \sum_{n\in\mathbb{Z}} \psi(n)ne^{2i\pi t n^2 z}$$

satisfies:

$$\theta_{\psi,t}^{-}\left(\frac{az+b}{cz+d}\right) = \psi(d)\left(\frac{t}{d}\right) \varepsilon_d^{-1}\left(\frac{c}{d}\right)(cz+d)^{3/2}\ \theta_{\psi,t}^{-}(z)$$

for any $\begin{pmatrix} a & b \\ c & d \end{pmatrix} \in \Gamma_0(4N^2t)$.

Remark: We recall here the theorem of Serre and Stark [28]. Let n be an integer and χ a character mod n. A modular form on $\Gamma_0(n)$ satisfying:

$$f(\gamma\cdot z) = \chi(\gamma)\, \varepsilon_d^{-1}\left(\frac{c}{d}\right)(cz + d)^{1/2} f(z)$$

for every $\gamma = \begin{pmatrix} a & b \\ c & d \end{pmatrix}$ in $\Gamma_0(n)$ is a linear combination of the θ-series $\theta^+_{\psi,t}$ for appropriate choices of (ψ,t) (ψ is a character mod N, t an integer such that $4N^2t$ divides n and $\chi(d) = \psi(d)\left(\frac{t}{d}\right)$.)

2.5. The Shale-Weil representation associated to a quadratic form.

Let (V,B) be a symplectic space. The representation $\tilde{R}$ of $\widehat{Sp}(B)$ is "almost" irreducible. It is a sum of two irreducible representations corresponding to even and odd functions. We will similarly study the decomposition into irreducible components of the k-tensor products of the representation $\hat{R}$. The principle is the following: Let (E,S) be an orthogonal vector space of dimension k, with a quadratic form S. Then the space $(V \otimes E, B \otimes S)$ with the bilinear form $B \otimes S$ is a symplectic space. The representation R of the group $Sp(B \times S)$ restrict to $Sp(B) \times O(S)$ in R_S. When S is positive definite, R_S, as a representation of $Sp(B)$, is the k-tensor product of R and the compact group $O(k)$ plays an analogous role for the decomposition of $\overset{k}{\otimes} R$ that the symmetric group $\mathfrak{S}_k$ for the representation of $GL(n,\mathbb{C})$ in $\overset{k}{\otimes}\mathbb{C}^n$: let us consider an irreducible representation λ of $O(k)$ occurring in R_S, then the restriction of the representation R_S to the space of vectors of type λ under $O(k)$ is jointly irreducible under $Sp(B) \times O(k)$. Thus the unitary representation of $Sp(B)$ occurring in R_S are in this way naturally parametrized by representations of the compact group $O(k)$. They are "lowest weight vectors" representations of $Sp(B)$ and their lowest weight vectors have potential interest for the construction of holomorphic modular forms on the Siegel generalized upper half-plane. This representation $\overset{k}{\otimes} R$ is studied in detail in [15]. When S is of type (p,q), the representation R_S is the tensor product of p copies of R with q copies of the conjugate representation of $\overline{R}$. The decomposition of this representation is not known in general, although the same feeling persists that the unitary representation of $Sp(n,\mathbb{R})$ "occurring" in R_S should be naturally parametrized by the unitary representations of $O(p,q)$ occurring in R_S. When $\dim V = 2$, i.e. $Sp(B) = SL(2,\mathbb{R})$, this is indeed true: we have $R_S = \int \sigma_s \otimes \tau_s \, ds$, where ds is a Borel measure on the dual space $\widehat{SL}(2,\mathbb{R}) \times \widehat{O}(p,q)$, $s = \sigma_s \otimes \tau_s$ with $\sigma_s \in \widehat{SL}(2,\mathbb{R})$, $\tau_s \in \widehat{O}(p,q)$ and for almost every s (with respect to ds), σ_s and τ_s determines each other [12]. The description of the discrete part of R_S was given by Rallis and Schiffmann [22]. This discrete part will be of interest for the construction of θ-series.

Our main goal is to give explicit formulas for the lowest weight vectors of these representations R_S of $Sp(B)$. We will do it referring to representation theory only as a background. We will start by the well-known case $SL(2,\mathbb{R}) \times O(k)$ (i.e. $\dim V = 2$, S positive definite), then we will describe explicitly the lowest weight vectors for $SL(2,\mathbb{R}) \times O(p,q)$. Finally we summarize some of the results for $Sp(n,\mathbb{R}) \times O(k)$.

2.5.1. Let E be a k-dimensional real vector space with a non-degenerate quadratic form S of signature (p,q) (with $p + q = k$). There exists a basis $e_1, e_2, \cdots, e_p, e_{p+1}, \cdots, e_{p+q}$ of E such that:

$$S(e_i, e_i) = 1 \quad \text{for} \quad i \leq p,$$

$$S(e_i, e_i) = -1 \quad \text{for} \quad i > p.$$

We will, for such a decomposition, write $E = E_+ \oplus E_-$ with

$$E_+ = \sum_{i=1}^{p} \mathbb{R}e_i, \quad E_- = \sum_{i=p+1}^{p+q} \mathbb{R}e_i.$$

We denote by $O(S)$ the orthogonal group of the form S. We have $O(S) \simeq O(p,q)$. The subgroup of $O(S)$ leaving stable the decomposition $E_+ \oplus E_-$ is naturally isomorphic to $O(p) \times O(q)$.

Let (V,B) be a symplectic vector space. We consider the space $V \otimes E$, with the bilinear form $B \otimes S$ given by $(B \otimes S)(x \otimes v, y \otimes v') = B(x,y)\, S(v,v')$. The form $(B \otimes S)$ is a non-degenerate alternate form on $V \otimes E$. Thus the space $V \otimes E$ with the bilinear form $B \otimes S$ is a symplectic space.

It is clear that the direct product $Sp(B) \times O(S)$ is naturally imbedded in $Sp(B \otimes S)$ by $(g_1, g_2) \to g_1 \otimes g_2$ acting in $V \otimes E$.

We will study the restriction of the Shale-Weil representation of $Sp(B \otimes S)$ to $Sp(B) \times O(S)$. Let ℓ be a Lagrangian subspace of (V,B), then $\ell \otimes E$ is a Lagrangian subspace of $(V \otimes E, B \otimes S)$. We recall that we have constructed the universal covering group $\widetilde{G}_\ell$ of the group $Sp(B) = G$ using the choice of ℓ as a Lagrangian space. As a set, $\widetilde{G}_\ell$ is equal to $G \times \mathbb{Z}$, the multiplicative law being given by: $(g_1,n_1)\cdot(g_2,n_2) = (g_1g_2, n_1 + n_2 + \tau(\ell,g_1\ell,g_1g_2\ell))$.

2.5.2. Lemma: The map $(g,n) \to (g \otimes 1,(p-q)n)$ is a homomorphism from $\widetilde{G}_\ell$ to $\widetilde{Sp(B \otimes S)}_{\ell\otimes E}$.

Proof: We have to verify that

$$\tau(\ell \otimes E,\ g_1\ell \otimes E,\ g_1g_2\ell \otimes E) = (p-q)\tau(\ell,g_1\ell,g_1g_2\ell),$$

which is immediate.

As the group $O(S)$ leaves stable the Lagrangian space $\ell \otimes E$, $O(S)$ is naturally imbedded in $\widetilde{Sp(B \otimes S)}_\ell$ by $g_2 \to (1 \otimes g_2, 0)$. We will study the restriction of the representation $\widetilde{R}$ of the group $\widetilde{Sp(B \otimes S)}$ to $\widetilde{Sp(B)} \times O(S)$. We note by $\widetilde{R}_S$ this restriction. We denote by R_S the restriction of the projective representation $R_{\ell\otimes E}$ to $Sp(B) \times O(S)$.

2.5.3. Lemma: If k is even, the projective representation R_S is equivalent to a true representation of the symplectic group $Sp(B)$.

Proof: The cocycle associated to the choice of $\ell \otimes E$ as Lagrangian subspace of $V \otimes B$ is given by:

$$c(g_1,g_2) = e^{-\frac{i\pi}{4}(p-q)\tau(\ell,g_1\ell,g_1g_2\ell)} .$$

As $p - q$ is even, this is a coboundary: let $s(g) = s(\ell^+,g\cdot\ell^+)$(1.7.7), we have

$$c(g_1,g_2) = s(g_1)^{-(\frac{p-q}{2})} s(g_2)^{-(\frac{p-q}{2})} s(g_1g_2)^{(\frac{p-q}{2})} .$$

Hence the function

$$R(g) = s(g)^{-(\frac{p-q}{2})} R_S(g)$$

defines a true representation of G, equivalent to the projective representation R_S.

2.5.4. Let us consider an orthogonal decomposition $E = E_1 \oplus E_2$ of the space E. We denote by S_1,S_2 the restriction of S to E_1,E_2. The corresponding decomposition $V \otimes E = V \otimes E_1 \oplus V \otimes E_2$ is stable under $Sp(B) \times O(S_1) \times O(S_2)$. Let us consider the representation $\widetilde{R}_{S_1}$ of $\widetilde{Sp}(B) \times O(S_1)$. $\widetilde{R}_{S_1}$ is realized in $H(\ell \otimes E_1)$, subspace of functions on the Heisenberg group associated to $(V \otimes E_1, B \otimes S_1)$. Similarly $\widetilde{R}_{S_2}$ is realized in $H(\ell \otimes E_2)$. It is clear that the map

$$(\varphi_1 \otimes \varphi_2)(\exp v_1 \otimes e_1 \exp v_2 \otimes e_2 \exp tE)$$
$$= \varphi_1(\exp v_1 \otimes e_1)\, \varphi_2(\exp v_2 \otimes e_2)\, e^{-2i\pi t}$$

establishes an isomorphism of $H(\ell \otimes E_1) \otimes H(\ell \otimes E_2)$ with $H(\ell \otimes (E_1 \oplus E_2))$. Therefore the restriction of the representation $\widetilde{R}_S$ to $\widetilde{Sp}(B) \times O(S_1) \times O(S_2)$ is given by

$$\widetilde{R}_S(g,\sigma_1,\sigma_2) = \widetilde{R}_{S_1}(g,\sigma_1) \otimes \widetilde{R}_{S_2}(g,\sigma_2),$$

i.e. $\check{R}_S$ as a representation of $\widetilde{Sp}(B)$ is the (inner) tensor product $\widetilde{R}_{S_1} \otimes \widetilde{R}_{S_2}$. As a representation of $O(S_1) \times O(S_2)$, it is the outer tensor product.

Let us remark that the representation $\widetilde{R}_{-S}$ is canonically equivalent to $\overline{\widehat{R}_S}$. In particular if S is of signature (p,q) our representation $\widetilde{R}_S$ as a representation of $\widetilde{Sp}(n;\mathbb{R}) = \widetilde{G}$ is isomorphic to $\overset{p}{\otimes} \widehat{R} \otimes \overset{q}{\otimes} \overline{\widehat{R}}$.

2.5.5. Let S be of signature $(1,1)$. The representation $\widetilde{R}_S$ of the group $Sp(B)$ is equivalent to $R_\ell \otimes \overline{R}_\ell$. Let us consider the natural representation U of $Sp(B)$ inside $L^2(V)$ given by $(U(g)f)(x) = f(g^{-1}x)$. The representation U is a unitary representation of G (if $g \in Sp(B)$, $\det g = 1$) in $L^2(V)$.

2.5.6. __Proposition:__ Let S be of signature $(1,1)$. The representation $\widehat{R}_S$ is equivalent to the representation U.

__Proof:__ We consider the vector space $E = \mathbb{R}e_1 \oplus \mathbb{R}e_2$ with the symmetric form $S(x_1e_1 + x_2e_2, y_1e_1 + y_2e_2) = x_1y_2 + x_2y_1$. The subspace $\mathbb{R}e_1$ is an isotropic subspace for S. Let $V = \ell \oplus \ell'$ be a decomposition of V in a sum of complementary Lagrangian subspaces. Then $\ell_E = \ell \otimes E$ and $V \otimes \mathbb{R}e_1 = \ell_1$ are both Lagrangian subspaces of $(V \otimes E, B \otimes S)$. The representations R_{ℓ_E} in $H(\ell_E)$ and R_{ℓ_1} in $H(\ell_1)$ of the group $Sp(B \otimes S)$ are unitary equivalent via the operator $\mathcal{F}_{\ell_1,\ell_E}$. We identify $H(\ell_1)$ with $L^2(V)$ via $\varphi(v) = \varphi(\exp v \otimes e_2)$. As $G = Sp(B)$

leaves stable the decomposition $V \otimes e_1 \oplus V \otimes e_2$ of $V \otimes E$ it is clear that $R_{\ell_1}(g)$ is given by $(U(g)f)(v) = f(g^{-1}v)$. This establishes the proposition.

We will, for later applications, give the formula for the intertwining operator $\mathcal{F}_{\ell_1, \ell\otimes E}$ between R_S and U. We identify $H(\ell_E)$ with $L^2(\ell' \otimes E)$ and write $\varphi(y \otimes e_1 + y' \otimes e_2)$ for $\varphi(\exp(y \otimes e_1 + y' \otimes e_2))$, $y, y' \in \ell'$. As

$$\ell_1/\ell_1 \cap \ell_E \simeq \ell' \otimes e_1,$$

we have

$$(\mathcal{F}_{\ell_1,\ell_E}\varphi)(v) = \int_{t\in\ell'} \varphi(\exp v \otimes e_2 \exp t \otimes e_1)\, dt.$$

Let us write $v = x + y$, where $x \in \ell$, $y \in \ell'$. Then:

2.5.7. $$(\mathcal{F}_{\ell_1,\ell_E}\varphi)(x + y) = \int \varphi(\exp y \otimes e_2 \exp x \otimes e_2 \exp t \otimes e_1)\, dt$$

$$= \int_{t\in\ell'} \varphi(\exp(t \otimes e_1 + y \otimes e_2))e^{-2i\pi B(x,t)} dt.$$

i.e. $\mathcal{F}_{\ell_1,\ell_E}$ is a partial Fourier transform with respect to the variable t.

Let (E, S) be an orthogonal vector space, with S of signature (p,q).

We will restrict now our attention to the case where V is our 2-dimensional canonical symplectic vector space $\mathbb{R}P \oplus \mathbb{R}Q$. We choose $\ell = \mathbb{R}P \otimes E$ as a Lagrangian subspace and $\ell' = \mathbb{R}Q \otimes E$ as a complementary Lagrangian space. Hence the space of $\tilde{R}_S$ is identified with $L^2(E)$ by $\varphi(v) = \varphi(\exp Q \otimes v)$. We consider R_S as representation of $\mathcal{G}^\sim \times O(S)$, where $\mathcal{G}^\sim$ is the universal covering group of $SL(2,\mathbb{R})$. We write also $\mathcal{G} = \tilde{S}L(2,\mathbb{R})$.

2.5.8. The formula for the representation R_S (the canonical projective representation) becomes on the set of generators of $SL(2,\mathbb{R})$

$$(R_S(g(a))f)(v) = |a|^{k/2}\, f(a.v), \text{ for } g(a) = \begin{pmatrix} a & 0 \\ 0 & a^{-1} \end{pmatrix}.$$

$$(R_S(u(t))f)(v) = e^{i\pi t S(v,v)} f(v), \text{ for } u(t) = \begin{pmatrix} 1 & t \\ 0 & 1 \end{pmatrix}.$$

$$(R_S\begin{pmatrix} 0 & 1 \\ -1 & 0 \end{pmatrix} f)(v) = \int e^{2i\pi S(v,w)} f(w)\, dw.$$

We have

$$\tilde{R}_S(g,n) = e^{\frac{i\pi}{4}(p-q)n} R_S(g)$$

for $(g,n) \in \mathcal{G} \simeq \breve{G}$ from Lemma 2.5.2.

As $O(S)$ leaves stable ℓ and ℓ' the representation $\tilde{R}_S$ of $O(S)$ is simply given by:

$$(\tilde{R}_S(u)f)(v) = f(u^{-1}v), \text{ for } u \in O(S).$$

2.5.9. Let us first restrict our attention to the case where S is a positive definite form defined on a space E of dimension p. We are interested to study the decomposition of $\tilde{R}_S$ under $\mathcal{G} \times O(p)$.

Let λ be an irreducible representation of the group $O(p)$. As $O(p)$ and $\mathcal{G}$ commutes, the isotypic component $L(\lambda)$ of type λ of $L^2(E)$ is stable by $\mathcal{G}$. If $\lambda \neq \mu$, $L(\lambda)$ and $L(\mu)$ are orthogonal subspaces of $L^2(E)$.

Let us consider the action of $O(p)$ on the space of polynomials $\mathcal{P}$ on E given by $(u \cdot P)(v) = P(u^{-1}v)$.

Similarly we denote by $\mathcal{P}(\lambda)$ the isotypic component of type λ in $\mathcal{P}$.

2.5.10. Lemma: The space $L(\lambda) \neq 0$ if and only if $\mathcal{P}(\lambda) \neq 0$.

Proof: The space of functions of the form $P(x)e^{-\pi S(x,x)}$, with $P \in \mathcal{P}$ is dense in $L^2(E)$. (As S is positive definite, these functions are in $L^2(E)$.) Let $P = \Sigma P_\lambda$ the decomposition of P in isotypic components of type λ. Then, as $S(x,x)$ is invariant under $O(S)$,

$$P(x)e^{-\pi S(x,x)} = \sum_\lambda (P_\lambda(x)e^{-\pi S(x,x)}) \text{ with } P_\lambda(x)e^{-\pi S(x,x)} \in L(\lambda).$$

For $f \in L(\lambda)$,

$$\langle f, P(x)e^{-\pi S(x,x)}\rangle = \langle f, P_\lambda(x)e^{-\pi S(x,x)}\rangle.$$

Hence $\mathcal{P}(\lambda) = 0$ implies $L(\lambda) = 0$.

We now analyze the possible λ's such that $\mathcal{P}(\lambda)$ is not zero. Let us consider the identification of E with E^* given by S. This extends to an identification of the space of polynomials on E with the space of constant coefficients differential operators on E. If $(x_1, x_2, \cdots, x_p)$ is an orthonormal system of coordinates, the differential operator corresponding to $P(x_1,x_2,\cdots,x_p)$ is $P(\frac{\partial}{\partial x_1},\frac{\partial}{\partial x_2},\cdots,\frac{\partial}{\partial x_p})$. The space of polynomials on E is provided with the hermitian inner product $\langle P,Q\rangle = (P(\frac{\partial}{\partial x})\cdot\overline{Q})(0)$. This inner product is invariant under the group $O(S)$.

We denote by Δ_S the Laplacian operator corresponding to S, i.e.

$$\Delta_S = \frac{\partial^2}{\partial x_1^2} + \cdots + \frac{\partial^2}{\partial x_p^2}$$

for an orthonormal system of coordinates. The operator Δ_S commutes with the action of $O(S)$.

We denote by $\mathcal{H}$ the space of harmonic polynomials on E, i.e. $\mathcal{H} = \{P \in \mathcal{P} ; \Delta_S P = 0\}$. $\mathcal{H}$ is invariant under $O(S)$. Let us remark that $\mathcal{H}$ is the orthogonal complement of the space $\mathcal{P} S = \{P(x)S(x,x)$, with $P \in \mathcal{P}\}$ for the inner product $\langle P,Q\rangle$: in fact, if $Q \in \mathcal{H}$, clearly

$$(P(\frac{\partial}{\partial x})\Delta_S \cdot \overline{Q})(0) = 0$$

as $\Delta_S \cdot \overline{Q} \equiv 0$ on E. Conversely if $((\frac{\partial}{\partial x_i})^n(\Delta_S\overline{Q}))(0) = 0$ for every i,n, all the derivatives of $\Delta_S\overline{Q}$ are equal to zero at the origin, hence $\Delta_S\overline{Q} \equiv 0$.

2.5.11. Lemma:

a) Every element $P \in \mathcal{P}$ can be written on the form $P = \Sigma\, S^i P_i$ where $P_i \in \mathcal{H}$ (i.e. $P(x) = \Sigma\, S(x,x)^i P_i(x)$).

b) $\mathcal{P}(\lambda) \neq 0$, if and only if $\mathcal{H}(\lambda) \neq 0$.

Proof:

We can write $P = P_0 + SP_1$ as $S\mathcal{P}$ and $\mathcal{H}$ are orthogonal with respect to the positive definite hermitian form $\langle P,Q\rangle$. By induction of deg.P a) follows. Now if

$P \in \mathcal{P}(\lambda)$, we have $P = \Sigma S^i P_{i,\lambda}$ where $P_{i,\lambda}$ is the isotypic component of type λ of P_i. Hence if $\mathcal{P}(\lambda) \neq 0$, $\mathcal{H}(\lambda) \neq 0$.

2.5.12. Let us consider the space $\mathcal{H}(n)$ of homogeneous harmonic polynomials of degree n. It is well known that the representation d_n of $O(S)$ into $\mathcal{H}(n)$ is an irreducible representation of $O(S)$. We have finally proven that $L(\lambda) \neq \{0\}$ if and only if λ is of the form d_n.

We denote by $L^2(d_n)$ the subspace of functions in $L^2(E)$ of type d_n with respect to the action of $O(S)$. The space $L^2(d_n)$ is stable by $\mathcal{G} \times O(S)$.

If $\dim E = 1$, we have $\mathcal{H} = \{P;\ (\frac{\partial}{\partial \xi})^2 P = 0\}$. Hence $P(\xi) = 1$ or $P(\xi) = \xi$, corresponding to the two different irreducible representations of $O(1) = \{\pm 1\}$. Hence $L^2 = L(0) \oplus L(1)$ where $L(0) = \{f;\ f(x) = f(-x)\}$ consists of the even functions $L(1) = \{f;\ f(x) = -f(-x)\}$ consists of the odd functions.

If $\dim E > 1$, then for every $n \geq 0$, $\mathcal{H}(n) \neq 0$. For example, if $S = x_1^2 + x_2^2 + \cdots + x_p^2$ in orthonormal coordinates, $(x_1 + ix_2)^n$ is such an element of $\mathcal{H}(n)$. Hence we have $L^2 = \bigoplus_{n \geq 0} L^2(d_n)$.

Let $\mathcal{S}(E)$ be the space of rapidly decreasing functions on E. Then, if f belongs to $\mathcal{S}(E)$, f is a C^∞ vector for the representation $\widehat{R}$. Let (H,X,Y) be the canonical basis of $s\ell(2)$. We have:

2.5.13. a) $d\widehat{R}_S(X) = i\pi S$, where $i\pi S$ denotes the multiplication operator $f(v) \to i\pi S(v,v) f(v)$.

b) $\widetilde{dR}_S(H) = \sum_{i=1}^{p} x_i \frac{\partial}{\partial x_i} + p/2 = \frac{1}{2} \sum_{i=1}^{p} (x_i \frac{\partial}{\partial x_i} + \frac{\partial}{\partial x_i} x_i)$.

c) $\widetilde{dR}_S(Y) = \frac{i}{4\pi} \Delta_S$, where $\Delta_S = \Sigma (\frac{\partial}{\partial x_i})^2$.

The formulas a) and b) follows by differentiation of 2.5.8. As $\sigma(X) = \sigma X \sigma^{-1} = -Y$, and σ acts by the Fourier transform it follows that $\widetilde{dR}_S(Y)$ is given by the Fourier transform with respect to S of the operator $i\pi S$, leading to c). We will still denote by H the operator

$$\sum_{i=1}^{p} x_i \frac{\partial}{\partial x_i} + p/2 .$$

Let us now consider the basis J_0, J^-, J^+ of $\mathfrak{sl}(2,\mathbb{C})$.

2.5.14. Lemma:

$$\widetilde{dR}(J_0)\cdot(f(\xi)e^{-\pi S(\xi,\xi)}) = (\frac{i}{4\pi} \Delta_S \cdot f - iH\cdot f)e^{-\pi S(\xi,\xi)}$$

$$2\widetilde{dR}(J^-)(f(\xi)e^{-\pi S(\xi,\xi)}) = \frac{i}{4\pi} (\Delta_S \cdot f)(\xi)e^{-\pi S(\xi,\xi)} .$$

Proof: As

$$\frac{\partial^2}{\partial x^2} (f(x)e^{-\pi x^2}) = (\frac{\partial^2 f}{\partial x^2} + (2\pi)^2 x^2 - 4\pi(x \frac{\partial f}{\partial x} + \frac{1}{2}))e^{-\pi x^2},$$

we obtain:

$$\widetilde{dR}_S(Y)(f(\xi)e^{-\pi S(\xi,\xi)}) = (\frac{i}{4\pi} \Delta_S \cdot f + i\pi Sf - iH\cdot f)e^{-\pi S(\xi,\xi)}.$$

Writing $J_0 = Y - X$, $2J^- = Y + X + iH$, we obtain the lemma.

2.5.15. Proposition: Let $P \in \mathcal{H}(n)$, then the function

$f_P(\xi) = P(\xi)e^{-\pi S(\xi,\xi)}$ satisfies:

a) $d\widetilde{R}_S(J_0)\cdot f_P = -i(\frac{p}{2} + n)f_P$

b) $d\widetilde{R}_S(J^-)\cdot f_P = 0.$

Proof: This follows immediately from 2.5.14.

Thus the functions $f_P(\xi)$ are lowest weight vectors for the representation $\widetilde{R}_S$. Our first aim is then reached for the case of a definite quadratic form.

2.5.16. Let us explicit the relation between the representation R_S and the representations $(T_k, \bar{O}(P^+))$ which are the "typical" representations of $SL(2,\mathbb{R})$ having lowest weight vectors.

We consider the subrepresentation $(\widetilde{R}_S, L^2(d_n))$ of $\mathcal{G}\times O(p)$ in the space of functions of type d_n with respect to the action of $O(p)$. The vector f_P is a lowest weight vector of weight $\alpha = p/2 + n$ in $L^2(d_n)$. As in 2.3.3, we thus consider $f_{P,z} = y^{-\alpha/2}(\widetilde{R}_S(b(z))\cdot f_P)$, i.e. explicitly

2.5.17. $$f_{P,z}(\xi) = P(\xi)e^{i\pi z S(\xi)}.$$

(We write $S(\xi)$ for $S(\xi,\xi)$.) This function $z \to f_{P,z}$ of P^+ in $L^2(d_n)$ satisfies the fundamental relation:

2.5.18. $$R_S(g)\cdot f_{P,z} = \text{"}(cz+d)\text{"}^{-\alpha} f_{P,g\cdot z}.$$

This relation follows from 2.5.15 a) and the discussion in 2.3.3. It could also be checked directly on the set of generators $\begin{pmatrix}1 & x\\ 0 & 1\end{pmatrix}$, $\begin{pmatrix}0 & 1\\ -1 & 0\end{pmatrix}$ of $SL(2,\mathbb{R})$: as P is harmonic, the Fourier transform of $P(\xi)e^{i\pi z S(\xi)}$ is proportional to $z^{-\alpha}P(\xi)e^{-i\pi z^{-1}S(\xi)}$.

Let us consider the operator $\bar{\mathcal{F}}_n: L^2(d_n) \to \bar{\mathcal{O}}(P^+, \mathcal{H}(n))$ from functions on E of type d_n to antiholomorphic functions on P^+ with values in $\mathcal{H}(n)$, given by:

$$\langle(\bar{\mathcal{F}}_n \varphi)(z), Q\rangle = \int_E e^{-i\pi \bar{z} S(\xi)} \varphi(\xi)\, Q(\xi)\, d\xi .$$

Writing $\bar{\mathcal{F}}_n\varphi$ as $\langle\varphi, f_{\bar{Q},z}\rangle_{L^2(E)}$, the relation 2.5.18 shows that $\bar{\mathcal{F}}_n \tilde{R}_S(g) = \bar{T}_{(p/2)+n}(g)\cdot \bar{\mathcal{F}}_n$, for $g \in \tilde{\mathcal{G}}$. Hence $\bar{\mathcal{F}}_n$ intertwines the representation $\tilde{R}_S$ of $\tilde{\mathcal{G}} \times O(p)$ on the space $L^2(d_n)$ with the representation $\bar{T}_{(p/2)+n} \otimes d_n$ of $\tilde{\mathcal{G}} \times O(p)$ on $\bar{\mathcal{O}}(P^+) \otimes \mathcal{H}(n)$.

The image of the lowest weight vector f_P in $(\tilde{R}_S, L^2(d_n))$ is proportional to the lowest weight vector $(\psi_1 \otimes P)(w) = (\bar{w}-i)^{-\alpha} \otimes P$ of the representation $(\bar{T}_\alpha \otimes d_n, \bar{\mathcal{O}}(P^+) \otimes \mathcal{H}(n))$. It is in fact easy to check directly that: $(\bar{\mathcal{F}}_n f_{P,z})(w)$ is proportional to $(w-z)^{-\alpha} \otimes P$: The relation to be proven is:

$$\int_E e^{i\pi(z-\bar{w})S(\xi)}\, P(\xi)Q(\xi)d\xi = c(\bar{w}-z)^{-\alpha}\, (P,Q),$$

whenever Q is harmonic of degree n. By homogeneity, it is enough to check this relation for $(z-\bar{w}) = i$, i.e. that

$$\int_E e^{-\pi S(\xi)} P(\xi)Q(\xi)d\xi = c(P,Q)$$

where c is a non-zero constant. Both members of this equality are inner product on $\mathcal{H}(n)$ invariant by $O(p)$. As d_n is irreducible, they are proportional. We summarize the preceding discussion in the:

2.5.19. Proposition: The operator $\mathcal{F}_n : L^2(d_n) \to \overline{\mathcal{O}}(P^+) \otimes \mathcal{H}(n)$ given by

$$\langle (\mathcal{F}_n\varphi)(z), Q\rangle = \int_E e^{-i\pi\overline{z}S(\xi)}\varphi(\xi)Q(\xi)d\xi$$

intertwines the representation $(R_S, L^2(d_n))$ of $\mathcal{G} \times O(p)$ in $L^2(d_n)$ with the representation $\overline{T}_{(p/2)+n} \otimes d_n$ of $\mathcal{G} \times O(p)$ in $\overline{\mathcal{O}}(P^+) \otimes \mathcal{H}(n)$. Furthermore $\mathcal{F}_n$ is an injective operator such that

$$\mathcal{F}_n f_{P,z} = c(\overline{w}-z)^{-\alpha} \otimes P\ , \quad \text{with } \alpha = p/2 + n$$

where c is a non-zero constant.

Proof: The only point of the proposition which doesn't follow from the preceding discussion is the injectivity of $\mathcal{F}_n$. However if $(\mathcal{F}_n\varphi)(\overline{z}) = 0$, by considering $(\frac{\partial}{\partial\overline{z}})^i(\mathcal{F}_n\varphi)(\overline{z})$ we see that φ is orthogonal to all the functions of the form $e^{-\pi S(\xi)}S^i(\xi)Q(\xi)$ with Q harmonic in $\mathcal{H}(n)$. As this space of functions is dense in $L^2(d_n)$, $\varphi = 0$.

2.5.20. Remark: The representations $(\overline{T}_k, \overline{\mathcal{O}}(P^+))$ of $SL(2,\mathbb{R})$ are unitarizable for $k > 0$: Let

$$H_k = \{f \in \overline{\mathcal{O}}(P^+)\ ;\ \int |f|^2\, y^{k-2}\, dxdy < \infty\}\ ,$$

then $H_k \neq \{0\}$ for $k > 1$ and it is easily checked that $\overline{T}_k$ acts unitarily and irreducibly on H_k. These representations, for $k > 1$, are the representations of the antiholomorphic (relative) discrete series of $\widetilde{SL}(2,\mathbb{R})$. If $k > 0$, we can define

H_k to be

$$H_k = \{f \in \bar{\mathcal{O}}(P^+) \; ; \; \int |f'(z)|^2 \, y^k \, dxdy < \infty$$
$$f \to 0 \text{ at } \infty\}$$

and it is not difficult to see that $(\overline{T}_k, H_k)$ is an irreducible unitary representation of $SL(2,\mathbb{R})$, for $k > 0$.

We can then write the decomposition of $\widetilde{R}_S$ in unitary irreducible representations of $\widetilde{SL}(2,\mathbb{R}) \times O(p)$, as:

$$\widetilde{R}_S = \bigoplus_n (\overline{T}_{(p/2)+n} \otimes d_n)$$

(when $p = 1$, n is restricted to be 0 or 1).

We now consider the case of a form S of signature (p,q) on the vector space E of dimension $k = p + q$. We will suppose that both p and q are non-zero. We will first give a somewhat sketchy procedure to obtain lowest weight vectors for the representation $\widetilde{R}_S$, then a group theoretical approach.

Let us choose a decomposition of our space E as an orthogonal sum $E = E_1 \oplus E_2$, where the restriction S_1 of S to E_1 is positive definite, the restriction $-S_2$ of S to E_2 is negative definite. We write $\Delta_S = \Delta_{S_1} - \Delta_{S_2}$ where Δ_{S_1} (resp. Δ_{S_2}) is the Laplacian associated to S_1 (resp. S_2) acting on E_1 (resp. E_2).

Let us consider the infinitesimal action of $\widetilde{R}_S$ on the space $\mathcal{S}(E)$. We have:

$$d\widetilde{R}_S(X) = i\pi(S_1 - S_2) = i\pi S$$

$$d\widetilde{R}_S(H) = (H_1 + H_2) = H$$

$$d\widetilde{R}_S(Y) = \frac{i}{4\pi}(\Delta_{S_1} - \Delta_{S_2}) = \frac{i}{4\pi}\Delta_S .$$

2.5.21. <u>Proposition</u>: If a function $P(\xi)e^{-\pi S(\xi,\xi)}$ on E is a lowest weight vector of weight λ, it satisfies the relations:

1) $P(\xi)e^{-\pi S(\xi,\xi)}$ is in $L^2(E)$
2) $\Delta_S \cdot P = 0$ in the weak sense
3) $H \cdot P = \lambda P$

(i.e. P is harmonic with respect to Δ_S and homogeneous of degree $\lambda - k/2$).

<u>Proof</u>: If $f = P(\xi)\, e^{-\pi S(\xi)}$ is a lowest weight vector of weight λ for the representation $\widetilde{R}_S$, we have, for every $\varphi \in \mathcal{S}(E)$,

$$\langle \widetilde{R}_S(\delta(\theta))\cdot f,\varphi\rangle_{L^2} = e^{-i\lambda\theta}\langle f,\varphi\rangle = \langle f,\widetilde{R}_S(\delta(\theta))^{-1}\varphi\rangle .$$

Differentiating the last two equalities and using Lemma 2.5.14, we obtain $\frac{i}{4\pi}\Delta_S\cdot P - iH\cdot P = -i\lambda\theta$ in the weak sense. Similarly, if f is a C^∞-vector for $\widetilde{R}_S$, we have

$$\langle d\widetilde{R}_S(X)\cdot f,\varphi\rangle_{L^2} = \langle f, d\widetilde{R}_S(-\overline{X})\cdot\varphi\rangle_{L^2} .$$

Using Lemma 2.5.14, we obtain the relation $\Delta_S \cdot P = 0$.

It is a delicate question to decide when a function satisfying the equation 1), 2), 3) is actually a C^∞-vector for the representation $\widetilde{R}_S$. We will first ignore this point, and construct vectors satisfying 1), 2), 3). As S is not positive definite, the function $e^{-\pi S(\xi)}$ is clearly far from being in $L^2(E)$. We will see however that there exists continuous functions $P(\rho)$, solutions in the weak sense of the equations $\Delta_S \cdot P = 0$, $H \cdot P = \lambda P$ and supported on the cone $S(\rho,\xi) > 0$ such that $P(\xi)e^{-\pi S(\rho,\xi)}$ is in $L^2(E)$. $P(\xi)$ will be on the form $S(\rho)^k u(\xi)$, with $k > 0$, where $u(\xi)$ is supported on the cone $S(\rho) > 0$, so $S(\rho)^k$ will cancel the singularities of the characteristic function of the set $\{\xi; S(\xi) > 0\}$. We now proceed to our construction:

We write Δ for Δ_S.

2.5.22. Lemma: For f differentiable, we have:
$\Delta S^\alpha f = S^\alpha \Delta f + 4\alpha S^{\alpha-1}(H + \alpha - 1)f$, on the set $S(x) \neq 0$.

Proof: It is easy to compute that, for
$S = (x_1^2 + x_2^2 + \cdots + x_p^2 - y_1^2 + \cdots + y_q^2)$

$$\left(\frac{\partial}{\partial x_i}\right)^2 S^\alpha f = 2\alpha S^{\alpha-1} f + 4\alpha(\alpha-1)x_i^2 S^{\alpha-2} f + 4\alpha S^{\alpha-1} x_i \frac{\partial}{\partial x_i} f + S^\alpha \frac{\partial^2}{\partial x_i^2} f$$

$$\left(\frac{\partial}{\partial y_j}\right)^2 S^\alpha f = -2\alpha S^{\alpha-1} f + 4\alpha(\alpha-1)y_j^2 S^{\alpha-2} f - 4\alpha S^{\alpha-1} y_j \frac{\partial}{\partial y_j} f + S^\alpha \frac{\partial^2}{\partial y_j^2} f$$

and the lemma follows.

2.5.23. Lemma: Let P_1 be a harmonic polynomial of degree n on E_1, P_2 a harmonic polynomial of degree m on E_2. Then, if:

1) $\frac{p+q}{2} + \gamma - 1 + 2\alpha + 2\beta + n + m = 0$

2) $\alpha(\frac{p}{2} + \alpha - 1 + n) = 0$

3) $\beta(\frac{q}{2} + \beta - 1 + m) = 0$

the function $\varphi = P_1 P_2 S_1^\alpha S_2^\beta S^\gamma$ satisfies $\Delta\varphi = 0$ on the set

$S(x) \neq 0$, $S_1(x) \neq 0$ $S_2(x) \neq 0$.

<u>Proof</u>: By Lemma 2.5.22, we have

$$\Delta_S(S^{\gamma} P_1 P_2 S_1^{\alpha} S_2^{\beta}) = S^{\gamma}((\Delta_1-\Delta_2) P_1 S_1^{\alpha} P_2 S_2^{\beta}) + 4\gamma S^{\gamma-1}((H+\gamma-1)\cdot P_1 P_2 S_1^{\alpha} S_2^{\beta})$$

But the condition 1) implies that $(H + (\gamma-1))\cdot(P_1 P_2 S_1^{\alpha} S_2^{\beta}) = 0$, as follows from the homogeneity degree. Similarly 2) and 3) implies

$$\Delta_1(S_1^{\alpha} P_1) = 0, \ \Delta_2(S_2^{\beta} P_2) = 0.$$

2.5.24. <u>Theorem</u>: Let P_1 be a harmonic polynomial of degree n on E_1 and P_2 a harmonic polynomial of degree m on E_2. Let

$$\psi_{P_1,P_2} = P_1 P_2 S_1^{-(n + \frac{p-2}{2})} S^{\frac{p-q}{2} + n - m - 1}$$

on $S(x,x) > 0$

$= 0$ on $S(x,x) < 0$

then

1) $\Delta\cdot\psi_{P_1,P_2} = 0$ on the set $S_1(x) \neq 0$, $S(x) \neq 0$

$H\cdot\psi_{P_1,P_2} = (n - m + \frac{1}{2}(p-q))\cdot\psi_{P_1,P_2}$.

2) If $\frac{p-q}{2} + (n - m) > 1$,

the function $\psi_{P_1,P_2}(x)e^{-\pi S(x,x)}$ belongs to $L^2(E)$.

3) If $n - m > q$, the function $\psi_{P_1,P_2}(x)e^{-\pi S(x,x)}$ belongs to $L^1(E)$.

4) If $p + q > 2$, and $n - m > q$ then ψ_{P_1,P_2} is in $L^1(E) \cap L^2(E)$ and is continuous.

<u>Proof</u>:

1) follows from 2.5.23. (It corresponds to the case $\beta = 0$.)

2) Let us write $x = u + v$ with $u \in E_1$, $v \in E_2$. We have to compute:

$$\int_{S_1(u)>S_2(v)} (S_1(u))^{-(p-2+2n)} (S_1(u) - S_2(v))^{p-q+2(n-m)-2} e^{-2\pi(S_1(u)-S_2(v))} |P_1(u)|^2 |P_2(v)|^2 \, dudv.$$

Let us consider polar coordinates on E_1 and E_2, i.e. we write $u = S_1(u)^{1/2}\sigma_1$, $v = S_2(u)^{1/2}\sigma_2$, where σ_1,σ_2 are points on the unit sphere in E_1,E_2. If $t_1 = S_1(u)$, $t_2 = S_2(v)$, then $du = t_1^{p-2/2} dt_1 \, d\sigma_1$, $dv = t_2^{q-2/2} dt_2 \, d\sigma_2$, for $d\sigma_1, d\sigma_2$ the surface measures on the unit sphere. Then, using the homogeneity property of P_1, P_2, we have to see when:

$$\int_{\substack{t_1>t_2 \\ t_1>0 \\ t_2>0}} t_1^{-(p-2+2n)} (t_1-t_2)^{p-q+2(n-m)-2} e^{-2\pi(t_1-t_2)} t_1^n t_2^m t_1^{(p-2)/2} t_2^{(q-2)/2} dt_1 dt_2 < \infty$$

Changing t_1 in $(t + t_2)$, we obtain that the preceding integral is equal to

$$\int_{\substack{t_2>0 \\ t>0}} (t + t_2)^{-((p-2)/2 + n)} t_2^{m + (q-2)/2} e^{-2\pi t} t^{p-q+2(n-m)-2} \, dtdt_2 .$$

But

$$\int_{t_2>0} \frac{t_2^{m+(q-2)/2}}{(t+t_2)^{(p-2)/2+n}}\, dt_2$$

is convergent, provided that $n - m + (p-q)/2 > 1$. In this case, changing t_2 in tt_2', this equals $t^{m-n+(q-p)/2+1}$. Thus the full integral is

$$\int e^{-2\pi t}\, t^{n-m+(p-q)/2-1}\, dt,$$

which is convergent.

The assertion 3) is proven in the same way. Using polar coordinates, we have to see when

$$\int t_1^{-((p-2)/2+n)} (t_1-t_2)^{(p-q)/2+n-m-1}\, t_1^{n/2} t_2^{m/2} e^{-\pi t} t_1^{(p-2)/2} t_2^{(q-2)/2} dt_1 dt_2 < \infty$$

This integral is equal to:

$$\int_{t>0} (t)^{(p-q)/2+(n-m-1)} e^{-\pi t} \left(\int_{t_2>0} \frac{t_2^{(m+q-2)/2}}{(t+t_2)^{n/2}}\, dt_2\right) dt\,.$$

As before, this is convergent, if and only if $n - m > q$, and is equal in this case to:

$$\int_{t>0} t^{(n-m)/2+p/2-1}\, e^{-\pi t}\, dt.$$

4) Let us remark that if $n - m > q$, the condition 3) implies $(p-q)/2 + (n-m) > (p+q)/2$. On the set $S_1(x) > S_2(x)$, the only singularity of $S_1(x)$ is for $S_1(x) = S_2(x) = 0$, i.e. at $x = 0$. The factor $S^{(p-q)/2+n-m-1}$ is then positive and cancels

the singularity on $S(x) = 0$ of the characteristic function $x \neq 0$ of the set $S(x) > 0$. The homogeneity degree being positive the function is continuous at 0.

We now give a group theoretical approach to the construction of the Rallis-Schiffmann functions $f_{P_1,P_2}(x) = \psi_{P_1,P_2}(x)e^{-\pi S(x)}$ based on the script of R. Howe [12].

We write, according to the orthogonal decomposition $E = E_1 \oplus E_2$, $L^2(E) = L^2(E_1) \otimes L^2(E_2)$.

Then $\widetilde{R}_S$, as a representation of $SL(2,\mathbb{R}) \times O(p) \times O(q)$, is isomorphic to $\bigoplus_{n,m} T_{(p/2)+n} \otimes T_{(q/2)+m} \otimes d_n \otimes d_m$. Let $d = (p/2) + n - ((q/2) + m)$. We will see here that, when $d > 1$, the representation $\overline{T}_{(p/2)+n} \times T_{(q/2)+m}$ contains $\overline{T}_d$. Let v_1 be the lowest weight vector of $\overline{T}_d$, $P_1 \in \mathcal{H}(n)$, $P_2 \in \mathcal{H}(m)$. The Rallis-Schiffmann function f_{P_1,P_2} corresponds to the (unique) vector $v_1 \otimes P_1 \otimes P_2$ of $\overline{T}_d \otimes d_n \otimes d_m \subset \widetilde{R}_S$. Hence we need only to describe explicitly the formula for $v_1 \otimes P_1 \otimes P_2$ in the given model $\widetilde{R}_S$.

2.5.25. Remark: The decomposition of $\overline{T}_\alpha \otimes T_\beta$ has been studied by Gutkin [9] and Repka [25]. In particular $\overline{T}_\alpha \otimes T_\beta$ $(\alpha \geq \beta)$ contains discretely the sum

$$\bigoplus_{\substack{j \text{ integer} \\ \alpha-\beta-2j>1}} T_{\alpha-\beta-2j} .$$

It is then not difficult to prove (see ([11])) the following:

2.5.26. Theorem: The discrete spectrum of the representation $\widetilde{R}_S$

is given as follows:

A) Let $p > 1$, $q > 1$, then

$$(\tilde{R}_S)_d = \bigoplus_{\alpha > 1} \overline{T}_\alpha \otimes V_\alpha \oplus \bigoplus_{\beta > 1} (T_\beta \otimes V_\beta)$$

where α, β runs over the integers, if $\frac{p-q}{2}$ is an integer or α, β runs over the $\frac{1}{2}$ integers, if $\frac{p-q}{2}$ is a half integer. The representation V_α (resp. V_β) is a irreducible representation of $O(p,q)$. Its restriction to $O(p) \times O(q)$ is

$$V_\alpha = \bigoplus_{n,m} d_n \otimes d_m \text{ with } n - m + \frac{p-q}{2} = \alpha + 2j, \ j \geq 0$$

(resp. $V_\beta = \oplus\, d_n \otimes d_m$ with $m - n + \frac{q-p}{2} = \beta + 2j$, $j \geq 0$).

B) Let $p > 1$, $q = 1$. Then $(R_S)_d = \bigoplus_{\alpha > 1} \overline{T}_\alpha \times V_\alpha$;

(with $V_\alpha = \bigoplus_{\substack{n \\ m=0,1}} d_n \otimes d_m$; $n - m + \frac{p-1}{2} = \alpha + 2j$, $j \geq 0$).

C) If $p = q = 1$, then $(\tilde{R}_S)_d = 0$.

In particular, it is possible to describe all the K-finite vectors of the representation $T_d \otimes V_d$ by differentiating our particular vector f_{P_1,P_2} with respect to the infinitesimal action of $SL(2,\mathbb{R}) \times O(p,q)$.

We proceed now to the explicit description of the vector $v_1 \otimes P_1 \otimes P_2$ in $L^2(E)$.

Let us consider the action of $O(S_1) \times O(S_2) = O(p) \times O(q)$ on $L^2(E)$. Let $L^2(d_n \otimes d_m)$ be the isotypic component of $L^2(E)$ of type $d_n \otimes d_m$. The operator $\bar{\mathcal{F}}_n \otimes \mathcal{F}_m$ intertwines the representation $\tilde{R}_S$ of $\widetilde{SL}(2,\mathbb{R}) \times O(p) \times O(q)$ restricted to $L^2(d_n \otimes d_m)$ with the representation $(\bar{T}_{(p/2)+n} \otimes T_{(q/2)+m}) \otimes d_n \otimes d_m$. The representation $\bar{T}_{(p/2)+n} \otimes T_{(q/2)+m}$ operates on the space of functions $F(\bar{z}_1, z_2)$ antiholomorphic in z_1, holomorphic in z_2, by:

$$((\bar{T}_{(p/2)+n} \otimes T_{(q/2)+m})(g^{-1})) \cdot F)(\bar{z}_1, z_2)$$
$$= (c\bar{z}_1+d)^{-((p/2)+n)}(cz_2+d)^{-((q/2)+m)} F(g\cdot\bar{z}_1, g\cdot z_2) .$$

Now, a function $F(\bar{z}_1, z_2)$ antiholomorphic in z_1 and holomorphic in z_2 is entirely determined by its restriction to the diagonal $(\bar{z}, z)$. Thus, if we consider the representation $\bar{T}_{(p/2)+n,(q/2)+m}$ of $\widetilde{SL}(2,\mathbb{R})$ acting on "all" functions on P^+ by:

$$(\bar{T}_{(p/2)+n,(q/2)+m}(g^{-1}) \cdot f)(u)$$
$$= (c\bar{u}+d)^{-((p/2)+n)}(cu+d)^{-((q/2)+m)} f(g\cdot u) \quad (u \in P^+)$$

the operator $\varphi \to ((\bar{\mathcal{F}}_n \otimes \mathcal{F}_m) \cdot \varphi)(\bar{z}, z)$ intertwines the representation $\tilde{R}_S | L^2(d_n \otimes d_m)$ with $\bar{T}_{(p/2)+n,(q/2)+m} \otimes d_n \otimes d_m$. We still denote this operator as $\bar{\mathcal{F}}_n \otimes \mathcal{F}_m$. Let $d = ((p/2)+n) - ((q/2)+m)$. The operator $(Mf)(u) = (\operatorname{Im} u)^{-((q/2)+m)} f(u)$ intertwines the representation $T_{d,0}$ acting on "all" functions on P^+ with the representation $\overline{T}_{(p/2)+n,(q/2)+m}$. The representation T_d (acting on $\mathcal{E}(P^+)$) is naturally contained in $T_{d,0}$, thus in $\overline{T}_{(p/2)+n,(q/2)+m}$. It follows from 2.3.5 that the function on P^+ given by:

$$\psi'_\tau(u) = (\mathrm{Im}\ u)^{-((q/2)+m))}(\overline{u}-\tau)^{-d}$$

verifies:

2.5.27. $$\overline{T}_{(p/2)+n,(q/2)+m}(g)\cdot\psi'_\tau = (c\tau+d)^{-d}\ \psi'_{g\cdot\tau}.$$

Our aim is to give an explicit formula of the function ψ_τ in $L^2(d_n \otimes d_m)$ such that $(\overline{\mathcal{F}}_n \otimes \mathcal{F}_m)\cdot\psi_\tau = \psi'_\tau$.

According to the decomposition $E = E_1 \oplus E_2$, we write an element of E as $u + v$ with $u \in E_1$ and $v \in E_2$.

Let $P_1(u)$ be a harmonic polynomial of degree n with respect to S_1 and $P_2(v)$ a harmonic polynomial of degree m with respect to S_2. Let $d = ((p/2)+n) - ((q/2)+m))$. We consider the Rallis-Schiffmann function of $\tau \in P^+$:

$$\psi_\tau(u,v) = P_1(u)(S_1(u))^{-(\frac{p-2}{2}+n)} P_2(v)S(u+v)^{d-1}e^{2i\pi\tau S(u+v)}$$

$$\text{if}\quad S(u+v) = S_1(u) - S_2(v) > 0$$

$$= 0 \quad \text{if}\quad S_1(u) - S_2(v) \leq 0.$$

We now prove the:

2.5.28. Theorem: If $d > 1$, ψ_τ belongs to $L^2(d_n \otimes d_m)$ and $(\overline{\mathcal{F}}_n \otimes \mathcal{F}_m)\cdot\psi_\tau = \psi'_\tau \otimes P_1 \otimes P_2$.

<u>Proof</u>: The equality to be verified is:

$$\int_{S_1(u)>S_2(v)} e^{-i\pi\bar{z}S_1(u)} e^{i\pi z S_2(v)} P_1(u)P_2(v)(S_1(u))^{-(\frac{p-2}{2}+n)}$$

$$(S_1(u)-S_2(v))^{d-1} e^{i\pi\tau(S_1(u)-S_2(v))} Q_1(u)Q_2(v)\, du dv$$

$$= y^{-((q/2)+n)}(\bar{z}-\tau)^{-d} \langle P_1,Q_1\rangle \langle P_2,Q_2\rangle .$$

We write $\bar{z} = z - 2iy$, $t_1 = S_1(u)$, $t_2 = S_2(v)$. Using polar coordinates separately on E_1 and E_2 and the relations

$$\int_{S_i} P_i(\sigma)\overline{Q_i(\sigma)}\, d\sigma = \langle P_i,Q_i\rangle$$

(S_i unit sphere of E_i $(i = 1,2)$), the integral to be calculated is:

$$\int_{\substack{t_1>t_2\\ t_2>0}} e^{i\pi(\bar{z}-\tau)(t_1-t_2)} e^{-2\pi y t_2} (t_1-t_2)^{(d-1)} t_2^{\frac{q-2}{2}+m}\, dt_1 dt_2 .$$

This separates in an integral on $(t_1-t_2) > 0$ and on $t_2 > 0$. By homogeneity the result is clearly proportional to $\psi'_\tau(z)$.

The first assertion follows from 2.5.28 2).

It then follows from 2.5.25, 2.5.26 and from the fact that the operator $\bar{\mathcal{F}}_n \otimes \mathcal{F}_m$ is injective on $L^2(d_n \otimes d_m)$ that we have the fundamental formula:

2.5.29. $$R_S(g)\cdot\psi_\tau = j(g,\tau)^{-d}\,\psi_{g\cdot\tau},$$

where $j(\begin{pmatrix} a & b \\ c & d\end{pmatrix},\tau) = (c\tau+d)$.

<u>Remark</u>: As ψ_{P_1,P_2} is a harmonic function with respect to Δ_S, our function $\psi_\tau = \psi_{\tau,P_1,P_2} = \psi_{P_1,P_2}e^{2i\pi\tau S}$ is similar to the function $f_{P,z}$ (2.5.17) (when $S \gg 0$). The striking fact about the Rallis-Schiffmann function ψ_τ is that ψ_τ is supported on the set $S \geq 0$ and is not a C^∞-function on E.

2.5.30. Let us consider $\widetilde{R}_S$ as a representation of $\mathcal{G} \times O(S)$. The choice of the decomposition $E = E_1 \oplus E_2$ of E as a orthogonal direct sum of subspaces where S_1,S_2 are definite, is equivalent to the choice of a maximal compact subgroup of $O(S) \simeq O(p,q)$, namely $O(S_1) \times O(S_2) \simeq O(p) \times O(q)$. Clearly under the representation $(\widetilde{R}_S(g)\varphi)(x) = \varphi(g^{-1}x)$ of the group $O(S)$ in $L^2(E)$, our given function ψ_{τ,P_1,P_2} transforms under $O(p) \times O(q)$ as does $P_1 \otimes P_2$, i.e. ψ_{τ,P_1,P_2} is of type $d_n \otimes d_m$ under $O(p) \times O(q)$.

Let us summarize the results of the preceding discussion:

a) The function ψ_{τ,P_1,P_2} associated to the harmonic polynomials P_1 and P_2 of degree n and m respectively is in $L^2(E)$ if $d = ((p/2)+n) - ((q/2)+m) > 1$. It is in $L^1(E) \cap L^2(E)$, if $n - m > q$ and $p + q > 2$.

b) Let $d > 1$, then ψ_{τ,P_1,P_2} depends holomorphically of τ and satisfies:

$$\widetilde{R}_S(g)\cdot\psi_{P_1,P_2,\tau} = j(g,\tau)^{-d}\,\psi_{P_1,P_2,g\cdot\tau}$$

where $j(g,\tau) = c\tau + d$, if $g = \begin{pmatrix} a & b \\ c & d \end{pmatrix} \in SL(2,\mathbb{R})$. (If d is not an integer, $\widetilde{R}_S$ is a representation of $\widetilde{G} = \{(g,\varphi)\}$ and $j((g,\varphi);\tau)^{-d} = e^{-d\varphi(\tau)}$).

c) $\psi_{P_1,P_2,\tau}$ is a vector of type $d_n \otimes d_m$ under the action of $O(S_1) \times O(S_2)$.

(As remarked before, these properties in fact characterize ψ_{τ,P_1,P_2} uniquely.)

2.5.31. We will be mainly interested in the case where S is of signature $(2,q)$ on E. Let $D = \{z \in E^{\mathbb{C}}$ such that $S(z,z) = 0,\ S(z,\bar{z}) > 0\}$. ($D$ is a line bundle over the hermitian symmetric space associated to the group $O(2,q)$ via the map $z \to \mathbb{C}z$.) Then the 2-dimensional plane $(\mathbb{C}z \oplus \mathbb{C}\bar{z}) \cap E$ is a positive definite plane E_1. There exists a basis e_1, e_2, f_1, f_2, $\cdots$, f_q on E such that $z = e_1 + ie_2$, $S(\Sigma\, x_i e_i + \Sigma\, y_j f_j) = x_1^2 + x_2^2 - \sum_{j=1}^{q} y_j^2$. On $E_1 = \mathbb{R}e_1 \oplus \mathbb{R}e_2$, a harmonic polynomial P_1 of degree n is given by $P_1 = (x_1 + ix_2)^n$ or $(x_1 - ix_2)^n$. Now for $P_1 = (x_1 - ix_2)^n$, $S_1^{-n}P_1 = (x_1^2 + x_2^2)^{-n}(x_1 - ix_2)^n = (x_1 + ix_2)^{-n} = S(x,z)^{-n}$. Hence the vector f_{P_1,P_2} associated with $P_1 = (x_1 - ix_2)^n$, $P_2 = 1$ is given by $S(x,z)^{-n}\, S(x)^{n-(q/2)}$ and depends holomorphically of the variable z in D. Hence we obtain:

2.5.32. <u>Theorem</u>: Let E be a space with a quadratic form S of signature $(2,q)$ (with $q \geq 1$). Let $z \in E^{\mathbb{C}}$ with $S(z,z) = 0$ and $S(z,\bar{z}) > 0$. Then

a) The function

$$\psi^n_{\tau,z}(x) = S(x,z)^{-n}\, S(x)^{n-(q/2)}\, e^{i\pi\tau S(x)}, \text{ on } S(x) > 0$$
$$= 0 \quad \text{on } S(x) < 0$$

is in $L^2(E)$ if $n > q/2$.

b) For $n > q$, $\psi^n_{z,\tau}$ is continuous and in $L^1(E) \cap L^2(E)$.

c) $\psi^n_{z,\tau}$ satisfies the fundamental relation:

$$R_S(g_1,g_2)\cdot\psi^n_{\tau,z} = j(g_1,\tau)^{-d}\, \psi_{g_1\cdot\tau,g_2\cdot z}$$

where $g_1 \in \widetilde{G}$, $g_2 \in O(S)$, $d = n + 1 - (q/2)$.
(Remark: In appropriate coordinates for the projective space associated to D by $z \to \mathbb{C}z$, an automorphy factor $j(g_2,\mathbb{C}z)$ will also appear in the variable g_2.)

For some appropriate linear functional θ on $L^2(E)$ semi-invariant under the action of discrete subgroups $\Gamma_1 \times \Gamma_2$ of $\widetilde{G} \times O(S)$, the coefficient $\theta(\tau,z) = \langle\theta,\psi^n_{\tau,z}\rangle$ will be an automorphic form with respect to τ and z. Some explicit examples will be given in Section 2.7, 2.8, 2.9 and will lead to kernels of important correspondences.

2.5.33. Let us comment now on the case $p = q = 1$. Let $E = E_1 \oplus E_2$ with $S(x) = x_1^2 - x_2^2$. The representation $\widetilde{R}_{S_1}$ is the direct sum of $T_{1/2}$ and $T_{3/2}$, corresponding to the decomposition of $L^2(E)$ in even and odd functions. The vector $w_1(x_1) = x_1 e^{-\pi x_1^2}$ is the lowest weight vector of the representation $T_{3/2}$. Similarly $R_{S_2} = T_{1/2} \oplus T_{3/2}$, and the vector $w_2(x_2) = e^{-\pi x_2^2}$ is the highest weight vector of the representation $T_{1/2}$. The decomposition of $R_S = \tilde{R}_{S_1} \otimes \tilde{R}_{S_2}$ in irreducible representations of $SL(2,\mathbb{R})$ is given by a direct integral, without any discrete spectrum, as it can be easily seen from the isomorphism $R_S \cong U$. However the representation $\overline{T}_1$ occurs in this direct integral decomposition, even if not discretely. Our vector $v(x) = \psi_{P_1,P_2}(x)$ corresponding to $P_1 = x_1$, $P_2 = 1$ is given by

$$\psi_{P_1,P_2}(x) = (\text{sign } x_1)\, e^{-\pi(x_1^2 - x_2^2)} \quad \text{if} \quad (x_1^2 - x_2^2) > 0$$

$$= 0 \quad \text{if} \quad (x_1^2 - x_2^2) \leq 0 ,$$

Although $v(x)$ is not in $L^2(E)$, we can consider it as a tempered distribution on E. Under the action of the representation R_S on tempered distributions, v spans a representation equivalent to $\overline{T}_1$.

Let us look at the action of the group $O(S) = O(1,1)$ on $v(x)$. As x_1 doesn't change sign on the connected component of the hyperboloid $x_1^2 - x_2^2 = k$ for $k > 0$, $v(x)$ is invariant under $SO(1,1)$. Thus $v(x)$ transforms under the character d_1 of $O(1,1)$ such that

$$d_1\begin{pmatrix} \varepsilon_1 & 0 \\ 0 & \varepsilon_2 \end{pmatrix} = \text{sign } \varepsilon_1 , \qquad (\varepsilon_1 = \pm 1) .$$

Thus in the "philosophical" correspondence between representations of $SL(2,\mathbb{R})$ and of $O(1,1)$ via R_S, $\overline{T}_1$ is associated to this character d_1 of $O(1,1)$.

The vector

$$(w_1 \otimes w_2)(x_1,x_2) = x_1 e^{-\pi(x_1^2 + x_2^2)}$$

is of the same weight than v under the action of

$$K = \begin{pmatrix} \cos\theta & \sin\theta \\ -\sin\theta & \cos\theta \end{pmatrix} \subset SL(2,\mathbb{R}) .$$

But v "belongs" to the isotypic component associated to the trivial representation of $SO(1,1)$. Hence it is natural to expect the formula

$$v = \int_{SO(1,1)} (R_S(g)\cdot w)\, dg ,$$

if this has a meaning.

2.5.34. __Lemma__: Let $(x_1^2 - x_2^2) \neq 0$, then

$$v(x_1,x_2) = \int_{SO(1,1)} (R_S(g)\cdot w)(x_1,x_2)\, dg .$$

__Proof__: We have to integrate the function $x_1 e^{i\pi(x_1^2+x_2^2)}$ over the connected component of the hyperbole $x_1^2 - x_2^2 = k$ with respect to the invariant measure under $SO(1,1)$. If $k < 0$, as (x_1,x_2) and $(-x_1,x_2)$ belongs then to the same connected component of the hyperbole $x_1^2 - x_2^2 = k$, the integral is zero, by antisymmetry. Now let $x_1^2 - x_2^2 = k^2$; we parametrize the branch of the hyperbole $x_1^2 - x_2^2 = k^2$, with $x_1 > 0$, by $x_1 + x_2 = k\alpha$, $x_1 - x_2 = k\alpha^{-1}$; $k > 0$, $\alpha > 0$. The integral to be calculated is:

$$\frac{1}{2}\int k(\alpha+\alpha^{-1})\, e^{-\pi\frac{k^2(\alpha^2+\alpha^{-2})}{2}}\, \frac{d\alpha}{\alpha}.$$

We choose $u = (\alpha - \alpha^{-1})$ as new variable. Then $du = (1+\alpha^{-2})\, d\alpha = (\alpha+\alpha^{-1})\frac{d\alpha}{\alpha}$. Thus this is

$$\frac{1}{2}\int k\, e^{-\pi\frac{k^2}{2}(u^2+2)}\, du,$$

and is clearly proportional to $e^{-\pi k^2} = e^{-\pi(x_1^2-x_2^2)}$.

Let us consider $w_\tau^1(x_1) = x_1 e^{i\pi\tau x_1^2}$, $w_\tau^2(x_2) = e^{-i\pi\bar{\tau}x_2^2}$ and

$$\begin{aligned} w_\tau(x_1,x_2) &= (\operatorname{Im}\tau)^{1/2}\, w_\tau^1(x_1)\, w_\tau^2(x_2) \\ &= (\operatorname{Im}\tau)^{1/2}\, x_1 e^{i\pi\tau x_1^2}\, e^{-i\pi\bar{\tau}x_2^2}. \end{aligned}$$

Then it follows from 2.5.18 and 2.3.4 that

2.5.35. $$R_S(\sigma)\cdot w_\tau = (c\tau+d)^{-1}\, w_{\sigma.\tau}.$$

(However w_τ is not holomorphic in τ.)

Now let us consider

$$\begin{aligned} v_\tau(x_1,x_2) &= (\operatorname{sign} x_1)\, e^{i\pi\tau(x_1^2-x_2^2)}, && \text{if } x_1^2 - x_2^2 > 0 \\ &= 0 && \text{on } x_1^2 - x_2^2 < 0. \end{aligned}$$

The calculation of the Lemma 2.5.34, proves that v_τ is the "projection" of w_τ on the trivial representation of $SO(1,1)$, i.e. we have the formula:

2.5.36. $$v_\tau(x_1,x_2) = \int_{g\in SO(1,1)} (R_S(g)\cdot w_\tau)(x_1,x_2)\, dg.$$

We will use this integral representation of v_τ in order to explain the behavior of the modular forms considered by Hecke, in the framework of the Weil representation.

2.5.37. We now summarize some of the results on the decomposition of R_S in irreducible components, in the case where V is an arbitrary symplectic space of dimension $2n$, and S is a positive definite quadratic form on E.

Let W be a real vector space of dimension k and W^* its dual vector space. We take as model of symplectic space $V = W \oplus W^*$, with $B(x_1 + f_1, x_2 + f_2) = f_2(x_1) - f_1(x_2)$. The space $\ell = W$ and $\ell' = W^*$ are complementary Lagrangian subspaces in (V,B).

Let E be a k-dimensional vector space, with a positive definite symmetric form S. We consider the decomposition $W \otimes E + W^* \otimes E$ of $V \otimes E$ in complementary Lagrangian subspaces. The Weil representation R_S associated to the Lagrangian subspace $\ell \otimes E$ is then realized in $L^2(W^* \otimes E)$. We identify $W^* \otimes E$ with $\mathrm{Hom}_{\mathbb{R}}(W,E)$. The action of the group $O(S)$ on $L^2(\mathrm{Hom}(W,E))$ is then simply given by $(\sigma.\varphi)(x) = \varphi(\sigma^{-1}x)$, for $x \in \mathrm{Hom}(W,E)$.

Let D be the Siegel upper half plane associated to (V,B), i.e. $D = \{Z\colon (W^*)^{\mathbb{C}} \to (W^{\mathbb{C}})$ such that ${}^tZ = Z$, $\mathrm{Im}\, Z \gg 0\}$. Let D' be the Siegel upper half plane associated to $((V \otimes E), B \otimes S)$,

$D' = \{Z' : (W^* \otimes E)^{\mathbb{C}} \to (W \otimes E)^{\mathbb{C}}$, such that ${}^tZ' = Z'$; $\operatorname{Im} Z' \gg 0\}$

where E is identified with E^*, via S.

As S is positive definite, the map $Z \to Z \otimes id_E$ is an injection of D in D' (with respect to an orthogonal set of coordinates of E, $Z \otimes id_E$ is represented by the matrix

$$\begin{pmatrix} Z & & & 0 \\ & Z & & \\ & & \ddots & \\ 0 & & & Z \end{pmatrix}.$$

Let us define for $Z \in D$ the function $v_Z^S(X) = v_{Z\otimes id}(X)$. For $X \in \operatorname{Hom}(W,E)$ and $Z \in D$, the matrix $Z\,{}^tXSX \in \operatorname{Hom}_{\mathbb{C}}(W^{\mathbb{C}},W^{\mathbb{C}})$. It is immediate to check:

2.5.38. Proposition:

a) $v_Z^S(X) = e^{i\pi \operatorname{Tr}(Z\,{}^tXSX)}$.

b) For $g \in Sp(V,B)$, $R_S(g)\cdot v_Z^S = m(g,Z)^k v_{g.Z}^S$.

c) v_Z^S is invariant under the action of $O(S)$.

The function v_Z^S is thus the analogue of the "lowest weight vector" $e^{i\pi z S(\xi)}$ in the case of $SL(2,\mathbb{R})$.

Let k be even, then the representation $R_S(g)$ is equivalent to a true representation $R_S'(g) = s(g)^{k/2}\, R_S(g)$ of $Sp(n,\mathbb{R})$. Hence we have:

$$R_S'(g)\cdot v_Z^S = (\det(CZ+D))^{-k/2}\, v_{g\cdot Z}^S$$

for $g = \left(\frac{A\,|\,B}{C\,|\,D}\right) \in Sp(n,\mathbb{R})$.

2.5.39. We now generalize the functions $f_{P,z}(\xi) = P(\xi)e^{i\pi z S(\xi)}$, P harmonic polynomial on E, discussed in the case of $SL(2,\mathbb{R})$.

Let us consider the space $\mathcal{P}$ of all complex valued polynomials on $Hom(W,E)$. This is isomorphic to the space of all complex polynomials on $Hom(W^{\mathbb{C}},E^{\mathbb{C}})$. We denote by $O(S,\mathbb{C})$ the subgroup of complex transformation of $E^{\mathbb{C}}$ leaving stable the form $S^{\mathbb{C}}$ on $E^{\mathbb{C}}$. The group $GL(n;\mathbb{C}) \times O(S,\mathbb{C})$ acts on $\mathcal{P}$ via $((A,\sigma)\cdot P)(X) = P(\sigma^{-1}XA)$, for $A \in GL(n,\mathbb{C})$ and $\sigma \in O(S,\mathbb{C})$.

For $X \in Hom_{\mathbb{C}}(W^{\mathbb{C}},E^{\mathbb{C}})$, let us consider the symmetric matrix ${}^tXSX$. The coefficients $({}^tXSX)_{i,j}$ generate the algebra of all $O(S,\mathbb{C})$ invariant polynomial functions on $Hom(W,E)$. Thus we can describe the algebra $\mathcal{D}_S$ of all $O(S,\mathbb{C})$-invariant constant coefficients differential operators on $Hom(W,E)$ as follows: we fix a basis of $W^{\mathbb{C}}$ and an orthogonal basis of $E^{\mathbb{C}}$. Writing X in $Hom(W^{\mathbb{C}},E^{\mathbb{C}})$ as:

$$X = \begin{pmatrix} x_{11}, & \cdots, & x_{n1} \\ x_{1k}, & \cdots, & x_{nk} \end{pmatrix},$$

the algebra $\mathcal{D}_S$ is generated by the operators:

$$\Delta_{i,j} = \sum_{\ell=1}^{n} \frac{\partial}{\partial x_{\ell i}} \frac{\partial}{\partial x_{\ell j}} .$$

Similarly to 2.5.11, we define the space of $O(S,\mathbb{C})$ harmonic polynomials by:

$$\mathcal{H} = \{P \in \mathcal{P} \text{ such that } \Delta_{ij} P = 0 , \text{ for all } i,j\}.$$

Let us consider the action of the group $O(k) = O(S)$ on $\mathcal{H}$.

We write $\mathcal{H} = \oplus\, \mathcal{H}(\lambda)$ for the decomposition of $\mathcal{H}$ in isotypic components under $O(k)$. We then have:

2.5.40. Theorem:

a) Let $\Sigma = \{\lambda \in O(k)^{\wedge}$ such that $\mathcal{H}(\lambda) \neq \{0\}\}$. The restriction of the representation $\tilde{R}_S$ of $Sp(B) \times O(k)$ on the isotypic component of type λ under $O(k)$ is an irreducible representation $W_\lambda \otimes \lambda$ of $Sp(B) \times O(k)$.

b) We have $\tilde{R}_S = \bigoplus_{\lambda \in \Sigma} W_\lambda \otimes \lambda$. The correspondence $\lambda \to W_\lambda$ is injective on Σ.

Remark: The representations W_λ are representations of $\tilde{Sp}(B)$ with lowest weight vectors. Our conjecture in [15] is that $\{W_\lambda\}$ exhausts the list of unitary representations of the metaplectic group with lowest weight vectors. When $k > 2n$, the representations W_λ are the members of the anti-holomorphic discrete series of $Sp(B)$. When k is small, these representations can be realized into a subspace of anti-holomorphic functions on the Siegel upper half plane solutions of a system of differential equations.

Let us now describe the lowest weight vectors of the representation R_S of $Sp(B)$. Let us consider the space $\mathcal{H}$ of $O(k)$ harmonic polynomials on $Hom(W,E)$. $\mathcal{H}$ is stable under $GL(n,\mathbb{C}) \times O(k,\mathbb{C})$. We have the following:

2.5.41. Theorem:

a) The isotypic component $\mathcal{H}(\lambda)$ of $\mathcal{H}$ of type λ under $O(S,\mathbb{C})$ is irreducible under $GL(n,\mathbb{C}) \times O(S,\mathbb{C})$.

b) The isotypic component $\mathcal{H}(\tau)$ of $\mathcal{H}$ of type τ under $GL(n,\mathbb{C})$ is irreducible under $GL(n,\mathbb{C}) \times O(S,\mathbb{C})$.

In other words, if $\Sigma = \{\lambda$, irreducible representations of $O(S,\mathbb{C})$, such that $\mathcal{H}(\lambda) \neq 0\}$, then, for $\lambda \in \Sigma$, there exists a unique irreducible representation $\tau = \tau(\lambda)$ of $GL(n;\mathbb{C})$ such that $\mathcal{H}(\tau \otimes \lambda) \neq \{0\}$. Furthermore, the map $\lambda \to \tau(\lambda)$ is injective on Σ. (It is possible to describe explicitly the set Σ and the correspondence $\lambda \to \tau(\lambda)$, see [15].)

Let now $P \in \mathcal{H}(\tau \otimes \lambda)$. We consider the vector

$$f_{P,Z}(X) = P(X)\, e^{i\pi Tr(Z\,{}^tXSX)} \quad \text{of} \quad L^2(Hom(W,E)) .$$

We have then the following:

2.5.42. <u>Theorem</u>: Let $P \in \mathcal{H}(\tau \otimes \lambda)$, then for $g = \left(\begin{array}{c|c} A & B \\ \hline C & D \end{array}\right) \in Sp(n,\mathbb{R})$

$$R_S(g)\cdot f_{P,Z} = m(g,Z)^k\, f_{\tau({}^t(CZ+D)^{-1})\cdot P,\, g\cdot Z} .$$

<u>Proof</u>: This relation can be checked on the set of generators $\left(\begin{array}{c|c} 1 & X \\ \hline 0 & 1 \end{array}\right)$, $\left(\begin{array}{c|c} 0 & 1 \\ \hline -1 & 0 \end{array}\right)$ of $Sp(n;\mathbb{R})$, or by infinitesimal methods as in the case of $SL(2,\mathbb{R})$.

2.5.43. We finally give a special example of $P \in \mathcal{H}(\tau \otimes \lambda)$. Let $\dim E = \dim W = n$. Then $Hom(W,E)$ is a space of $n \times n$ matrices and the function $X \to \det X$ is well defined, up to a scalar depending of the choice of basis of W and E. Let us consider an orthogonal basis of E with respect to S. Then it is easy to verify in these coordinates that $\det X$ is an $O(S)$ harmonic polynomial. Clearly P is of type $\lambda(\sigma) = \det \sigma$, $\tau(A) = \det A$ with respect to $O(S) \times GL(n,\mathbb{C})$.

2.5.44. <u>Corollary</u>: Let, for $\nu = 0,1$ and $k = n$

$$f_{Z,\nu}(X) = (\det X)^{\nu} e^{i\pi \mathrm{Tr}(Z^t XSX)}$$

then $R_S(g)\cdot f_{\nu,Z} = m(g,Z)^k \, (\det(CZ+D))^{-1} \, f_{\nu,g\cdot Z}$.

2.6. θ-series associated to quadratic forms.

Let (E,S) be a k-dimensional vector space with a non-degenerate symmetric form of signature (p,q). We first consider the two-dimensional canonical symplectic space $V = \mathbb{R}P \oplus \mathbb{R}Q$ and the associated symplectic space $(E \otimes V, S \otimes B)$. The group $O(S) \times SL(2,\mathbb{R})$ is naturally imbedded in $Sp(S \otimes B)$. The space $\ell = E \otimes P$ is a Lagrangian subspace of $E \otimes V$.

Let L be a lattice in E. Let $L^* = \{\xi \in E,\ S(\xi,L) \in \mathbb{Z}\}$ the dual lattice of L with respect to the form S. The lattice $r = L^* \otimes P + L \otimes Q$ is a self-dual lattice in $E \otimes V = E \otimes P \oplus E \otimes Q$. We consider the character χ of $\exp(r \oplus \mathbb{R}E) = R$ given by $\chi(\exp(\ell^* \otimes P + \ell \otimes Q + tE)) = (-1)^{S(\ell,\ell^*)}e^{2i\pi t}$. We assume $L \subset L^*$ (i.e. $S(L,L) \subset \mathbb{Z}$). We denote by n_L the level of L, i.e. n_L is the smallest integer such that $n_L\ell^* \in L$ for every $\ell^* \in L^*$.

Let us consider the action of $SL(2,\mathbb{Z})$ on $E \otimes V$. We have

$$\begin{pmatrix} a & b \\ c & d \end{pmatrix} \cdot (\ell^* \otimes P + \ell \otimes Q) = (a\ell^*+b\ell) \otimes P + (c\ell^*+d\ell) \otimes Q.$$

Hence the lattice $L^* \otimes P \oplus L \otimes Q = r$ is stable under

$$\Gamma_0(n_L) = \{\begin{pmatrix} a & b \\ c & d \end{pmatrix},\ c \equiv 0 \bmod n_L\}.$$

The pair (r,χ) is stable under

2.6.1. $$\Gamma_0(n_L,\chi) = \{\begin{pmatrix} a & b \\ c & d \end{pmatrix},\ c \equiv 0 \bmod n_L,\ \begin{matrix} acS(\ell^*,\ell^*) \equiv 0 \text{ Mod } 2 \\ bdS(\ell,\ell) \equiv 0 \text{ Mod } 2 \end{matrix}\}.$$

In particular if $L = L^*$ and L is even $(S(\ell,\ell) \in 2\mathbb{Z},\ \ell \in L)$

then $\Gamma_0(n_L,\chi)$ is the full modular group $SL(2,\mathbb{Z})$. In general $\Gamma_0(n_L,\chi)$ contains the congruence subgroup $\Gamma_0(2n_L,2)$.

Let us consider the representation $W = W(r,\chi)$ of the Heisenberg group N in $H(r,\chi)$. It is clear that the operator $(A_r(\gamma)\varphi)(n) = \varphi(\gamma^{-1}\cdot n)$ is an unitary operator on $H(r,\chi)$ satisfying $A_r(\gamma)W(n)A_r(\gamma)^{-1} = W(\gamma\cdot n)$. Hence $A_r(\gamma)$ is proportional to the operator $R_\ell(\gamma)$ of the canonical projective Weil representation. There exists a scalar $\alpha(\gamma)$ such that the following diagram is commutative:

2.6.2.

$$\begin{array}{ccc} H(\ell) & \xrightarrow{\alpha(\gamma)\, R_\ell(\gamma)} & H(\ell) \\ {\scriptstyle \theta^\chi_{r,\ell}}\Big\downarrow & & \Big\downarrow{\scriptstyle \theta^\chi_{r,\ell}} \\ H(r,\chi) & \xrightarrow{A_r(\gamma)} & H(r,\chi) \end{array}$$

As in 2.2.30, we have $\alpha(\gamma) = b(\gamma\ell,\ell;(r,\chi)) = b(\ell,\gamma^{-1}\ell;(r,\chi))$.

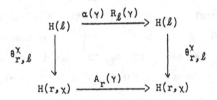

Let $(e_1, e_2, \cdots, e_k)$ be a $\mathbb{Z}$-basis of L, then $L^* = \mathbb{Z}\, e_1^* \oplus \cdots \oplus \mathbb{Z}\, e_k^*$ where e_i^* is the dual basis. Let $S = (S(e_i,e_j))$ the matrix of S with respect to the basis e_i. The number $D = (\det S)$ is independent of the choice of the $\mathbb{Z}$-basis (e_i), and is called the discriminant of L. We have $e_i = \sum_j S(e_i,e_j)e_j^*$, i.e. $L^* = S^{-1}\mathbb{Z}^k$, if we have identified L to $\mathbb{Z}^k$ inside $\mathbb{R}^k = E$ (S is the matrix $S = (S(e_i,e_j))$). Consequently, the number n_L is the smallest integer such that $n_L S^{-1}$ have integral coefficients. In particular n_L is a divisor of (det S); the equation $n_L = S\cdot(n_L S^{-1})$ implies that $n_L^k = (\det S)(\det(n_L S^{-1}))$. Hence n_L contains all the prime factors occurring in det S.

The elements $P_i = e_i^* \otimes P$, $Q_j = e_j \otimes Q$ form a symplectic basis of $E \otimes V$ such that

$$r = \sum_{i=1}^{k} \mathbb{Z} P_i \oplus \sum_{i=1}^{k} \mathbb{Z} Q_j .$$

For $\gamma = \begin{pmatrix} a & b \\ c & d \end{pmatrix}$ in $SL(2,\mathbb{R})$ the symplectic transformation $\gamma(e \otimes v) = e \otimes \gamma v$ is expressed in the basis (P_i, Q_j) by the matrix:

$$\left(\begin{array}{c|c} a & bS \\ \hline cS^{-1} & d \end{array}\right).$$

Hence for $\gamma \in \Gamma_0(n_L, \gamma)$,

$$\alpha(\gamma) = b(\ell, \gamma^{-1}\ell) = b(d, -cS^{-1})$$

$$= b(cS^{-1}, d)e^{-\frac{i\pi}{4}\operatorname{sign}(dS^{-1})} = b(cS^{-1}, d)e^{-\frac{i\pi}{4}(\operatorname{sign}(dc)(\operatorname{sign} S)}$$

from 2.2.29.

We now have to calculate:

$$b(cS^{-1}, d) = d^{-\frac{k}{2}} \sum_{\xi \in \mathbb{Z}^k/d\mathbb{Z}^k} e^{\frac{i\pi\langle cS^{-1}\xi, \xi\rangle}{d}}.$$

2.6.4. <u>Proposition</u>: Let Q be a $(k \times k)$ symmetric matrix with integral coefficients and even diagonal. Let d be an integer prime to det Q. Let

$$b(Q, d) = d^{-k/2} \sum_{\xi \in \mathbb{Z}^k/d\mathbb{Z}^k} e^{i\pi\frac{Q(\xi,\xi)}{d}}.$$

Then, if d is an odd positive integer:

$$b(Q,d) = \left(\frac{\det Q}{d}\right)\left(\frac{2}{d}\right)^k \varepsilon_d^k .$$

<u>Proof</u>: We proceed as in 2.2.18.

α) Let $d = p$ be a odd positive prime. The even symmetric form (Q mod p) on the vector space $\mathbb{F}_p^k = (\mathbb{Z}/p\mathbb{Z})^k$ is diagonalizable, i.e. there exists a $(k \times k)$ matrix A with integral coefficients such that $AQ{}^tA = \begin{pmatrix} 2q_1 & 0 \\ 0 & 2q_k \end{pmatrix}$ mod p Changing ξ in $A\xi$, our sum becomes

$$p^{-k/2}\Big(\sum_{\xi\in(\mathbb{Z}/p\mathbb{Z})^k} e^{2i\pi\frac{\Sigma q_j\xi_j^2}{p}}\Big) = \prod_{\alpha=1}^{k} \Big(p^{-1/2} \sum_{\xi\in\mathbb{Z}/p\mathbb{Z}} e^{\frac{2i\pi q_\alpha\xi^2}{p}}\Big)$$

$$= \prod_{\alpha=1}^{k} \left[\left(\frac{q_\alpha}{p}\right)\varepsilon_p\right] = \left(\frac{\pi q_\alpha}{p}\right)\varepsilon_p^k$$

by 2.2.14. But $(\det Q)(\det A)^2 \equiv (\pi q_\alpha)2^k$ mod p. Hence $\left(\frac{\pi q_\alpha}{p}\right) = \left(\frac{\det Q}{p}\right)\left(\frac{2}{p}\right)^k$.

β) Let $d = p^r$. Let $\xi_0 \in \mathbb{Z}^k/p^r\mathbb{Z}^k$. We consider the elements of $\mathbb{Z}^k/p^r\mathbb{Z}^k$ of the form $\xi_0 + p^{r-1}\ell$ where ℓ varies over $\mathbb{Z}^k/p\mathbb{Z}^k$. We have

$$\sum_{\ell\in\mathbb{Z}^k/p\mathbb{Z}^k} e^{\frac{i\pi\, Q(\xi_0+p^{r-1}\ell,\,\xi_0+p^{r-1}\ell)}{p^r}}$$

$$= e^{\frac{i\pi\, Q(\xi_0,\xi_0)}{p^r}} \sum_{\ell\in\mathbb{Z}^k/p\mathbb{Z}^k} e^{2i\pi\frac{Q(\xi_0,\ell)}{p}}$$

$$= e^{\frac{i\pi\, Q(\xi_0,\xi_0)}{p^r}} \sum_{\ell\in\mathbb{Z}^k/p\mathbb{Z}^k} e^{4i\pi\sum_{\alpha=1}^{k}\frac{\xi_0^\alpha\ell^\alpha q_\alpha}{p}}$$

using the diagonal form of Q over $(\mathbb{Z}/p\mathbb{Z})^k$ (with q_α prime to p). As the sum over all p-roots of unity is zero, we find that our partial sum is zero except if $\xi_0^\alpha \equiv 0 \bmod p$, i.e. if $\xi_0 \in p\mathbb{Z}^k$. Hence in our Gauss sum

$$\sum_{\xi_0 \in \mathbb{Z}^k/p^r\mathbb{Z}^k} e^{\frac{i\pi Q(\xi_0,\xi_0)}{p^r}},$$

we need only to consider the element $\xi_0 \in p\mathbb{Z}^k$. We obtain as in 2.2.18

$$b(Q,p^r) = b(Q,p^{r-2}) = \left(\frac{\det Q}{p^r}\right)(\varepsilon_{p^r})^k\left(\frac{2}{p^r}\right)^k,$$

by induction hypothesis.

γ) Finally if $d = d_1d_2$, with d_1 and d_2 relatively prime, we write any element ξ of $\mathbb{Z}^k/d\mathbb{Z}^k$ as $\xi = d_2\xi_1 + d_1\xi_2$ with $\xi_1 \in \mathbb{Z}^k/d_1\mathbb{Z}^k$, $\xi_2 \in \mathbb{Z}^k/d_2\mathbb{Z}^k$, and we obtain (following 2.2.18) our proposition.

2.6.5. Remark: Let Q be an even form such that $\det Q$ is odd. Then

a) $\dim E = k$ is even and $(-1)^{k/2}(\det Q) \equiv 1 \text{ Mod } 4$.

b) $b(Q,2) = \left(\frac{2}{\det Q}\right)$.

Proof: Let us first consider the case where $k = 2$. Then $Q = \begin{pmatrix} 2a & c \\ c & 2b \end{pmatrix}$, with c odd. Hence $\det Q = -c^2 \text{ Mod } 4 \equiv -1 \text{ Mod } 4$. Now to prove b), we have to calculate

$$\frac{1}{2}\sum_{\substack{\xi_1 \in \mathbb{Z}/2\mathbb{Z} \\ \xi_2 \in \mathbb{Z}/2\mathbb{Z}}} e^{i\pi(a\xi_1^2 + c\xi_1\xi_2 + b\xi_2^2)}.$$

This depends only of (a,b,c) mod 2.

If $a \equiv 0$ mod 2, we can write $\xi_1\xi_2 + b\xi_2^2 = \xi_2(\xi_1+b\xi_2)$. Changing of coordinates, we have:

$$b(Q,2) = \frac{1}{2} \sum_{\substack{\xi_1=0,1 \\ \xi_2=0,1}} e^{i\pi\xi_1\xi_2} = 1 .$$

As $\det Q = -c^2$ (c odd) is congruent to 1 or -1 mod 8, we have $b(Q,2) = 1 = (\frac{2}{\det Q})$.

If $a \neq 0$, $b \neq 0$, $c \neq 0$ mod 2, we have to calculate

$$\frac{1}{2} \sum_{\substack{\xi_1=0,1 \\ \xi_2=0,1}} e^{i\pi(\xi_1^2+\xi_1\xi_2+\xi_2^2)} = -1 .$$

But $\det Q = 4ab - c^2$ is congruent to 3 or -3 mod 8, and we have again $b(Q,2) = -1 = (\frac{2}{\det Q})$.

Now let $k > 2$. As $\det Q$ is odd, we can find a coefficient $q_{ij} = Q(e_i,e_j)$ of Q which is odd. We can suppose (e_i,e_j) are the first elements of the basis (e_1,e_2). The restriction of Q to the space $\mathbb{Z}e_1 + \mathbb{Z}e_2$ has matrix $\begin{pmatrix} 2a & c \\ c & 2b \end{pmatrix}$. Now for $i > 2$, we can find $x_i, y_i \in \mathbb{Z}$ such that

$$Q(e_i - x_ie_1 - y_ie_2, e_1) = 0 \mod 8$$

$$Q(e_i - x_ie_1 - y_ie_1, e_2) = 0 \mod 8 ,$$

as $\det \begin{pmatrix} 2a & c \\ c & 2b \end{pmatrix}$ is invertible mod 8. Thus there exists a unipotent matrix U such that

$$UQ{}^tU = \begin{pmatrix} Q_1 & & 0 \\ & Q_2 & \\ 0 & & \ddots Q_\ell \end{pmatrix} \mod 8$$

where Q_α are 2×2 matrix. Now $\det(UQ^tU) = \det Q =$ $(\det Q_1)\cdot(\det Q_2) \cdots \cdot(\det Q_\ell) \bmod 8$. This implies a) and b) as

$$b(Q,2) = \pi\, b(Q_i,2) = \pi\left(\frac{2}{\det Q_i}\right) = \left(\frac{2}{\det Q}\right).$$

If Q is an even symmetric form, such that $\det Q$ is odd, we can compute $b(Q,d)$ for d prime to $(\det Q)$, using the recurrence formulas:

If $d = d_1 2^r$ with d_1 odd:

$$b(Q,d_1 2^r) = b(Q2^r,d_1)\, b(Qd_1,2^r)$$

$$b(Q,2^r) = b(Q,2^{r-2})$$

$$b(Q,2) = \left(\frac{2}{\det Q}\right).$$

2.6.7. Let $\gamma = \begin{pmatrix} a & b \\ c & d \end{pmatrix}$ belonging to $\Gamma_0(n_L,\chi)$. Then cS^{-1} is an even symmetric form. Now from the equation $\det(cS^{-1})(\det S) = c^k$, $\det(cS^{-1})$ is a divisor of c^k. As $ad - bc = 1$, d and $\det(cS^{-1})$ are relatively prime. Hence we obtain:

2.6.8. <u>Proposition</u>: Let $\gamma = \begin{pmatrix} a & b \\ c & d \end{pmatrix}$ belonging to $\Gamma_0(n_L,\chi)$ then the diagram 2.6.2 is commutative, with

$$a(\gamma) = e^{-\frac{i\pi}{4}(\text{sign } cd)\text{sign } S}\, b(cS^{-1},d).$$

For d odd positive:

$$b(cS^{-1},d) = \left(\frac{c}{d}\right)^k \left(\frac{2}{d}\right)^k \left(\frac{\det S}{d}\right) \varepsilon_d^k.$$

(for d even, we can use the Remark 2.6.6 to give an explicit expression of $b(cS^{-1},d)$).

We now can give uniform formulas for the representation $\widetilde{R}_S$ on the congruence subgroup $\Gamma_0(2n_L,2)$.

Let us consider the universal covering group $\widetilde{G}$ of $SL(2;\mathbb{R})$ and the true representation $\widetilde{R}_S$ of $\widetilde{G}$ in $H(\ell)$ defined in 2.5.

2.6.9. Theorem. Let $\gamma \in \Gamma_0(2n_L,2)$, then

$$\widetilde{R}_S(\widetilde{\gamma}) = \lambda(\widetilde{\gamma})\, A_r(\gamma)$$

with $\lambda(\widetilde{\gamma}) = \varepsilon_d^{-k} \left(\frac{2c}{d}\right)^k \left(\frac{D}{d}\right)$.

<u>Proof</u>: We recall (2.5.2) that $\widetilde{R}_S(\widetilde{\gamma}) = e^{\frac{i\pi}{4}m} R_S(\gamma)$ where for $\gamma = \begin{pmatrix} a & b \\ c & d \end{pmatrix}$, $m = -(p-q)$ sign c if $c \neq 0$. Then from 2.6.2, we obtain

$$\begin{aligned}
\widetilde{R}(\widetilde{\gamma}) &= \alpha(\gamma)^{-1} A_r(\gamma)\, e^{-\frac{i\pi}{4}(\text{sign } c)(p-q)} \\
&= \alpha(\gamma)^{-1} A_r(\gamma)\, e^{-\frac{i\pi}{4}(\text{sign } c)(\text{sign } S)} \\
&= e^{\frac{i\pi}{4}\text{sign}(cd)(\text{sign } S)}\, b(cS^{-1},d)^{-1} e^{-\frac{i\pi}{4}(\text{sign } c)(\text{sign } S)} A_r(\gamma).
\end{aligned}$$

Hence if d is positive, we obtain:

$$R(\gamma) = \varepsilon_d^{-k} \left(\frac{2c}{d}\right)^k \left(\frac{D}{d}\right) A_r(\gamma) \qquad \text{by 2.6.4.}$$

If d is negative, we proceed in a similar manner.

2.6.10. For applications in Section 2.8, we will analyze a slightly more general situation.

Let Λ be a lattice in (E,S) of level n_Λ and determinant D

Let $\gamma = \begin{pmatrix} a & b \\ c & d \end{pmatrix}$ be an element of $SL(2,\mathbb{Z})$ such that $abS(x) \equiv cdS(y) \equiv 0 \mod 2$, for every $x,y \in \Lambda$. Let $h \in \Lambda^*$, we denote by $\theta(\Lambda,h)$ the distribution on E given by

$$(\theta(\Lambda,h),f) = \sum_{x\in\Lambda} f(h+x) .$$

2.6.11. Proposition:

$$R_S(\gamma)\cdot\theta(\Lambda,h) = D^{-1/2}c^{-k/2} \sum_{k\in\Lambda^*} c(h,k)\ \theta(\Lambda,k)$$

with

$$c(h,k) = \sum_{y\in\Lambda/c\Lambda} e^{-i\pi \frac{d}{c} S(h+y)} e^{\frac{2i\pi}{c} S(k,h+y)} e^{-i\pi \frac{a}{c} S(k)}$$

Proof: For a change, we will give a direct proof using the explicit formula for $R_S(\gamma)$. By definition

$$\begin{aligned}(R_S(\gamma^{-1})\cdot\varphi)(\exp y \otimes Q) &= (\mathcal{F}_{\ell,\gamma^{-1}\ell}\, A(\gamma^{-1})\varphi)(\exp y \otimes Q)\\ &= u \int (A(\gamma^{-1})\varphi)(\exp y \otimes Q \exp x \otimes P)\, dx\\ &= u \int \varphi(\exp y \otimes \gamma Q \exp x \otimes \gamma P)\, dx\end{aligned}$$

where u is the positive constant such that this operator is unitary. We write

$$\exp y \otimes \gamma Q \exp x \otimes \gamma P = \exp \alpha \otimes Q \exp \beta \otimes P \exp tE$$

and we obtain

$$(R_S(\gamma^{-1})\varphi)(\exp y \otimes Q)) = u' \int \varphi(\exp \alpha \otimes Q)\, e^{-2i\pi t}\, d\alpha ,$$

as $\varphi \in H(\ell)$. The explicit calculation of t in function of y and α leads to the formula

$$(R_S(\gamma^{-1})\cdot\varphi)(\exp y \otimes Q)$$
$$= u' \int \varphi(\exp \alpha \otimes Q)\, e^{-i\pi \frac{a}{c} S(\alpha)}\, e^{\frac{2i\pi S(\alpha,y)}{c}}\, e^{-i\pi \frac{d}{c} S(y)}\, d\alpha .$$

Let $Q_1, Q_2, \cdots, Q_k$ be a basis of E, and suppose that $H(\ell)$ is identified with $L^2(E)$ with the measure $dy_1\, dy_2 \cdots dy_k$ derived from this basis, then if $\det S$ is the determinant of the matrix $S(Q_i, Q_j)$, the normalization of u' in order that $R_S(\gamma^{-1})$ is unitary is

$$u' = |\det S|^{1/2}\, c^{-n/2} .$$

Let Λ be our lattice, we choose $d\alpha$ as such that the volume of Λ is 1. Hence $|\det S|$ with respect to this measure $d\alpha$ is equal to the discriminant D of the lattice Λ. Thus we obtain the formula for $R_S(\gamma^{-1})$

2.6.12 $$(R_S(\gamma^{-1})\varphi)(\exp y \otimes Q)$$
$$= |D|^{1/2} c^{-k/2} \int \varphi(\exp \alpha \otimes Q)\, e^{-i\pi \frac{a}{c} S(\alpha)}\, e^{\frac{2i\pi S(\alpha,y)}{c}}\, e^{-i\pi \frac{d}{c} S(y)}$$

which is a special case of 1.6.21 3).

We have to compute:

$$F_h = \sum_{\xi \in \Lambda} e^{-i\pi \frac{d}{c} S(h+\xi)} \int \varphi(\exp \alpha \otimes Q)\, e^{-i\pi \frac{a}{c} S(\alpha)}\, e^{\frac{2i\pi}{c} S(\alpha, \xi+h)}\, d\alpha .$$

Let us write $\xi \in \Lambda$ as $\xi = y + c\ell$, where y describes a system of representatives of $\Lambda/c\Lambda$ and ℓ describes Λ. Then as $cdS(\ell) \equiv 0 \mod 2$, $\ell \in \Lambda$, we have

$$F_h = \sum_{y\in\Lambda/c\Lambda} e^{-i\pi\frac{d}{c}S(h+y)} \sum_{\ell\in\Lambda} \int \varphi(\exp \alpha\otimes Q)\, e^{-i\pi\frac{a}{c}S(\alpha)}\, e^{\frac{2i\pi}{c}S(\alpha,h+y)}\, e^{2i\pi S(\alpha,\ell)} d\alpha$$

Now the dual lattice of Λ with respect to S is Λ^*. Then $\mathrm{vol}\,\Lambda^* = (\mathrm{card}\,\Lambda^*/\Lambda)^{-1} = D^{-1}$. By Poisson formula we obtain

$$F_h = D^{-1} \sum_{y\in\Lambda/c\Lambda} e^{-i\pi\frac{d}{c}S(h+y)} \sum_{k\in\Lambda^*} \varphi(\exp k\otimes Q)\, e^{-i\pi\frac{a}{c}S(k)}\, e^{\frac{2i\pi}{c}S(k,h+y)}$$

and

$$(R_S(\gamma)\cdot\theta(\Lambda,h),\varphi)$$

$$= c^{-k/2}D^{-1/2} \sum_{\substack{k\in\Lambda^*\\ y\in\Lambda/c\Lambda}} \varphi(\exp k\otimes Q)\, e^{-i\pi\frac{a}{c}S(k)}\, e^{\frac{2i\pi}{c}S(k,h+y)}\, e^{-i\pi\frac{d}{c}S(h+y)} .$$

Now let us remark that the function

$$c(h,k) = \sum_{y\in\Lambda/c\Lambda} e^{-i\pi\frac{d}{c}S(h+y)}\, e^{\frac{2i\pi}{c}S(k,h+y)}\, e^{-i\pi\frac{a}{c}S(k)}$$

is invariant by the translation $k \to k + \ell$, with $\ell \in \Lambda$ as we have the equality

$$e^{-i\pi\frac{d}{c}S(h+y)}\, e^{\frac{2i\pi}{c}S(k+\ell,h+y)}\, e^{-i\pi\frac{a}{c}S(k+\ell)}$$

$$= e^{-i\pi\frac{d}{c}S(h+(y-a\ell))}\, e^{\frac{2i\pi}{c}S(k,h+(y-a\ell))}\, e^{-i\pi\frac{a}{c}S(k)}$$

($ad - bc = 1$ and $abS(\ell,\ell) \equiv 0 \mod 2$). Thus we obtain our proposition.

2.6.13. Let us consider S of signature (p,q). We choose a decomposition of E of the form $E = E_1 \oplus E_2$ such that (E_1,E_2) are orthogonal, the restriction S_1 of S to E_1

is positive definite and the restriction of S to E_2 is negative definite. Let P_1 be a homogeneous polynomial of degree n on E_1 harmonic with respect to S_1 and P_2 a homogeneous polynomial of degree m on E_2 harmonic with respect to S_2. We consider the function:

$$\psi_{P_1,P_2} = S_1^{-(\frac{p-2}{2}+n)} P_1 P_2 \, S^{\frac{p-q}{2}+(n-m-1)} \quad \text{for } S(x) > 0$$

$$= 0 \quad \text{if } S(x) < 0.$$

and

$$f_{P_1,P_2}(x) = \psi_{P_1,P_2}(x)\, e^{-\pi S(x,x)}.$$

Let us suppose that $n - m > q$, $p > 1$, $q \geq 1$, then the function f_{P_1,P_2} is continuous, in $L^1(E) \cap L^2(E)$ and is a lowest weight vector of the representation $\widetilde{R}_S$.

Let L be a lattice in E of discriminant D and level n_L. We can form the coefficient $(\delta_L, \widetilde{R}_S(\sigma) f_{P_1,P_2})$. We obtain

2.6.14. Theorem: The function $\theta_{P_1,P_2}(\tau) = \sum_{\substack{\xi \in L \\ S(\xi)>0}} \psi_{P_1,P_2}(\xi) e^{i\pi S(\xi)\tau}$ is a holomorphic function of τ on P^+ and satisfies

$$\theta_{P_1,P_2}\left(\frac{a\tau+b}{c\tau+d}\right) = \left(\frac{D}{d}\right)\left(\frac{2c}{d}\right)^k \varepsilon_d^{-k} (c\tau+d)^{\frac{p-q}{2}+n-m} \theta_{P_1,P_2}(\tau)$$

for $\gamma = \begin{pmatrix} a & b \\ c & d \end{pmatrix}$ in $\Gamma_0(2n_L, 2)$, which $k = p + q$.

<u>Remark</u>: As $\phi_{P_1,P_2}(\xi)$ is supported on $S(\xi) > 0$, we indeed sum only over the part of the lattice L in the cone $S(\xi) > 0$. This is clearly a necessary condition for the sum to be convergent Our condition $n - m > q$, $p > 1$, $q \geq 1$, assures that the function f_{P_1,P_2} is continuous and in L^1, hence the sum converges absolutely and defines a holomorphic function of τ.

2.6.15. Let us consider the case $p = q = 1$. Let L be a lattice in (E,S), where $S(x_1,x_2) = x_1^2 - x_2^2$. Let $P_1 = x_1$, $P_2 = 1$. We consider, as in 2.5.33, the function

$$v_\tau(x_1,x_2) = (\text{sign } x_1)\, e^{i\pi\tau(x_1^2-x_2^2)}, \quad \text{if } x_1^2 - x_2^2 > 0$$
$$= 0 \quad\quad , \quad \text{if } x_1^2 - x_2^2 \leq 0 .$$

Let us consider the group $G = SO(1,1)$. Let G_0 be the subgroup of G leaving L stable and such that G_0 acts by the identity on L^*/L. For $h \in L^*$, the set $h + L$ is invariant under G_0. We denote by $h+L/G_0$ the set of orbits of G_0 on $h + L$. As G_0 is contained in $SO(1,1)$ the function $(\xi_1,\xi_2) \to \text{sign } \xi_1$ is constant on an orbit of G_0. We thus can form

$$\theta_S(\tau,h) = \sum_{\xi\in h+L/G_0} (\text{sign } \xi_1)\, e^{i\pi\tau(\xi_1^2-\xi_2^2)} .$$

2.6.16. <u>Theorem</u>: Let L be a lattice in (E,S) such that $S(\xi) \neq 0$ if $\xi \in L$, $\xi \neq 0$. Let N be the smallest integer such that NS^{-1} is an even integral form. Then for $\gamma = \begin{pmatrix} a & b \\ c & d \end{pmatrix}$, $a \equiv d \equiv 1 \text{ Mod } N$, $b \equiv c \equiv 0 \text{ Mod } N$, $\theta_S(\gamma\cdot\tau,h) = (c\tau+d)\, \theta_S(\tau,h)$.

Proof: Let us consider the distribution

$$(\theta_{L,h},\varphi) = \sum_{\xi\in L} \varphi(h+\xi) .$$

It follows from our study that $\theta_{L,h}$ is a semi-invariant distribution under the subgroup $\Gamma(N) = \{\gamma \equiv 1 \text{ Mod } N\}$. The group $SO(1,1)$ acts naturally on E and clearly $\theta_{L,h}$ is invariant under G_0. Let us consider the function

$$\omega_\tau(x_1,x_2) = (\mathrm{Im}\ \tau)^{1/2}\, x_1 e^{i\pi\tau x_1^2} e^{-i\pi\tau x_2^2} .$$

The function ω_τ is rapidly decreasing on E. We recall (2.5.35), (2.5.36) that

$$v_\tau(x_1,x_2) = \int_{SO(1,1)} (g\cdot\omega_\tau)(x_1,x_2)\, dg .$$

The function $g \to (\theta_{L,h},R(g)\cdot\omega_\tau)$ is invariant by left translation of G_0. Furthermore the double integral

$$\int_{G_0\backslash SO(1,1)} \Big(\sum_{\xi\in L+h} (R(g)\cdot\omega_\tau)(\xi_1,\xi_2)\Big)\, d\dot g$$

is absolutely convergent. Interchanging the order of summation, this is

$$\sum_{\xi\in L+h/G_0} \Big(\int_{SO(1,1)} (R(g)\cdot\omega_\tau)(\xi_1,\xi_2)dg\Big) = \theta_S(\tau,h) .$$

But ω_τ verifies $R_S(\sigma)\cdot\omega_\tau = (c\tau+d)^{-1}\, \omega_{\sigma\cdot\tau}$, for $\sigma \in SL(2,\mathbb{R})$, thus, for any $g \in SO(1,1)$, the function

$$(\theta_{L,h},R(g)\cdot\omega_\tau) = \theta_S(g,\tau,h)$$

verifies

$$\theta_S(g,\gamma\cdot\tau,h) = (c\tau+d)\ \theta_S(g,\tau,h)\ .$$

From the integral expression

$$\theta(\tau,h) = \int_{G_0\backslash SO(1,1)} \theta_S(g,\tau,h)\ d\dot{g}\ ,$$

we obtain our theorem.

<u>Remark</u>: Let $K = \mathbb{Q}(\sqrt{D})$ a real quadratic field. Let $\mathcal{O}$ be the ring of integers of K. We may identify $\mathcal{O}$ (or an ideal of $\mathcal{O}$) to a lattice L in $\mathbb{R}^2$ via $u \to (u,u')$. Let $S(x,y) = x^2 - Dy^2$, then $S(u,u') = N(u)$. The dual lattice is $\frac{\mathcal{O}}{\sqrt{D}}$ and is of level D, for $D \equiv 1 \text{ Mod } 4$. The group G_0 is the group U_0 of units ε of $\mathcal{O}$ such that $\varepsilon \equiv 1 \bmod \mathcal{O}\sqrt{D}$. The corresponding θ-series $\sum\limits_{\frac{u}{\sqrt{D}} \equiv \frac{\alpha}{\sqrt{D}} \bmod \mathcal{O}} (\text{sign } u)\ q^{N(u)}$ has been considered by Hecke. They also appear in character formulas for the highest weight representations of the Kac-Moody Lie algebra $\hat{s\ell}_2$, as discovered by D. Petersor and V. Kac.

2.6.17. We similarly explicit now the transformation properties of the θ-series on the Siegel upper half plane associated to a even number of variables.

We consider (V,B) a symplectic space of dimension $2n$ with the fixed self dual lattice $r = \oplus\, \mathbb{Z} P_1 \oplus \cdots \oplus \mathbb{Z} P_n \oplus \mathbb{Z} Q_1 \oplus \cdots \oplus \mathbb{Z} Q_n$, and the decomposition $V = \ell \oplus \ell'$, with $\ell = \bigoplus\limits_i \mathbb{R}P_i$, $\ell' = \bigoplus\limits_j \mathbb{R}Q_j$. We write $r = r_1 \oplus r_2$ with $r_1 = r \cap \ell$, $r_2 = r \cap \ell'$.

Let (E,S) be an orthogonal vector space, with a lattice L such that $S(L,L) \subset \mathbb{Z}$. Let L^* be its dual lattice. Let $(e_1, e_2, \cdots, e_k)$ be a $\mathbb{Z}$-basis of L, then $L^* = \mathbb{Z} e_1^* \oplus \cdots \oplus \mathbb{Z} e_k^*$

is a $\mathbb{Z}$-basis of L^*. The lattice $r_L = r_1 \otimes L^* + r_2 \otimes L$ is a self-dual lattice in $(V \otimes E, B \otimes S)$. Let χ_L be the quasi-character of r_L associated to the decomposition $r_L = r_1 \otimes L^* \oplus r_2 \otimes L$, i.e.

$$\chi_L(\exp(x_1 \otimes v^* + x_2 \otimes v)) = e^{i\pi B(x_1,x_2)S(v^*,v)}$$

for $x_1 \in r_1$, $x_2 \in r_2$, $v^* \in \ell^*$, $v \in \ell$, and let $\Gamma(r_L,\chi_L)$ be the associated θ-group in $Sp(B \otimes S)$. The basis $(e_i^* \otimes P_j, e_\ell \otimes Q_K)$ is a symplectic basis of (V,B) which is a $\mathbb{Z}$-basis of r_L over $\mathbb{Z}$. With respect to this basis, the matrix representing the image $g \otimes \mathrm{Id}$ of the transformation $g = \left(\begin{array}{c|c} A & B \\ \hline C & D \end{array}\right)$ of $Sp(B)$ is $Sp(B \otimes S)$ is the matrix

$$\left(\begin{array}{c|c} A \otimes \mathrm{id} & B \otimes S \\ \hline C \otimes S^{-1} & D \otimes \mathrm{id} \end{array}\right).$$

In particular, we have the:

2.6.18. <u>Lemma</u>:

a) Let S be even and q be the smallest integer such that qS^{-1} is integral and with even diagonal coefficients. Let $\Gamma_0^{(n)}(q) = \{\left(\begin{array}{c|c} A & B \\ \hline C & D \end{array}\right) ; C \in qM_n(\mathbb{Z})\}$. Then if $\gamma \in \Gamma_0^{(n)}(q)$, $\gamma \otimes \mathrm{id} \in \Gamma(r_L,\chi_L)$.

b) Let S be arbitrary, let n_S be the smallest integer such that $n_S S^{-1}$ has integral coefficients. Let

$$\Gamma_0^{(n)}(2n_S,2) = \{\left(\begin{array}{c|c} A & B \\ \hline C & D \end{array}\right); C \in 2M_n(\mathbb{Z}), D \in 2n_S M_n(\mathbb{Z})\}.$$

then if $\gamma \in \Gamma_0^{(n)}(2n_S,2)$, $\gamma \otimes \mathrm{id} \in \Gamma(r_L,\chi_L)$.

2.6.16. We now suppose k even. We will thus explicit the multiplicator of the θ-function on the congruence subgroup described in the lemma.

In the case where k is even and S is of signature (p,q) we have seen that the projective representation R_S is equivalent to a true representation of $Sp(n,\mathbb{R})$. In fact, if we define $R'(g) = s(g)^{-(p-q/2)} R_S(g)$, then $R'(g)$ is a true representation of $G = Sp(n,\mathbb{R})$.

We consider θ_{r_L} the intertwing operator between $H(\ell \otimes E) = L^2(\ell' \otimes E)$ and $H(r_L,\chi_L)$, in particular $\theta_{r_L}\varphi$ is a function on $V \otimes E$ such that

$$(\theta_{r_L}\varphi)(0) = \sum_{\xi \in r_2 \otimes L} \varphi(\xi) .$$

Let us consider S even and let $\gamma \in \Gamma_0^{(n)}(q)$. Then the image $\gamma \otimes \text{id}$ of γ belongs to $\Gamma_0(r_L,\chi_L)$. Thus the operator $A_{r_L}(\gamma \times \text{id})$ operates on $H(r_L,\chi_L)$ naturally, by conjugation.

2.6.19. Let L be an even lattice in (E,S) with $\dim E = 2k$. Let $(\det S)$ be the determinant of S over a $\mathbb{Z}$-basis of E and q the smallest integer such that qS^{-1} is even integral. The function on $\mathbb{Z}$ defined by $\chi_S(n) = \left(\frac{(-1)^{k/2} \det S}{n}\right)$ for n odd defines a Dirichlet character mod q.

We now prove the:

2.6.20. Theorem: Let L be an even lattice in the orthogonal vector space (E,S) where $\dim E = k$ is even. Then

$$R_S^!(\gamma)\theta_{r_L} = \chi(\gamma)\, A_{r_L}(\gamma)\, \theta_{r_L} ,$$

with $\chi(\nu) = \chi_S(\det D)$, for $\nu = \left(\frac{A|B}{C|D}\right) \in \Gamma_0^{(n)}(q)$.

Proof: We know that $R_S(\gamma)\theta_{r_L} = b(\gamma \otimes \mathrm{id})\, A_{r_L}(\gamma)\theta_{r_L}$ with $b(\gamma \times \mathrm{id})$ given in (2.2.26). Thus we obtain

$$R_S^!(\nu)\theta_{r_L} = s(\gamma)^{((p-q)/2)}\, b(\gamma \otimes \mathrm{id})\, A_{r_L}(\nu)\, \theta_{r_L}$$

$$= u(\nu)\, A_{r_L}(\nu)\, \theta_{r_L} .$$

As $R_S^!(\nu)$ and $A_{r_L}(\gamma)$ are true representations of $\Gamma_0^{(n)}(q)$, $u(\gamma)$ is a character of $\Gamma_0^{(n)}(q)$. We thus have to verify that $u(\gamma)$ coincides with $\chi(\gamma)$ on a set of generators of $\Gamma_0^{(n)}(q)$. Let

$$\gamma = \begin{pmatrix} a & b \\ c & d \end{pmatrix} \in \Gamma_0^{(1)}(q) \subset SL(2,\mathbb{Z}) \quad (\text{i.e. } c \in q\mathbb{Z}).$$

We consider $\tilde{\gamma}_i$ the element of $Sp(V)$ which operates by $\begin{pmatrix} a & b \\ c & d \end{pmatrix}$ on $(\mathbb{R}P_i \oplus \mathbb{R}Q_i)$ and by id on $\sum_{j\neq i} (\mathbb{R}P_j \oplus \mathbb{R}Q_j)$. It then follows from the calculation of $b(\gamma \times \mathrm{id})$ in one-dimension, that $u(\tilde{\gamma}_i) = \chi(\tilde{\gamma}_i)$.

Let us consider now the following elements of $\Gamma_0(q)$:

$$g(A) = \left(\frac{A \mid 0}{0 \mid {}^tA^{-1}}\right), \text{ with } A \in SL(n,\mathbb{Z})$$

$$u(X) = \left(\frac{1 \mid X}{0 \mid 1}\right), \text{ with } X = {}^tX \in M_{n,n}(\mathbb{Z})$$

$$g'(T) = \left(\frac{1 \mid 0}{qT \mid 1}\right), \text{ with } T = {}^tT \in M_{n,n}(\mathbb{Z}) .$$

It follows from [8], that the elements $\{\{\widehat{\gamma_i}\}, g(A), u(x), g'(T)\}$

generates $\Gamma_0^{(n)}(q)$. To complete the proof of the Theorem, it is then sufficient to verify that $u(\gamma) = 1$ for $\gamma = g(A)$, $u(X)$ or $g'(T)$. For $\gamma = g(A)$ or $u(X)$ this is clear, as $b(\gamma \otimes \mathrm{id}) = s(\gamma) = 1$.

Now let $g'(T) \otimes \mathrm{id} = \left(\begin{array}{c|c} 1 & 0 \\ \hline qT\otimes S^{-1} & 1 \end{array}\right)$. Then

$$b(g'(T) \otimes \mathrm{id}) = b(1, qT \otimes S^{-1}) = b(qT \otimes S^{-1}, 1)^{-1} e^{\frac{i\pi}{4}(\mathrm{sign}\ T)(\mathrm{sign}\ S)}$$

by (2.2.29)

$s(g'(T)) = i^n \,\mathrm{sign}\,(\det T)$ if T is invertible.

Hence we have to verify that:

$$e^{-\frac{i\pi}{4}(\mathrm{sign}\ T)(p-q)} \cdot \left(e^{\frac{i\pi}{2}n}\ \mathrm{sign}\ (\det T)\right)^{\frac{p-q}{2}} = 1 .$$

If $\mathrm{sign}\ T = a-b$, then $\mathrm{sign}\ (\det T) = (-1)^b$ and $a+b = n$ (if T is invertible). Then we have

$$e^{-\frac{i\pi}{4}(\mathrm{sign}\ T)(p-q)} = e^{-\frac{i\pi}{4}(n-2b)(p-q)} = e^{-\frac{i\pi}{2}n(\frac{p-q}{2})}\, e^{i\pi b(\frac{p-q}{2})}$$

and the equality is satisfied.

2.6.21. We now give some standard applications of the Theorem 2.6.20 together with the Theorem 2.5.41 to transformation properties of θ-functions.

Let S be a positive definite quadratic form on E. As in 2.5.36, we write $V = W \oplus W^*$ and realize the representation R_S in $L^2(\mathrm{Hom}(W,E))$. Let $\mathbb{Z} e_1 \oplus \cdots \oplus \mathbb{Z} e_k$ a $\mathbb{Z}$-basis of L in E. Then the lattice $r_2 \otimes L$ in $W^* \otimes E$ is identified with the lattice $M_{n,k}(\mathbb{Z})$ of $n \times k$ matrices with integral coefficients.

We consider $r_L = r_1 \otimes L^* + r_2 \otimes L$. Let $N(B \otimes S)$ be the Heisenberg group associated to $(V \otimes E, B \otimes S)$. We identify elements of $\mathrm{Hom}(W \otimes E) = W^* \otimes E \subset V \otimes E$ to elements of $N(B \otimes S)$ by the exponential map. The operator θ_{r_L, χ_L} from $L^2(\mathrm{Hom}(W,E))$ to $H(r_L, \chi_L)$ has the form:

$$(\theta_{r_L, \chi_L} f)(n) = \sum_{X \in M_{n,k}(\mathbb{Z})} f(n \exp X) \; ; \quad n \in N(B \otimes S).$$

Let S be even on the lattice L.

We define $\theta_S(Z) = \sum_{X \in M_{n,k}(\mathbb{Z})} e^{i\pi \mathrm{Tr}(Z\,{}^t XSX)}$.

2.6.22. Theorem:

a) Let k be even and q the smallest integer such that qS^{-1} is integral and with even diagonal coefficients, then for every $g = \left(\frac{A|B}{C|D}\right) \in \Gamma_0^{(n)}(q)$,

$$\theta_S((AZ + B)(CZ + D)^{-1}) = \chi_S(\det D)(\det(CZ + D))^{k/2}\, \theta_S(Z).$$

b) Let $n = k$ be even, Q an integer and χ a Dirichlet character mod Q.

We define, for $\nu = 0$ or 1,

$$\theta_S^{\nu,\chi}(Z) = \sum_{X \in M_{n,n}(\mathbb{Z})} \chi(\det X)(\det X)^{\nu} e^{i\pi \mathrm{Tr}(Z\,{}^t XSX)} .$$

Then, for every $g = \left(\frac{A|B}{C|D}\right) \in \Gamma_0^{(n)}(qQ^2)$,

$$\theta_S^{\nu,\chi}((AZ+B)(CZ+D)^{-1}) = \chi(\det D)(\det(CZ+D))^{k/2+\nu}\, \theta_S^{\nu,\chi}(Z) .$$

Proof:

a) As usual, we write $\theta_S(Z) = (\theta_{r_L, \chi_L} . v_Z^S)(0)$, where v_Z^S

is given in 2.5.32. Then a) follows immediately from 2.6.20 and 2.5.38.

b) The function $M \to \chi(\det M)$ is constant on the cosets $M + QX$ of $M_{n,n}(\mathbb{Z})/QM_{n,n}(\mathbb{Z})$.

Let us thus introduce the lattice QL. As $(QL^*) = \frac{1}{Q} L^*$ its level is qQ^2. Applying the Theorem 2.6.18 to the lattice QL, we have that for $\gamma \in \Gamma_0^{(n)}(qQ^2)$,

$$R_S'(\gamma)\theta_{r_{QL}} = \chi_S(\det D)\, A_{r_{QL}}(\gamma)\cdot\theta_{r_{QL}} .$$

But, now for $f_{\nu,Z}$ given in 2.5.42:

$$\theta_S^{\nu,\chi}(Z) = \sum_{M\in M_{n,n}(\mathbb{Z})/QM_{n,n}(\mathbb{Z})} \chi(\det M)(\theta_{r_{QL}}\cdot f_{\nu,Z})(\exp M).$$

Thus for $g \in \Gamma_0^{(n)}(qQ^2)$:

$$\theta_S^{\nu,\chi}(g\cdot Z) = \sum_{M\in M_{n,n}(\mathbb{Z})/QM_{n,n}(\mathbb{Z})} \chi(\det M)(\theta_{r_{QL}}\cdot f_{\nu,g\cdot Z})(\exp M)$$

$$= (\det(CZ+D))^{k/2+\nu} \sum_{M\in M_{n,n}(\mathbb{Z})/QM_{n,n}(\mathbb{Z})} \chi(\det M)(\theta_{r_{QL}}\cdot R_S(g)\cdot f_{\nu,Z})(\exp M)$$

by 2.5.42

$$= \chi_S(\det D)(\det(CZ+D))^{k/2+\nu} \sum_{M\in M_{n,n}(\mathbb{Z})/QM_{n,n}(\mathbb{Z})} \chi(\det M)(A_{r_{QL}}(g)\cdot\theta_{r_{QL}} f_{\nu,Z})(\exp$$

by 2.6.20

Now, if $M \in M_{n,n}(\mathbb{Z})$ and $g = \left(\frac{A|B}{C|D}\right) \in \Gamma_0(qQ^2)$ it is easy to see that $g^{-1}\cdot M = MA + u$, with $u \in r_{QL} = r_2 \otimes L^*/Q + r_1 \otimes QL$, and $(B \otimes S)(MA,u) \in 2\mathbb{Z}$. Thus, for $\varphi \in H(r_{QL},\chi_{QL})$, $\varphi(g^{-1}M) = \varphi(MA)$. Thus b) follows by changing in the last equality M in MA and remarking that, as $(\det A)(\det D) \equiv 1 \mod Q$, $\chi(\det A)^{-1} = \chi(\det D)$.

2.7. The Shimura correspondence.

2.7.1. Let us consider the vector space E of real 2×2 symmetric matrices x with the quadratic form $S = -2 \det x$.

If

$$x = \begin{pmatrix} x_1 & x_3 \\ x_3 & x_2 \end{pmatrix} \quad \text{and} \quad y = \begin{pmatrix} y_1 & y_3 \\ y_3 & y_2 \end{pmatrix},$$

the associated bilinear form $S(x,y)$ is given by $S(x,y) = 2x_3y_3 - x_1y_2 - x_2y_1$. If

$$x = \begin{pmatrix} v_1+u_1 & u_2 \\ u_2 & v_1-u_1 \end{pmatrix}, \quad S(x) = 2(u_1^2 + u_2^2 - v_1^2).$$

In particular S is of signature $(2,1)$.

The group $SL(2,\mathbb{R})$ acts on E by $g\cdot x = gx{}^tg$. This action leaves $S(x)$ stable. Hence we obtain a map from $SL(2,\mathbb{R})$ to $O(2,1)$. It is easy to see that this map is surjective on the connected component of $O(2,1)$ and that its kernel consists of $\begin{pmatrix} \pm 1 & 0 \\ 0 & \pm 1 \end{pmatrix}$. The group $SL(2,\mathbb{R})/\{\pm 1\}$ is denoted by $PSL_2(\mathbb{R})$.

Let us consider the symplectic vector space $(\mathbb{R}P \oplus \mathbb{R}Q) \otimes E$ and the imbedding of $\mathcal{O} \times O(2,1)$ into $\widetilde{Sp}(B \otimes S)$ defined in 2.5. The corresponding representation $\widetilde{R}_S$ gives us a representation of $G_2 \times O(2,1)$, where G_2 is the two fold covering of $SL(2,\mathbb{R})$. The formula for the action of $\mathcal{O}$ on $L^2(E) \simeq L^2(E \otimes \mathbb{R}Q)$ are given in 2.5.8. The action of $O(2,1)$ on $L^2(E)$ is simply given by $(g\cdot f)(x) = f(g^{-1}x{}^tg^{-1})$.

As in 2.5.24, we consider $D = \{v \in E^{\mathbb{C}};\ S(v,v) = 0,\ S(v,\bar{v}) > 0\}$
A basepoint v_0 of D is

$$v_0 = \begin{pmatrix} -1 & i \\ i & 1 \end{pmatrix}.$$

Let us consider, for $z \in P^+$, the unique element

$$b(z) = \begin{pmatrix} y^{1/2} & xy^{-1/2} \\ 0 & y^{-1/2} \end{pmatrix}$$

of B_0 such that $b(z)\cdot i = z$. We consider the action of $SL(2,\mathbb{R})$ on $E^{\mathbb{C}}$ given by $g\cdot x = gx\,{}^tg$. For this action, we have the:

2.7.2. Lemma:

a) $u(\theta)\cdot v_0 = e^{2i\theta}v_0$, for $u(\theta) = \begin{pmatrix} \cos\theta & -\sin\theta \\ \sin\theta & \cos\theta \end{pmatrix}$.

b) $J^-\cdot v_0 = 0$

c) $b(z)\cdot v_0 = y^{-1}\begin{pmatrix} z^2 & z \\ z & 1 \end{pmatrix}$, for $z \in P^+$.

Proof: a) follows from direct computation of

$$u(\theta)\cdot v_0 = \begin{pmatrix} \cos\theta & -\sin\theta \\ \sin\theta & \cos\theta \end{pmatrix}\begin{pmatrix} -1 & i \\ i & 1 \end{pmatrix}\begin{pmatrix} \cos\theta & \sin\theta \\ -\sin\theta & \cos\theta \end{pmatrix}.$$

b) For $X \in \mathfrak{g}$, $X\cdot v_0 = \frac{d}{d\varepsilon}(\exp \varepsilon X\cdot v_0)_{|\varepsilon=0}$

$$= \frac{d}{d\varepsilon}(\exp \varepsilon X\, v_0 + v_0\, {}^t(\exp \varepsilon X))_{|\varepsilon=0}$$

$$= Xv_0 + v_0\,{}^tX.$$

Thus, as $J^- = \frac{1}{2}\begin{pmatrix} i & 1 \\ 1 & -i \end{pmatrix}$, $J^- v_0 + v_0\,{}^t(J^-) =$

$$\frac{1}{2}\left(\begin{pmatrix} i & 1 \\ 1 & -i \end{pmatrix}\begin{pmatrix} -1 & i \\ i & 1 \end{pmatrix} + \begin{pmatrix} -1 & i \\ i & 1 \end{pmatrix}\begin{pmatrix} i & 1 \\ 1 & -i \end{pmatrix}\right) = 0$$

c) follows from the computation

$$b(z)\cdot v_0 = \begin{pmatrix} y^{1/2} & xy^{-1/2} \\ 0 & y^{-1/2} \end{pmatrix}\begin{pmatrix} -1 & i \\ i & 1 \end{pmatrix}\begin{pmatrix} y^{1/2} & 0 \\ xy^{-1/2} & y^{-1/2} \end{pmatrix}.$$

As in 2.5.31, we consider the function:

$$(\psi^n_{v_0})(x) = S(x,v_0)^{-n}\, S(x)^{n-1/2}\, e^{-\pi S(x)} \quad \text{on} \quad S(x) > 0$$

$$= 0 \qquad \text{on} \quad S(x) \leq 0\,.$$

This function is continuous and in $L^1(E) \cap L^2(E)$, if $n > 1$.

2.7.3. <u>Proposition</u>: Let $n > 1$, the function $\psi^n_{v_0}$ is a lowest weight vector of weight $(n + (1/2), 2n)$ for the action of $\widetilde{G} \times PSL_2(\mathbb{R})$.

<u>Proof</u>: The fact that $\psi^n_{v_0}$ is a lowest weight vector of weight $n + 1/2$ for the first factor $\widetilde{G}$ has already been established in 2.5.31. Let us check the corresponding assertion for $PSL_2(\mathbb{R})$. The fact that $\psi^n_{v_0}$ is an eigenvector for the action of $u(\theta)$ follows from 2.7.2. As the action of the Lie algebra of $PSL_2(\mathbb{R})$ is given by linear vector fields, it follows from 2.7.2. b) that $J^-\cdot\psi^n_{v_0} = 0$, as $J^-\cdot S(v_0,x) = 0$, and the other factors are invariant under the full group $O(2,1)$.

From the Remark 2.5.25, we obtain:

2.7.4. <u>Theorem</u>: Let $n \geq 1$. The representation $\mathbb{T}_{n+(1/2)} \otimes \mathbb{T}_{2n}$ of $G_2 \times PSL_2(\mathbb{R})$ is contained as a discrete subspace in $L^2(E)$ with multiplicity one. The vector $\psi^n_{v_0}$ is the lowest weight vector of this representation.

Let $\tau = \alpha + i\beta$, $z = x + iy$ be given points in $P^+ \times P^+$ and $Q(z) = \begin{pmatrix} z^2 & z \\ z & 1 \end{pmatrix}$, then we have the formula: $(n > 1)$.

2.7.5. $$(\widetilde{R}_S(b(\tau) \times b(z)) \cdot \psi^n_{v_0})(x)$$

$$= \beta^{(n+(1/2))/2}\, y^n\, S(x,Q(z))^{-n}\, S(x)^{n-(1/2)}\, e^{i\pi S(x)\tau}$$

$$\text{on } S(x) > 0$$

$$= 0 \quad \text{on} \quad S(x) \leq 0.$$

2.7.6. For $g = \begin{pmatrix} a & b \\ c & d \end{pmatrix}$, and $g \cdot z = \dfrac{az + b}{cz + d}$,

$$Q(g \cdot z) = (cz + d)^{-2} g\, Q(z)\, {}^t g.$$

This formula follows from direct computations.

2.7.7. As shown in 2.3.3 the fact that $\psi^n_{v_0}$ is a vector of weight $(n + (1/2), 2n)$ for $\widetilde{R}_S$ can be translated as follows: For $(\tau, z) \in P^+ \times P^+$, we denote by $\psi^n(\tau, z)$ the function in $L^2(E)$ given by

$$\psi^n(\tau,z)(x) = S(x,Q(z))^{-n}\, S(x)^{n-(1/2)}\, e^{i\pi S(x)\tau}, \text{ on } S(x) > 0$$

$$= 0 \quad \text{on} \quad S(x) \leq 0,$$

then, for $(\sigma, g) \in G_2 \times PSL_2(\mathbb{R})$,

$$\widetilde{R}_S(\sigma,g)\cdot\psi^n(\tau,z) = j(\sigma,\tau)^{-(n+(1/2))}\, j(g,z)^{-2n}\, \psi^n(\sigma\cdot\tau, g\cdot z)$$

where $j((\begin{smallmatrix} a & b \\ c & d \end{smallmatrix}),z) = cz + d$, and more generally, for $\sigma = ((\begin{smallmatrix} a & b \\ c & d \end{smallmatrix}),\varphi) \in \mathcal{O}$ (i.e. $e^{\varphi(z)} = cz + d$),

$$j(\sigma,\tau)^{\alpha} = e^{\alpha\varphi(\tau)} .$$

Remark: For $g \in PSL_2(\mathbb{R}) \subset O(2,1)$, this formula is immediately derived from 2.7.6.

For $\sigma \in G_2$, this is a deeper property.

As $n > 1$ will be fixed in the following, we will often suppress the index n and write $\psi(\tau,z)$ instead of $\psi^n(\tau,z)$.

2.7.8. We will now construct semi-invariant distributions associated to the lattice $L = \{(\begin{smallmatrix} x_1 & x_3 \\ x_3 & x_2 \end{smallmatrix});\ x_i \in \mathbb{Z}\}$. Let us consider the orthogonal decomposition $E = E_{12} \oplus E_3$ of our space E, with $E_{12} = \{(\begin{smallmatrix} x_1 & 0 \\ 0 & x_2 \end{smallmatrix})\}$ and $E_3 = \{(\begin{smallmatrix} 0 & x_3 \\ x_3 & 0 \end{smallmatrix})\}$. We write

$$e_1 = (\begin{smallmatrix} 1 & 0 \\ 0 & 0 \end{smallmatrix}),\quad e_2 = (\begin{smallmatrix} 0 & 0 \\ 0 & 1 \end{smallmatrix}) \quad\text{and}\quad e_3 = (\begin{smallmatrix} 0 & 1 \\ 1 & 0 \end{smallmatrix}).$$

With respect to the decomposition $L^2(E) = L^2(E_{12}) \otimes L^2(E_3)$, our representation R_S is written as $R_{12} \otimes R_3$.

2.7.9. We restrict first our attention to R_{12}. As the restriction S_{12} of S to E_{12} is of signature $(1,1)$, $R_{12} = R_{S_{12}}$ is a true representation of $SL(2,\mathbb{R})$ equivalent to the natural representation U (2.5.5).

2.7.10. Let us consider the lattice $\mathbb{Z}e_1 \oplus \mathbb{Z}e_2$ in E_{12} and the δ-distributions $(\delta_{x_1,x_2},\varphi) = \varphi(x_1e_1 + x_2e_2)$. Let ψ be a character mod N and u a function on $\mathbb{Z}/N\mathbb{Z}$ such that $u(a\cdot j) = \psi(a)u(j)$, for a invertible mod $\mathbb{Z}$. We define

$$(\delta_u,\varphi) = \sum_{x_1,x_2\in\mathbb{Z}} u(x_1)\delta_{x_1,x_2}.$$

We consider δ_u as a distribution on $H(E_{12} \otimes \mathbb{R}P)$ via

$$(\delta_u,\varphi) = \sum_{x_1,x_2\in\mathbb{Z}} u(x_1)\ \varphi(\exp((x_1e_1 + x_2e_2) \otimes Q)).$$

Let $\Gamma_0(N) = \{\begin{pmatrix} a & b \\ c & d\end{pmatrix};\ c \equiv 0 \bmod N\}$. We still denote by ψ the character

$$\gamma = \begin{pmatrix} a & b \\ c & d\end{pmatrix} \to \psi(d)$$

of $\Gamma_0(N)$.

2.7.11. <u>Proposition:</u> For $\gamma \in \Gamma_0(N)$, $R_{12}(\gamma)\cdot\delta_u = \psi(\gamma)^{-1}\delta_u$.

<u>Proof:</u> We could use the results of Section 2.6. However we will give an alternate description of δ_u which will be fundamental for our applications. Let us consider the Lagrangian subspace $\ell = E_{12} \otimes \mathbb{R}P$ and $\ell_1 = \mathbb{R}e_1 \otimes (\mathbb{R}P \oplus \mathbb{R}Q)$. As seen in 2.5.5, the representation R_{12} of $SL(2,\mathbb{R})$ in $H(\ell) \simeq L^2(E_{12} \otimes Q)$ is equivalent to the natural representation U of $SL(2,\mathbb{R})$ in $L^2(\mathbb{R}P \oplus \mathbb{R}Q) = L^2(\mathbb{R}P \oplus \mathbb{R}Q) \otimes e_2) \simeq H(\ell_1)$ via the operator:

$$\begin{aligned}(\mathcal{F}_{\ell_1,\ell}\varphi)(\exp(xP + yQ) \otimes e_2) &= \int \varphi(\exp(xP + yQ) \otimes e_2 \exp te_1 \otimes Q)\ dt \\ &= \int \varphi(\exp(te_1 + ye_2) \otimes Q)e^{2i\pi tx}\ dt.\end{aligned}$$

i.e. $\mathcal{F}_{\ell_1,\ell}$ is the Partial Fourier Transform with respect to the variable t.

We write, as u is periodic mod N,

$$(\delta_u,\varphi) = \sum_{h\in\mathbb{Z}/N\mathbb{Z}} u(h) \sum_{m,n\in\mathbb{Z}} \varphi(h + mN,n).$$

Thus, applying Poisson summation formula in the first variable, we obtain:

$$(\delta_u,\varphi) = \frac{1}{N} \sum_{h\in\mathbb{Z}/N\mathbb{Z}} u(h) \sum_{m,n\in\mathbb{Z}} (\bar{\mathcal{F}}_{\ell_1,\ell}\varphi)((\frac{mP}{N} + nQ) \otimes e_2)\, e^{-\frac{2i\pi mh}{N}}$$

2.7.12. We define, for u a function on $\mathbb{Z}/N\mathbb{Z}$, the Fourier transform of u by

$$\hat{u}(m) = \sum_{h\in\mathbb{Z}/N\mathbb{Z}} u(h)\, e^{-\frac{2i\pi mh}{N}} .$$

It is clear that, if u satisfies $u(ah) = \psi(a)u(h)$ for ψ a character mod N, a invertible mod N, then $\hat{u}$ satisfies:

$$\hat{u}(am) = \psi(a)^{-1}\hat{u}(m) .$$

Using this definition, we thus have the:

2.7.13. <u>Formula:</u>

$$(\delta_u,\varphi) = \frac{1}{N} \sum_{m,n\in\mathbb{Z}} \hat{u}(m)(\bar{\mathcal{F}}_{\ell_1,\ell}\varphi)(\frac{mP + nNQ}{N} \otimes e_2) .$$

Let us define the distribution $\delta'_{\hat{u}}$ on the space $\mathbb{R}P \oplus \mathbb{R}Q$ by

$$(\delta'_{\hat{u}},\varphi) = \frac{1}{N} \sum_{m,n} \hat{u}(m)\, \varphi(\frac{mP + nNQ}{N}) .$$

As we have

$$\begin{pmatrix} a & b \\ Nc & d \end{pmatrix}\begin{pmatrix} m \\ nN \end{pmatrix} = \begin{pmatrix} am+nbN \\ N(cm+dn) \end{pmatrix},$$

we have $U(\gamma)\cdot\delta'_{\hat{u}} = \psi(\gamma)^{-1}\delta'_{\hat{u}}$. As δ_u is transformed to $\delta'_{\hat{u}}$ by $\mathcal{F}_{\ell_1,\ell}$, we get our proposition.

2.7.14. Let us consider the space E_3 and the representation R_3 of $\mathcal{G}$ in $L^2(E_3)$ associated to the quadratic form $S_3(x_3) = 2x_3^2$. Let $(\theta,\varphi) = \sum_{x_3\in\mathbb{Z}} \varphi(x_3)$ the θ distribution on E_3. We denote by $\widetilde{\Gamma}_0(4)$ the reciproc image of $\Gamma_0(4)$ in $\mathcal{G}$. The group $\Gamma_0(4)$ is conjugated to $\Gamma_0(2,2)$ by the element

$$g(\sqrt{2}) = \begin{pmatrix} \sqrt{2} & 0 \\ 0 & \frac{1}{\sqrt{2}} \end{pmatrix}.$$

Thus it follows from the Theorem 2.4.9 and 2.4.15 that $\widehat{R}(\sigma)\cdot\theta = \lambda(\sigma)^{-1}\cdot\theta$, where, for $\gamma\in\Gamma_0(4)$

$$\gamma = \begin{pmatrix} a & b \\ c & d \end{pmatrix} \ (c \equiv 0 \bmod 4),\ \lambda(\widetilde{\gamma}) = \varepsilon_d^{-1}\left(\frac{c}{d}\right).$$

Let ψ be a character mod $4N$ and u a function on $\mathbb{Z}/4N\mathbb{Z}$ satisfying $u(aj) = \psi(a)u(j)$. Let us consider our space

$$E = \{\begin{pmatrix} x_1 & x_3 \\ x_3 & x_2 \end{pmatrix}\}$$

and the distribution on the space E given by:

$$(\theta_u,\varphi) = \sum_{x_1,x_2,x_3\in\mathbb{Z}} u(x_1)\ \varphi(x_1,x_2,x_3).$$

We denote by $\widetilde{\Gamma}_0(4N)$ the inverse image of $\Gamma_0(4N)$ in $\mathcal{G}$. On $\widetilde{\Gamma}_0(4N)$ we consider the character $\lambda\psi$ where λ is the special character of $\widetilde{\Gamma}_0(4)$ associated to θ and ψ is just given by $\psi\begin{pmatrix} a & b \\ c & d \end{pmatrix} = \psi(d)$.

Let us consider the subgroup $\Gamma_0(0,2N) = \{\begin{pmatrix} a & b \\ c & d \end{pmatrix},\ b \equiv 0 \text{ Mod } 2N\}$ The map $\begin{pmatrix} a & b \\ c & d \end{pmatrix} \to \psi(d)^2$ defines a character ψ^2 of

$$\overline{\Gamma}_0(0,2N) = \Gamma_0(0,2N)/\{\pm 1\} .$$

We consider the representation $\widetilde{R}_S$ of $G_2 \times PSL_2(\mathbb{R})$.

We now state:

2.7.15. <u>Proposition</u>:

a) $\widetilde{R}_S(\sigma)\theta_u = \lambda(\sigma)^{-1}\ \psi(\sigma)^{-1}\ \theta_u$, for $\sigma \in \widetilde{\Gamma}_0(4N) \subset G_2$.

b) $\widetilde{R}_S(\gamma)\cdot\theta_u = \psi(\gamma)^2\ \theta_u$, for $\gamma \in \overline{\Gamma_0}(0,2N) \subset PSL_2(\mathbb{R})$.

<u>Proof</u>: a) follows immediately from the Proposition 2.7.11 and the fact that $\widetilde{R} \simeq R_{12} \otimes \widetilde{R}_3$ as a representation of G_2.

b) We have:

$$\begin{pmatrix} a & b \\ c & d \end{pmatrix}\begin{pmatrix} x_1 & x_3 \\ x_3 & x_2 \end{pmatrix}\begin{pmatrix} a & c \\ b & d \end{pmatrix} = \begin{pmatrix} x_1' & x_3' \\ x_3' & x_2' \end{pmatrix},$$

with $x_1' = a^2x_1 + 2abx_3 + b^2x_2$, thus $x_1' \equiv a^2x_1 \bmod 4N$ proving b).

2.7.16. Let $n > 1$ and let us consider the function $\psi_{v_0}^n$ which is in $L^1(E) \cap L^2(E)$, continuous and supported on $S(x) > 0$. Thus the coefficient $(\theta_u, \widetilde{R}(\sigma,g)\cdot\psi_{v_0}^n) = r_n(\sigma,g) =$ $\sum_{x_1,x_2,x_3\in\mathbb{Z}} u(x_1)(\widetilde{R}(\sigma,g)\cdot\psi_{v_0}^n)(x)$ is given by an absolutely convergent serie. It follows from 2.7.15, 2.7.7 that $r_n(\sigma,g)$ is an automorphic form with respect to both variables (σ,g) of weight $(n + \frac{1}{2}, 2n)$. The corresponding holomorphic function

$r_n(\tau,z)$ on $P^+ \times P^+$ is given by:

$$\sum_{\substack{x_1,x_2,x_3\in\mathbb{Z}\\ x_3^2>x_1x_2}} u(x_1)(2x_3z-x_1-x_2z^2)^{-n}(2(x_3^2-x_1x_2))^{n-(1/2)}e^{2i\pi(x_3^2-x_1x_2)\tau}$$

Hence we get:

2.7.17. <u>Theorem</u>: Let ψ be a character mod $4N$, u a function on $\mathbb{Z}/N\mathbb{Z}$ satisfying $u(aj) = \psi(a)u(j)$. Let $n > 1$, the function:

$$\Omega_u(\tau,z) = \sum_{\substack{x_1,x_2,x_3\in\mathbb{Z}\\ x_3^2>x_1x_2}} u(x_1)(2x_3z-x_1-x_2z^2)^{-n}(x_3^2-x_1x_2)^{n-(1/2)}e^{2i\pi(x_3^2-x_1x_2)\tau}$$

is a holomorphic function of (τ,z), which is:

- modular in τ with respect to $\widehat{\Gamma}_0(4N)$, with character $\lambda\psi$, of weight $n+(1/2)$
- modular in z, with respect to $\Gamma_0(0,2N)$, with character ψ^{-2}, of weight $2n$.

Let

$$\omega_{q,u}(z) = \sum_{\substack{x_1,x_2,x_3\in\mathbb{Z}\\ x_3^2-x_1x_2=q}} u(x_1)(2x_3z-x_1-x_2z^2)^{-n}$$

then we can reexpress $\Omega_u(\tau,z)$ as

2.7.18. $$\Omega_u(\tau,z) = \sum_{q=1}^{\infty} \omega_{q,u}(z)(q)^{n-(1/2)}e^{2i\pi q\tau}.$$

2.7.19. Let $n > 1$ and ψ a character mod $4N$. We denote by $S_{n+1/2}(\Gamma_0(4N),\lambda\psi)$ the space of holomorphic functions on P^+ satisfying:

a) $f(\frac{a\tau+b}{c\tau+d}) = \lambda(\tilde{\gamma})\psi(\gamma)(c\tau+d)^{n+(1/2)}f(\tau)$, for every

$$\gamma = \begin{pmatrix} a & b \\ c & d \end{pmatrix} \in \Gamma_0(4N).$$

b) $f(\tau) = \sum_{m>0} a_m e^{2i\pi m\tau}$.

Let $\mathcal{F}$ be a fundamental domain for $\Gamma_0(4N)\backslash P^+$. We can then form for $f \in S_{n+(1/2)}(\Gamma_0(4N),\lambda\psi)$ the Petersson inner product

$$F_u(z) = \int_{\mathcal{F}} \Omega_u^n(z,\tau)\overline{f(\tau)}(\operatorname{Im}\tau)^{n+(1/2)-2}\,|d\tau d\bar{\tau}|.$$

The resulting function is an automorphic form of weight $2n$ under $\Gamma_0(0,N/2)$ and character ψ^2.

We will now consider some special function u_0 and prove that the map $f \to F_{u_0}$ gives the Shimura correspondence.

2.7.20. We recall that if $n > 2$, we can define the Poincaré series associated to the cusp $P = \infty$, by

$$G^q(\tau) = \frac{1}{2} \sum_{\sigma\in\widetilde{\Gamma}_0(4N)/\widetilde{\Gamma}_\infty} \lambda(\sigma)\,\psi(\sigma)\,j(\sigma^{-1},\tau)^{-(n+(1/2))}e^{2i\pi q(\sigma^{-1}\cdot\tau)}$$

where $\Gamma_\infty = \begin{pmatrix} 1 & \mathbb{Z} \\ 0 & 1 \end{pmatrix}$, and $\widetilde{\Gamma}_\infty$ is the reciproc image of Γ_∞ in $\widetilde{G}$.

2.7.21. Let us denote by ψ, for ψ a character mod $4N$, the function on $\mathbb{Z}$ such that $\psi(a) = 0$ if $(a,4N) \neq 1$ and $\psi(a) = \psi(a \text{ Mod } N)$ if a is invertible mod $4N$. We will see that is u_0 is such that $\hat{u}_0 = \bar{\psi}$, the automorphic form Ω_{u_0} is naturally expressed as a sum of Poincare series $G^q(\tau)$ (associated to the cusp ∞). (If u was arbitrary, we would have to use several cusps of $\Gamma_0(4N)\backslash P^+$: a similar example will be treated in 2.8.)

2.7.22. Let ψ be an even character mod 4N. For u_0 such that $\hat{u} = \psi^{-1} = \bar{\psi}$, the distribution δ_{u_0} is transformed by the operator $\bar{\mathcal{F}}_{\ell_1,\ell}$ into the distribution

$$(\delta'_\psi, \varphi) = \frac{1}{4N} \sum_{\substack{m,n\in\mathbb{Z}\\(m,4N)=1}} \bar{\psi}(m)\, \varphi\!\left(\frac{mP + 4NnQ}{4N}\right).$$

In particular δ'_ψ is supported on the set $\{(\frac{mP + 4NnQ}{4N}\ ;\ (m,4N) = 1\}$. We analyze now the orbits of $\Gamma_0(4N)$ on this set. We denote the point $xP + yQ$ by $\binom{x}{y}$. The stabilizer of the point $\binom{1}{0}$ for the natural action of $SL(2,\mathbb{R})$ on $\mathbb{R}P \oplus \mathbb{R}Q$ is the subgroup $N = \begin{pmatrix}1 & t\\ 0 & 1\end{pmatrix}, (t \in \mathbb{R})$. As $\begin{pmatrix}a & b\\ c & d\end{pmatrix}\cdot\binom{1}{0} = \binom{a}{c}$ the orbit of $\binom{1}{0}$ under $SL(2,\mathbb{Z})$ is the subset (m,n) of integers (m,n) relatively prime. Hence the image under $\Gamma_0(4N)$ of $\binom{1}{0}$ is the subset $A_1 = \{\binom{m}{4Nn}$, with $(m,4Nn) = 1\}$. Thus $A_1 \simeq \Gamma_0(4n)/\Gamma_\infty$. For j invertible mod 4N, we consider $A_j = jA_1 = \{j(mP + 4NnQ);\ (m,4Nn) = 1\}$. We have:

2.7.23. <u>Lemma</u>:

a) $A = \bigcup_{j \text{ inv.mod } 4N} A_j$.

b) For j invertible mod 4N, the map $\gamma \to \gamma\binom{j}{0}$ defines a bijection of $\Gamma_0(4N)/\Gamma_\infty$ with A_j. We have $A_j = A_{j'}$ if and only if $j = \pm j'$.

<u>Proof:</u> Let $(mP + 4NnQ) \in A$ and let j be the greatest common divisor of $(m,4Nn)$. As $(m,4N) = 1$, j is prime to $4N$. Hence $(m,4Nn) = j(m',4Nn')$, with $(m',4Nn') = 1$. The assertion b) is immediate.

For φ a function on the space E_{12}, we thus have

$$(\delta_{u_0},\varphi) = \frac{1}{4N} \sum_{m,n\in\mathbb{Z}} \overline{\psi(m)} \left((\mathcal{F}_{\ell_1,\ell}\varphi)\left(\frac{mP+4NnQ}{4N} \otimes e_2\right)\right)$$

$$= \frac{1}{4N}\frac{1}{2} \left(\sum_{\substack{j\in\mathbb{Z} \\ (j,4N)=1 \\ \gamma\in\Gamma_0(4N)/\Gamma_\infty}} \overline{\psi(j)}\psi(\gamma)(U(\gamma)^{-1}\mathcal{F}_{\ell_1,\ell}\varphi)\left(\frac{j}{4N},0\right)\right)$$

$$= \frac{1}{4N}\frac{1}{2} \sum_{\gamma\in\Gamma_0(4N)/\Gamma_\infty} \psi(\gamma) \sum_{j\in\mathbb{Z}} \overline{\psi(j)}(\mathcal{F}_{\ell_1,\ell}(R_{12}(\gamma)^{-1}\varphi)\left(\frac{j}{4N},0\right)$$

But we can again apply the Poisson summation formula to the function $(R_{12}(\gamma)^{-1}\varphi)$. As $\mathcal{F}_{\ell_1,\ell}$ is the partial Fourier transform with respect to the first variable, we obtain:

2.7.24. $$(\delta_{u_0},\varphi) = \frac{1}{2} \sum_{\gamma\in\Gamma_0(4N)/\Gamma_\infty} \psi(\gamma) \sum_{j\in\mathbb{Z}} u(j)(R_{12}(\gamma)^{-1}\varphi)(j,0)$$

(The factor $\frac{1}{2}$ comes from the fact that $\Gamma_0(4N)\binom{j}{0} = \Gamma_0(4N)\binom{-j}{0}$.)

We now express $\Omega_{u_0}(\tau,z)$ in fuction of the Poincare series $G^q(\tau)$.

2.7.25. <u>Theorem:</u> Let us define

$$d_m^{u_0}(z) = \sum_{j\in\mathbb{Z}} u_0(j)\left(\frac{1}{2mz - j}\right)^n .$$

Then we have:

$$\Omega_{u_0}(\tau,z) = 2^{n-1/2} \sum_{m=1}^{\infty} d_m^{u_0}(z)\, m^{2n-1}\, G^{m^2}(\tau) .$$

<u>Proof</u>: We have to compute:

$$(\theta_{u_0}, \psi(\tau,z)) = \sum_{x_1,x_2,x_3\in\mathbb{Z}} u_0(x_1)\, \psi(\tau,z)(x_1,x_2,x_3).$$

Let us define

$$\psi_\theta(\tau,z)(x_1,x_2) = \sum_{x_3\in\mathbb{Z}} \psi(\tau,z)\begin{pmatrix} x_1 & x_3 \\ x_3 & x_2 \end{pmatrix},$$

as a function of (x_1,x_2). We then have to calculate: $(\delta_{u_0}, \psi_\theta(\tau,z))$. Using the formula 2.7.22, this is

$$\frac{1}{2} \sum_{\gamma\in\Gamma_0(4N)/\Gamma_\infty} \psi(\gamma) \sum_{j\in\mathbb{Z}} u(j)(R_{12}(\gamma)^{-1}.\psi_\theta(\tau,z))(j,0).$$

As R_{12} operates on the variable (x_1,x_2), we have:

$$\sum_{j\in\mathbb{Z}} u(j)\, R_{12}(\gamma)^{-1}\, \psi_\theta(\tau,z)(j,0)$$

$$= \sum_{\substack{j\in\mathbb{Z}\\ m\in\mathbb{Z}}} u(j)(R_{12}(\gamma)^{-1}.\psi(\tau,z))\begin{pmatrix} j & m \\ m & 0 \end{pmatrix}.$$

Let us write $R_S(\sigma) = R_{12}(\sigma)R_3(\sigma)$, i.e. $R_{12}(\sigma)^{-1} = R_3(\sigma)R_S(\sigma)^{-1}$. In particular, as $R_3(\sigma)$ for $\sigma \in \Gamma_0(4N)$ leaves semi-invariant the distribution θ, we have:

$$\sum_{m\in\mathbb{Z}} (R_{12}(\gamma)^{-1}.\psi(\tau,z))\begin{pmatrix} j & m \\ m & 0 \end{pmatrix}$$

$$= \sum_{m\in\mathbb{Z}} \lambda(\tilde{\gamma})(R_S(\tilde{\gamma})^{-1}.\psi(\tau,z))\begin{pmatrix} j & m \\ m & 0 \end{pmatrix}.$$

We recall (2.7.7)

$$R_S(\tilde{\gamma})^{-1}.\psi(\tau,z) = j(\tilde{\gamma}^{-1},\tau)^{-(n+(1/2))}\,\psi(\gamma^{-1}.\tau,z).$$

Now

$$\psi(\gamma^{-1}\tau,z)\begin{pmatrix} j & m \\ m & 0 \end{pmatrix} = \left(\frac{1}{2mz-j}\right)^n (2m^2)^{(n-(1/2))} e^{2i\pi m^2\gamma^{-1}\tau} \quad \text{for} \quad m \geq 1$$

$$= 0 \quad \text{for} \quad m = 0 .$$

Using now the order of summation:

$$\sum_{\gamma\in\Gamma_0(4N)/\Gamma_\infty} \quad \sum_{m,j\in\mathbb{Z}} = \sum_{m\in\mathbb{Z}} \quad \sum_{\gamma\in\Gamma_0(4N)/\Gamma_\infty} \quad \sum_{j\in\mathbb{Z}}$$

we obtain our theorem.

We now compute the development of $d_m^{u_0}(z)$ in Fourier series.

2.7.26. <u>Lemma</u>: For $z \in P^+$, we have:

$$d_m^{u_0}(z) = c(n,N) \sum_{r=1}^{\infty} r^{n-1}\,\overline{\psi(r)}\, e^{\frac{2i\pi rmz}{2N}} .$$

<u>Proof</u>: We write

$$d_m^{u_0}(z) = \sum_{h\in(\mathbb{Z}/4N\mathbb{Z})} u_0(h) \sum_{j\in\mathbb{Z}} \left(\frac{1}{2mz-(h+4Nj)}\right)^n .$$

Let us consider the function on $\mathbb{R}$ defined by:

$$x \to \left(\frac{1}{\left(\frac{2mz-h}{4N}\right) + x}\right)^n, \quad n > 1 .$$

From the integral formula

$$(\frac{1}{z})^n = c_n \int_{\xi>0} e^{2i\pi\xi z} \xi^{n-1} d\xi \quad (z \in P^+),$$

we see that, for $n > 1$, this function of x is the Fourier transform of the continuous function of ξ, supported on $\xi \geq 0$ given by

$$\xi \to \xi^{n-1} e^{\frac{2i\pi\xi(2mz-h)}{4N}} .$$

Thus, applying the Poisson summation formula, we obtain:

$$\sum_{j\in\mathbb{Z}} \left(\frac{1}{\frac{2mz-h}{4N} + j}\right)^n = c_n \sum_{r=1}^{\infty} e^{\frac{2i\pi r(2mz-h)}{4N}} r^{n-1}$$

and finally, up to a multiplicative constant depending only of n and N:

$$d_m^{u_0}(z) = \sum_{h\in\mathbb{Z}/4N\mathbb{Z}} u_0(h) \sum_{r=1}^{\infty} e^{-\frac{2i\pi rh}{4N}} r^{n-1} e^{\frac{2i\pi rmz}{2N}} .$$

As, by definition of u_0,

$$\sum_{h\in\mathbb{Z}/4N\mathbb{Z}} u_0(h) e^{-\frac{2i\pi rh}{4N}} = \overline{\psi(r)}$$

we obtain our lemma.

2.7.27. <u>Theorem</u>: Let $n > 1$, ψ a character mod $4N$,

$$f = \sum_{m>0} a(m) e^{2i\pi m\tau}$$

a function in $S_{n+(1/2)}(\Gamma_0(4N), \lambda\psi)$,

then

$$(S\ f)(z) = \sum_{\substack{r=1\\m=1}}^{\infty\ \infty} r^{n-1}\psi(r)\ a(m^2)\ e^{2i\pi rmz}$$

is an automorphic form with respect to the congruence subgroup $\Gamma_0(2N)$, with character ψ^2.

Proof: Let us compute:

$$(\overline{S}f)(z) = \int \overline{\Omega_{u_0}(\tau,z)}\ f(\tau)\ d\mu_n(\tau)\ .$$

From the characteristic property of Poincare series (2.3.23) and Theorem 2.7.25, we obtain (up to a multiplicative constant)

$$(\overline{S}f)(z) = \sum_{m=1}^{\infty} \overline{d_m^{u_0}(z)}\ a(m^2)$$

$$= \sum_{\substack{r=1\\m=1}}^{\infty\ \infty} r^{n-1}\ \psi(r)\ a(m^2)\ e^{-\frac{2i\pi mr\overline{z}}{2N}}\ .$$

From (2.7.17), $\Omega_{u_0}(\tau,z)$ is modular in z with respect to $\Gamma_0(0,2N)$ with character $\overline{\psi}^2$. Thus $\int \overline{\Omega_{u_0}(\tau,z)}\ f(\tau)\ d\mu_n(\tau) = (\overline{S}f)(z)$ is antiholomorphic in z and satisfies the relation:

$$(\overline{S}f)(\gamma\cdot p) = \psi(\gamma)^2\ \overline{J(\gamma,p)}^{2n}\quad (\overline{S}f)(p)$$

for $p \in P^+$, $\gamma \in \Gamma_0(0,2N)$. We have $(Sf)(z) = (\overline{S}f)(-2N\overline{z})$. From the fact that $(-2N)(\overline{\gamma\cdot p}) = \gamma'\cdot(-2N\overline{p})$ with

$$\gamma' = \begin{pmatrix} a & -2Nb \\ \frac{c}{-2N} & d \end{pmatrix},$$

we see that $(Sf)(z)$ is holomorphic in z, and automorphic with respect to the congruence subgroup $\Gamma_0(2N)$ with character ψ^2, Q.E.D.

2.7.28. Remark: The Dirichlet series corresponding to Sf (S like Shimura) is

$$\begin{aligned} &\Sigma\, r^{n-1}\, \psi(r)\, a(m^2)\, (mr)^{-s} \\ &= \left(\sum_{r=1}^{\infty} \psi(r)\, r^{n-1-s}\right)\cdot\left(\sum_{m=1}^{\infty} a(m^2) m^{-s}\right) \end{aligned}$$

Shimura proved originally Theorem 2.7.25 using Weil characterization of automorphic forms via functional equations of Dirichlet series.

2.8. Zagier modular forms and the Doi-Naganuma correspondence.

2.8.1. Let us consider the vector space E of all 2×2 matrices x provided with the quadratic form $S(x) = -2 \det x$.

If

$$x = \begin{pmatrix} x_1 & x_3 \\ x_4 & x_2 \end{pmatrix}, \quad S(x) = 2x_3x_4 - 2x_1x_2 .$$

If

$$x = \begin{pmatrix} x_1 & x_3 \\ x_4 & x_2 \end{pmatrix} \text{ and } y = \begin{pmatrix} y_1 & y_3 \\ y_4 & y_2 \end{pmatrix},$$

the value of the associated symmetric form $S(x,y)$ is

$$S(x,y) = x_3y_4 + x_4y_3 - (x_1y_2 + x_2y_1) .$$

We write also

$$x = \begin{pmatrix} u_1+v_1 & u_2-v_2 \\ u_2+v_2 & v_1-u_1 \end{pmatrix} .$$

In these coordinates

$$S(x) = 2(u_1^2 + u_2^2 - (v_1^2 + v_2^2)) .$$

In particular S is of signature $(2,2)$.

The group $SL(2,\mathbb{R}) \times SL(2,\mathbb{R})$ acts on E via $(g_1,g_2)\cdot x = g_1xg_2^{-1}$. This action leaves stable $S(x)$. Thus we obtain a map from $SL(2,\mathbb{R}) \times SL(2,\mathbb{R})$ into $O(2,2)$. It is easy to see that this map is surjective on the connected component of $O(2,2)$. Its kernel consists of

$$\begin{pmatrix}-1 & 0\\ 0 & -1\end{pmatrix} \times \begin{pmatrix}-1 & 0\\ 0 & -1\end{pmatrix}.$$

We consider the representation $\tilde{R}_S$ of $\tilde{\mathcal{G}} \times O(2,2)$. As S is of signature $(2,2)$, the restriction of $\tilde{R}_S$ to $\tilde{\mathcal{G}}$ defines indeed a true representation of $SL(2,\mathbb{R})$ which coincides with R_S (2.5.8). The formula for the action of $\tilde{\mathcal{G}}$ on $L^2(E)$ are given by 2.5.8. The action of $O(2,2)$ on $L^2(E)$ is simply given by $(g.f)(x) = f(g^{-1}.x)$. In particular, we obtain a representation of $SL(2,\mathbb{R}) \times (SL(2,\mathbb{R}) \times SL(2,\mathbb{R}))$ on $L^2(E)$. We will denote by σ an element of the first factor (we will write indifferently $\sigma \in \tilde{\mathcal{G}}$ or $\sigma \in SL(2,\mathbb{R})$ for the first factor) and by (g_1,g_2) an element of the second factor. The corresponding variables in P^+ are denoted by (τ,z_1,z_2). Following 2.5.31, we consider $D = \{v \in E^{\mathbb{C}},\ S(v,v) = 0,\ S(v,\bar{v}) > 0\}$. A base point v_0 of D is

$$v_0 = \begin{pmatrix}-i & -1\\ -1 & i\end{pmatrix}.$$

2.8.2. Lemma: For the action of $SL(2,\mathbb{R}) \times SL(2,\mathbb{R})$ on E given by $(g_1,g_2)\cdot x = g_1 x g_2^{-1}$, we have:

a) $(u(\theta_1),u(\theta_2))\cdot v_0 = e^{i\theta_1} e^{i\theta_2} v_0$.

b) $(J_1^-,J_2^-)\cdot v_0 = 0$.

c) $(b(z_1),b(z_2))\cdot v_0 = y_1^{-1/2}\, y_2^{-1/2} \begin{pmatrix}-z_1 & z_1 z_2\\ -1 & z_2\end{pmatrix}.$

<u>Proof</u>: This is proven in the same way than 2.7.2. As in 2.5.31, we consider the function:

$$(\psi^k_{v_0})(x) = S(x,v_0)^{-k}\, S(x)^{k-1}\, e^{-\pi S(x)}, \text{ on } S(x) > 0$$
$$= 0 \quad \text{if} \quad S(x) \leq 0.$$

This function is in $L^2(E)$ if $k > 1$, and in $L^1(E) \cap L^2(E)$ if $k > 2$. We thus obtain (as in 2.7.3):

2.8.3. Proposition: Let $k > 1$, the function $\psi^k_{v_0}$ is a lowest weight vector of weight (k,k,k) for the action R_S of $SL(2,\mathbb{R}) \times (SL(2,\mathbb{R}) \times SL(2,\mathbb{R}))$ on $L^2(E)$.

Remark: The following theorem follows easily from the Remark 2.5.25.

2.8.4. Theorem: For $k > 1$, the representation $T_k \otimes T_k \otimes T_k$ of $SL(2,\mathbb{R}) \times SL(2;\mathbb{R}) \times SL(2,\mathbb{R})$ is contained in $L^2(E)$ with multiplicity one. The vector $\psi^k_{v_0}$ is the lowest weight vector of this representation.

Let $\tau = \alpha + i\beta$, $z_1 = x_1 + iy_1$ and $z_2 = x_2 + iy_2$ be given points in $P^+ \times P^+ \times P^+$ and define

$$Q(z_1,z_2) = \begin{pmatrix} -z_1 & z_1z_2 \\ -1 & z_2 \end{pmatrix},$$

then:

2.8.5.
$$(R_S(b(\tau) \times b(z_1) \times b(z_2))\cdot\psi^k_{v_0})(x)$$
$$= \beta^{k/2}\, y_1^{k/2}\, y_2^{k/2}\, S(x,Q(z_1,z_2))^{-k}\, S(x)^{k-1}\, e^{i\pi S(x)\tau}, \quad \text{on } S(x) > 0$$
$$= 0 \quad \text{on } S(x) \leq 0.$$

2.8.6. As shown in (2.3.3), the fact that $\psi^k_{v_0}$ is a vector of weight (k,k,k) for R_S can be translated as follows: For $(\tau,z_1,z_2) \in P^+ \times P^+ \times P^+$, we denote by $\psi^k(\tau;z_1,z_2)$ the function in $L^2(E)$ such that

$$\psi^k(\tau,z_1,z_2)(x) = S(x,Q(z_1,z_2))^{-k}\ S(x)^{k-1}\ e^{i\pi S(x)\tau}, \text{ on } S(x) > 0$$

$$= 0 \quad \text{on } S(x) \leq 0\ .$$

Then for $(\sigma,g_1,g_2) \in \mathcal{G} \times (SL(2,\mathbb{R}) \times SL(2,\mathbb{R}))$

$$R_S(\sigma,g_1,g_2)\cdot\psi^k(\tau,z_1,z_2)$$
$$= j(\sigma,\tau)^{-k} j(g_1,z_1)^{-k} j(g_2,z_2)^{-k} \psi^k(\sigma\cdot\tau,g_1\cdot z_1,g_2\cdot z_2)$$

where

$$j(\begin{pmatrix} a & b \\ c & d \end{pmatrix},z) = cz + d\ .$$

<u>Remark</u>: For $(g_1,g_2) \in SL(2,\mathbb{R}) \times SL(2,\mathbb{R})$, this formula is immediately deduced from the relation:

$$Q(g_1\cdot z_1,g_2\cdot z_2) = j(g_1,z_1)^{-1} j(g_2,z_2)^{-1} g_1 Q(z_1,z_2) g_2^{-1}\ .$$

For $\sigma \in SL(2,\mathbb{R})$, this is a deeper property.

As k will be a fixed integer > 2, we will often suppress the index k, for example we will write $\psi(\tau;z_1,z_2)$ instead of $\psi^k(\tau,z_1,z_2)$.

2.8.7. Let $K = \mathbb{Q}(\sqrt{D})$, with $D > 0$ a square-free integer, a real quadratic field. We denote by $\lambda \to \lambda'$ the conjugation in K with respect to $\mathbb{Q}$, i.e. $(a + b\sqrt{D})' = (a - b\sqrt{D})$, and by $N(\lambda) = \lambda\lambda' = a^2 - Db^2$ the norm. Let $\begin{pmatrix} a & b \\ c & d \end{pmatrix}$ a matrix with coefficients in K, then $\begin{pmatrix} a & b \\ c & d \end{pmatrix}'$ denotes the matrix $\begin{pmatrix} a' & b' \\ c' & d' \end{pmatrix}$.

Let $\mathcal{O}$ be the ring of integers of $Q(\sqrt{D})$. λ is in $\mathcal{O}$ if and only if $\lambda + \lambda' \in \mathbb{Z}$, $\lambda\lambda' \in \mathbb{Z}$. If $D \not\equiv 1$ Mod 4 a basis of $\mathcal{O}$ over $\mathbb{Z}$ is thus given by $(1, \sqrt{D})$. If $D \equiv 1$ Mod 4, a basis of $\mathcal{O}$ over $\mathbb{Z}$ is given by $(1, \frac{1+\sqrt{D}}{2})$.

2.8.8. Let $SL(2,\mathcal{O}) = \{\begin{pmatrix} a & b \\ c & d \end{pmatrix} \in SL(2;\mathbb{R})\ ;\ a, b, c, d \in \mathcal{O}\}$. We consider the Hilbert modular subgroup $\Gamma_{\mathcal{O}}$ of $SL(2,\mathbb{R}) \times SL(2,\mathbb{R})$ defined by $\Gamma_{\mathcal{O}} = \{(g,g');\ g \in SL(2,\mathcal{O})\}$.

Let L be the lattice given as in [37], by:

$$L = \{\ell,\ 2 \times 2 \text{ Matrices with coefficients in } \mathcal{O} \text{ such that } \ell\ell' = (\det \ell)\ \mathrm{Id}\}.$$

Thus $\begin{pmatrix} a & b \\ c & d \end{pmatrix} \in L$, if and only if $\begin{pmatrix} a' & b' \\ c' & d' \end{pmatrix} = \begin{pmatrix} d & -b \\ -c & a \end{pmatrix}$, i.e.

$$L = \{\begin{pmatrix} \lambda & \sqrt{D}\,x_3 \\ \sqrt{D}\,x_4 & \lambda' \end{pmatrix}, \text{ with } \lambda \in \mathcal{O},\ x_3, x_4 \in \mathbb{Z}\}.$$

2.8.9. <u>Lemma</u>: The lattice L is stable under the action of $\Gamma_{\mathcal{O}}$.

<u>Proof</u>: This is clear as

$$(gxg'^{-1})(gxg'^{-1})' = gxx'g^{-1} = (\det x)\ \mathrm{Id}.$$

2.8.10. <u>Remark</u>: We have the formula:

$$\begin{pmatrix} a & b \\ c & d \end{pmatrix}\begin{pmatrix} x_1 & x_3 \\ x_4 & x_2 \end{pmatrix}\begin{pmatrix} d' & -b' \\ -c' & a' \end{pmatrix} = \begin{pmatrix} y_1 & y_3 \\ y_4 & y_2 \end{pmatrix}$$

with:

$$y_1 = ad'x_1 - bc'x_2 - ac'x_3 + bd'x_4$$

$$y_2 = -b'c\,x_1 + a'dx_2 + a'cx_3 - b'dx_4$$

$$y_3 = -ab'x_1 + a'bx_2 + aa'x_3 - bb'x_4$$

$$y_4 = cd'x_1 - c'dx_2 + dd'x_4 - cc'x_3 .$$

We see that if $\mathfrak{a}$ is an ideal of $\mathcal{O}$ containing $\mathcal{O}\sqrt{D}$ the lattice

$$L_{\mathfrak{a}} = \{\begin{pmatrix} \lambda & \sqrt{D}\,x_3 \\ \sqrt{D}\,x_4 & \lambda' \end{pmatrix} ; \lambda \in \mathfrak{a}\} \text{ is invariant under } SL(2,\mathcal{O}).$$

Let D_1 be a divisor of D. We obtain all such ideals as $\mathfrak{a}_{D_1} = D_1\mathcal{O} + \mathcal{O}\sqrt{D}$, and we denote the corresponding lattice by L_{D_1}. For $D_1 = 1$ this corresponds to our lattice L; for $D_1 = D$ we obtain the lattice considered by Kudla [17].

Let us consider the distribution δ_L given by $(\delta_L,\varphi) = \sum_{u\in L} \varphi(u)$. This is an invariant distribution under $\Gamma_{\mathcal{O}}$. We now investigate the transformation properties of δ_L with respect to the action of $\mathcal{G}$. We will suppose for the rest of this chapter that $D \equiv 1$ Mod 4.

2.8.11. <u>Proposition</u>: Let

$$\gamma = \begin{pmatrix} a & b \\ c & d \end{pmatrix} \in \Gamma_0(D) \subset \mathcal{G}, \text{ then } R_S(\gamma)\cdot\delta_L = \left(\frac{d}{D}\right)\delta_L .$$

<u>Proof</u>: We write $E = E_{12} \oplus E_{34}$, where $E_{12} = \{\begin{pmatrix} x_1 & 0 \\ 0 & x_2 \end{pmatrix}, x_i \in \mathbb{R}\}$ and $E_{34} = \{\begin{pmatrix} 0 & x_3 \\ x_4 & 0 \end{pmatrix}\}$. According to this decomposition, we write

our lattice $L = \Lambda \oplus \mathcal{L}$, where

$$\Lambda = \{\begin{pmatrix} \lambda & 0 \\ 0 & \lambda' \end{pmatrix};\ \lambda \in \mathcal{O}\} \quad \text{and} \quad \mathcal{L} = \{\begin{pmatrix} 0 & \sqrt{D}\,x_3 \\ \sqrt{D}\,x_4 & 0 \end{pmatrix},\ x_3, x_4 \in \mathbb{Z}\}.$$

We consider first the distribution $\theta_{\mathcal{L}}$ on the space E_{34} given by $(\theta_{\mathcal{L}}, \varphi) = \sum_{\xi \in \mathcal{L}} \varphi(\xi)$. It would follow from the formulas in Section 2.6. , that $\theta_{\mathcal{L}}$ is invariant under $\Gamma_0(D)$. However we will give an alternate proof which will be fundamental for the remaining of the section.

Let us consider the Weil representation R_{34} attached to the restriction S_{34} of $S = -2 \det x$ to the space $E_{34} = \mathbb{R}\, e_3 \oplus \mathbb{R}\, e_4$, $(e_3 = \begin{pmatrix} 0 & 1 \\ 0 & 0 \end{pmatrix}$, $e_4 = \begin{pmatrix} 0 & 0 \\ 1 & 0 \end{pmatrix})$. Our model is associated to the choice of the Lagrangian space $\ell = E_{34} \otimes P$. Let us choose $\ell_1 = \mathbb{R}\, e_3 \otimes (\mathbb{R}P + \mathbb{R}Q)$. We identify $H(\ell_1)$ with $L^2(\mathbb{R}P + \mathbb{R}Q)$ via $\varphi(xP + yQ) = \varphi\,(\exp(xP + yQ) \otimes e_4)$. The operator:

$$(\mathcal{F}_{\ell_1,\ell}\varphi)(\exp(xP+yQ) \otimes e_4) = \int \varphi(\exp(xP+yQ) \otimes e_4 \exp t(Q \otimes e_3))\, dt$$

intertwines the representation R_{34} with the natural representation U of $SL(2,\mathbb{R})$ in $L^2(\mathbb{R}P + \mathbb{R}Q)$ (2.5.6). We have:

$$(\mathcal{F}_{\ell_1,\ell}\varphi)(\exp(xP+yQ) \otimes e_4) = \int \varphi(\exp(te_3+ye_4) \otimes Q)\, e^{-2i\pi xt}\, dt$$

i.e. $\mathcal{F}_{\ell_1,\ell}$ is the Partial Fourier Transform with respect to the variable t. Hence, by Poisson summation formula,

$$(\theta_{\mathcal{L}}, \varphi) = \sum_{m,n \in \mathbb{Z}} \varphi(\exp(m\sqrt{D}\, e_3 + n\sqrt{D}\, e_4) \otimes Q)$$

is transformed in:

2.8.12. $(\theta_{\mathcal{L}}, \mathcal{F}_{\ell,\ell_1}\varphi) = (\sqrt{D})^{-1} \sum_{m,n\in\mathbb{Z}} \varphi(\exp(\frac{m}{\sqrt{D}} P + n\sqrt{D}\, Q) \otimes e_4)$.

Let us consider the lattice $R_D = \mathbb{Z}P \oplus \mathbb{Z}DQ$. As

$$\begin{pmatrix} a & b \\ Dc & d \end{pmatrix}\begin{pmatrix} m \\ Dn \end{pmatrix} = \begin{pmatrix} am+bDn \\ D(cm+dn) \end{pmatrix},$$

this lattice is invariant under the natural action of $\Gamma_0(D)$ on $\mathbb{R}P \oplus \mathbb{R}Q$. As our distribution $\theta_{\mathcal{L}}$ is transformed into the Poisson distribution associated to the $\Gamma_0(D)$-invariant lattice $\frac{1}{\sqrt{D}} R_D$, $\theta_{\mathcal{L}}$ is invariant under $\Gamma_0(D)$.

Let us now analyze the distribution $(\theta_\Lambda,\varphi) = \sum_{\xi\in\Lambda} \varphi(\xi)$. We consider the restriction S_{12} of S to E_{12}. We employ here the formula 2.6.8. The lattice

$$\Lambda^* = \{\begin{pmatrix} x & 0 \\ 0 & y \end{pmatrix} ; \ x\lambda + y\lambda' \in \mathbb{Z}, \text{ for every } \lambda \in \mathcal{O}\}$$

is easily seen to be

$$\Lambda^* = \{\begin{pmatrix} \frac{\mu}{\sqrt{D}} & 0 \\ 0 & \frac{-\mu'}{\sqrt{D}} \end{pmatrix}, \text{ with } \mu \in \mathcal{O}\}.$$

Its level n_Λ is D.

The lattice Λ has the $\mathbb{Z}$-basis

$$f_1 = \begin{pmatrix} 1 & 0 \\ 0 & 1 \end{pmatrix}, \ f_2 = \begin{pmatrix} \frac{1+\sqrt{D}}{2} & 0 \\ 0 & \frac{1-\sqrt{D}}{2} \end{pmatrix}.$$

Hence $S_\Lambda = \begin{pmatrix} -2 & -1 \\ -1 & \frac{D-1}{2} \end{pmatrix}$ and $\det S_\Lambda = -D$.

As Λ is an even lattice with respect to S_{12} and D is odd, we have $\Gamma_0(n_\Lambda,\chi) = \Gamma_0(D)$. As sign $S = 0$, $k = 2$, we get from 2.6.8, for d odd:

$$\alpha(\nu) = (\tfrac{-D}{d})\, \varepsilon_d^2 = (\tfrac{D}{d})(\tfrac{-1}{d})\, \varepsilon_d^2 = (\tfrac{D}{d}) = (\tfrac{d}{D})$$

as D is congruent to 1 Mod 4. Using the remarks 2.6.6, we see similarly that we have $\alpha(\gamma) = (\frac{d}{D})$, for every d. Q.E.D.

Let us consider the coefficient $R_k(\sigma, g_1, g_2) = (\delta_L, R_S(\sigma \times (g_1, g_2)) \cdot \psi^k_{v_0})$. It then follows from 2.8.11, 2.8.9 and 2.8.3 that the function $R_k(\sigma, g_1, g_2)$ is an automorphic form in σ with respect to the subgroup $\Gamma_0(D)$ with character $\begin{pmatrix} a & b \\ c & d \end{pmatrix} = (\frac{d}{D})$, and in (g_1, g_2) with respect to the Hilbert modular subgroup $\Gamma_{\mathcal{O}}$ of $SL(2,\mathbb{R}) \times SL(2,\mathbb{R})$. From 2.8.6, we explicit this as:

2.8.13. <u>Theorem:</u> (Zagier) Let $k > 2$, then the function:

$$\Omega_k(\tau, z_1, z_2) = \sum_{\xi \in L} \psi(\tau, z_1, z_2)(\xi)$$

$$= \sum_{\substack{\xi \in L \\ S(\xi) > 0}} S(\xi, Q(z_1, z_2))^{-k}\, S(\xi)^{k-1}\, e^{i\pi S(\xi)\tau}$$

is an automorphic function on $P^+ \times (P^+ \times P^+)$, which is:

- modular in (z_1, z_2) of weight (k,k) for the Hilbert modular group $\Gamma_{\mathcal{O}}$,
- modular in τ of weight k for the congruence subgroup $\Gamma_0(D)$ with character ε .

We explicit this function:

Let us write $\xi = \sqrt{D}\begin{pmatrix} u & a \\ -b & -u' \end{pmatrix}$ with $a,b \in \mathbb{Z}$ and $u \in \frac{\mathcal{O}}{\sqrt{D}} = \delta^{-1}$ (where $\delta = \mathcal{O}\sqrt{D}$ is the different of $K = \mathbb{Q}(\sqrt{D})$) then $-\det \xi = D(uu' - ab)$

$$S(\xi, Q(z_1,z_2)) = -(uz_2 + u'z_1 + a + bz_1z_2).$$

Let

$$\omega_m(z_1,z_2) = \sum_{\substack{u\in\delta^{-1} \\ a,b\in\mathbb{Z} \\ N(u)-ab=\frac{m}{D}}} \left(\frac{1}{uz_1 + u'z_2 + a + bz_1z_2}\right)^k .$$

Clearly each ω is a Hilbert modular form of weight (k,k) we have:

2.8.14. $$\Omega_k(\tau,z_1,z_2) = (\sqrt{D})^{-k}\, 2^{k-1} \sum_{m=1}^{\infty} \omega_m(z_1,z_2)\, m^{k-1}\, e^{2i\pi m\tau}.$$

We will now express $\Omega_k(z_1,z_2,\tau)$ as a linear combination of Poincare series attached to the cusps of $\Gamma_0(D)$. We recall (2.3.18) that a cusp is an equivalence class of $\mathbb{Q} \cup \{\infty\}$ under the action of $\Gamma_0(D)$. Let $\frac{x}{y}, \frac{x'}{y'} \in \mathbb{Q} \cup \{\infty\}$ with $(x,y) = 1$, $(x',y') = 1$. The relations

$$\frac{x'}{y'} = \frac{ax + by}{cx + dy}$$

for $\begin{pmatrix} a & b \\ c & d \end{pmatrix} \in \Gamma_0(D)$ implies $y' \equiv dy \text{ Mod } D$. Thus, as d is prime to D, we have $(y',D) = (y,D)$.

Let D_1 a divisor of D. We write $D = D_1D_2$. As D is square free, $(D_1,D_2) = 1$. We thus can find integers

$(p,q) \in \mathbb{Z}$ such that $pD_1 + qD_2 = 1$. If the cusp P is the equivalence class of $(\frac{-q}{D_1})$ we can choose $A_P = \begin{pmatrix} D_2 & -p \\ D_1 & q \end{pmatrix}$ as $A_P \cdot P = \infty$ (2.3.18).

Let us consider the natural action of $SL(2,\mathbb{R})$ on $\mathbb{R}P \oplus \mathbb{R}Q$. We write $\binom{x}{y}$ for the vector $xP + yQ$.

2.8.15. Lemma: Let (x,y) two integers, with $(x,y) = 1$ and $(y,D) = D_1$. There exists an element $\gamma \in \Gamma_0(D)$ such that

$$\binom{x}{y} = \gamma\, A_P^{-1} \cdot \binom{1}{0}, \text{ for } P = (\tfrac{-q}{D_1}) .$$

Proof: Let $D = D_1D_2$. We have $y = D_1y'$, and $(y,D_2) = 1$. As $(x,y) = 1$, y is prime to xD_2. Thus there exists $u,v \in \mathbb{Z}$ such that $xD_2u + vy = 1$. We write

$$\binom{x}{y} = \begin{pmatrix} x & -v \\ y & D_2u \end{pmatrix}\binom{1}{0} .$$

But

$$\begin{pmatrix} x & -v \\ y & D_2u \end{pmatrix} \cdot A_P = \begin{pmatrix} a & b \\ c & d \end{pmatrix},$$

with $c = D_2y + D_1D_2\alpha \equiv 0 \text{ Mod } D$. Hence $\binom{x}{y} = \gamma\, A_P^{-1}\binom{1}{0}$, with $\gamma \in \Gamma_0(D)$.

From the preceding lemma, it follows that the cusps of $\Gamma_0(D)$ are in one-to-one correspondence with the divisors D_1 of D. We write $P = (D_1)$ for the corresponding equivalence class of $(\frac{-q}{D_1})$.

As

$$A_{(D_1)}^{-1}\begin{pmatrix} 1 & n\omega_P \\ 0 & 1 \end{pmatrix} A_{(D_1)} = \begin{pmatrix} * & * \\ -n\omega_P D_1^2 & * \end{pmatrix},$$

the width ω_p of the cusp (D_1) is $D_2 = D/D_1$. We denote by $S_{(D_1)}$ the subgroup

$$A^{-1}_{(D_1)} \begin{pmatrix} 1 & \mathbb{Z} D_2 \\ 0 & 1 \end{pmatrix} A_{(D_1)} .$$

We consider, as in 2.3.19, the function $e^q_{(D_1)}(\tau) = e^{2i\pi q\tau/D_2}$ and we form, if $k > 2$, the Poincare series associated to the cusp (D_1) defined by:

$$G^q_{(D_1)}(\tau) = \frac{1}{2} \sum_{\gamma\in\Gamma_0(D)/S(D_1)} \mathcal{E}(\gamma)(T_k(\gamma)T_k(A^{-1}_{(D_1)})\, e^q_{(D_1)})(\tau) .$$

We explicit this as:

$$G^q_{(D_1)}(\tau) = \frac{1}{2} \sum_{\gamma\in\Gamma_0(D)/S(D_1)} \mathcal{E}(\gamma) j(A_{(D_1)}\gamma^{-1},\tau)^{-k} e^{2i\pi q(A_{(D_1)}\gamma^{-1})\cdot\tau/D_2}$$

For f a cusp form in $M(\Gamma_0(D), \mathcal{E}, k)$, we have

$$(f, G^q_{(D_1)}) = D_2^k \frac{(k-2)!}{(4\pi q)^{k-1}} a^q_{(D_1)}(f)$$

$$\left(\text{where } (T_k(A_{(D_1)})\cdot f)(\tau) = \sum_{q>0} a^q_{(D_1)}(f)\, e^{2i\pi q\tau/D_2}\right).$$

We now define, following Zagier:

$$G^n(\tau) = \sum_{\substack{D_2 \\ D_1D_2=D \\ D_2|n}} \psi(D_2)\, D_2^{-k}\, G^{n/D_2}_{(D_1)}(\tau)$$

where

$$\begin{aligned} \psi(D_2) &= \left(\frac{D_1}{D_2}\right)\sqrt{D_2}\ , \text{ if } D_2 \equiv 1 \text{ Mod } 4 \\ &= -i\left(\frac{D_1}{D_2}\right)\sqrt{D_2}\ , \text{ if } D_2 \equiv 3 \text{ Mod } 4 . \end{aligned}$$

We will prove the Zagier identity.

2.8.16. <u>Theorem</u>: For all $(z_1, z_2, \tau) \in P^+ \times P^+ \times P^+$, we have:

$$\sum_{m=1}^{\infty} \omega_m(z_1,z_2) m^{k-1} e^{2i\pi m\tau} = \sum_{n=1}^{\infty} \omega_n^0(z_1,z_2) n^{k-1} G^n(\tau)$$

where

$$\omega_n^0(z_1,z_2) = \sum_{\substack{u\in\delta^{-1} \\ a\in\mathbb{Z} \\ N(u)=n/D}} \left(\frac{1}{uz_1+u'z_2+a}\right)^k$$

<u>Proof</u>: Our method will be similar to the one in Section 2.7. We write $L = \Lambda \oplus \mathcal{L}$ in $E_{12} \oplus E_{34}$. Our representation R_S in $L^2(E) = L^2(E_{12}) \otimes L^2(E_{34})$ is written as $R_{12} \otimes R_{34}$. We will need two lemmas on the representation R_{12} that we will prove later:

2.8.17. <u>Lemma</u>: Let $D = D_1 D_2$, then:

$$D\Lambda^* + D_2\Lambda = \{\xi \in \Lambda;\ S_{12}(\xi,\xi) \in 2D_2\mathbb{Z}\}.$$

Let us consider the lattice $\Lambda + D_1\Lambda^* = \frac{1}{D_2}(D_2\Lambda + D\Lambda^*)$.

Then

$$\Lambda + D_1\Lambda^* = \{\xi \in \frac{\Lambda}{D_2};\ S_{12}(\xi,\xi) \in \frac{2\mathbb{Z}}{D_2}\}.$$

We define

$$(\theta_{D_1\Lambda^*+\Lambda}, \varphi) = \sum_{\ell\in D_1\Lambda^*+\Lambda} \varphi(\ell).$$

2.8.18. <u>Lemma</u>: $R_{12}(A_{(D_1)})\cdot\theta_\Lambda = s_{D_1}\theta_{D_1\Lambda^*+\Lambda}$, with $s_{D_1} = D_2^{-1}\psi(D_2)$.

2.8.19. We now analyze the Poisson distribution with respect to the lattice $\mathcal{L}$. From the proof of 2.8.11, the distribution $\theta_{\mathcal{L}}$ is transformed by the operator $\mathcal{F}_{\ell_1,\ell}$ in

$$(\theta_{\mathcal{L}}, \mathcal{F}_{\ell,\ell_1}\varphi) = (\sqrt{D})^{-1} \sum_{m,n\in\mathbb{Z}} \varphi(\exp(\tfrac{m}{\sqrt{D}}P + n\sqrt{D}Q) \otimes e_4).$$

We now analyze the orbits of $\Gamma_0(D)$ acting on the lattice $R_D = \mathbb{Z}P + \mathbb{Z}DQ$.

2.8.20. Lemma: Each element of R_D is conjugated under $\Gamma_0(D)$ to an element $jA^{-1}_{(D_1)}\binom{1}{0}$, where D_1 is a positive divisor of D and j a multiple of $D_2 = D/D_1$. Two elements $jA^{-1}_{(D_1)}\binom{1}{0}$ and $j'A^{-1}_{D_1'}\binom{1}{0}$ are conjugated if and only if $D_1 = D_1'$, $j = \pm j'$.

Proof: Let $m = jx$, $Dn = jy$ with $(x,y) = 1$ $(y,D) = D_1$, then j is determined up to sign and has to be a multiple of D_2. So the lemma follows from 2.8.14.

2.8.21. As the stabilizer of $A^{-1}_{D_1}\binom{j}{0}$ in $\Gamma_0(D)$, for $j \neq 0$, is

$$A^{-1}_{(D_1)}\begin{pmatrix}1 & \mathbb{Z}\\ 0 & 1\end{pmatrix} A_{(D_1)} \cap \Gamma_0(D) = S_{(D_1)},$$

we obtain

$$(\theta_{\mathcal{L}},\varphi)$$
$$= (\sqrt{D})^{-1} \sum_{\substack{D_1|D\\ j\in\mathbb{Z}-\{0\}\\ \gamma\in\Gamma_0(D)/S_{(D_1)}}} \tfrac{1}{2}(U(A_{(D_1)})U(\gamma)^{-1}\mathcal{F}_{\ell_1,\ell}\varphi)(\tfrac{D_2 j}{\sqrt{D}},0)$$
$$+ (\mathcal{F}_{\ell_1,\ell}\varphi)(0,0).$$

(The factor $\frac{1}{2}$ comes from the fact that the orbits of $A^{-1}_{D_1}\binom{j}{0}$

and $A_{D_1}^{-1}\binom{-j}{0}$ coincides). Let us now prove the Theorem 2.8.16.

The first member of the equality 2.8.16 is up to a constant:

$$(\theta_L, \psi(\tau,z_1,z_2)) = \sum_{\xi \in L} \psi(\tau,z_1,z_2)(\xi)$$

$$= \sum_{\lambda \in \mathcal{O}} \sum_{a,b \in \mathbb{Z}} \psi(\tau,z_1,z_2)\begin{pmatrix} \lambda & \sqrt{D}\,a \\ \sqrt{D}\,b & \lambda' \end{pmatrix}$$

Let us write $\psi_\lambda(x_3,x_4) = \psi(\tau,z_1,z_2)\begin{pmatrix} \lambda & x_3 \\ x_4 & \lambda' \end{pmatrix}$. We have to calculate $(\theta_{\mathcal{L}}, \psi_\lambda)$.

2.8.22. Lemma: $(\mathcal{F}_{\ell_1,\ell}\cdot\psi_\lambda)(0,0) = 0$.

Proof: We have to prove that

$$\int \left(\frac{1}{x_3+\lambda' z_1-\lambda z_2}\right)^k dx_3 = 0 ,$$

where λ is such that $N(\lambda) < 0$. As λ and λ' are of different signs, the element $\lambda' z_1 - \lambda z_2$ is not on the real axis. Thus we have

$$\int_{\mathrm{Im}\, z = c \neq 0} \left(\frac{1}{z}\right)^k dz = 0 \quad (\text{for } k > 1) ,$$

and the lemma is proven.

We then rewrite the formula 2.8.21 as:

$$(\theta_{\mathcal{L}}, \psi_\lambda)$$
$$= (\sqrt{D})^{-1} \sum_{\substack{D_1 | D \\ j \in \mathbb{Z} - \{0\} \\ \gamma \in \Gamma_0(D)/S(D_1)}} \frac{1}{2}\,(U(A_{(D_1)})U(\gamma)^{-1}\mathcal{F}_{\ell_1,\ell}\psi_\lambda)\left(\frac{D_2 j}{\sqrt{D}}, 0\right).$$

Let γ be fixed in $SL(2,\mathbb{R})$, then:

$$(\sqrt{D})^{-1} \sum_{j\in\mathbb{Z}-\{0\}} (U(\gamma)^{-1}.\mathcal{F}_{\ell_1,\ell}\psi_\lambda)(\frac{D_2 j}{\sqrt{D}},0)$$
$$= (\sqrt{D})^{-1} \sum_{j\in\mathbb{Z}} (U(\gamma)^{-1}\, \mathcal{F}_{\ell_1,\ell}\psi_\lambda)(\frac{D_2 j}{\sqrt{D}},0)$$

as

$$(U(\gamma)^{-1}.\mathcal{F}_{\ell_1,\ell}\psi_\lambda)(0,0) = (\mathcal{F}_{\ell_1,\ell}\psi_\lambda)(0,0) = 0\,.$$

Applying Poisson formula, this is:

$$D_2^{-1} \sum_{j\in\mathbb{Z}} (R_{34}(\gamma)^{-1}.\psi_\lambda)(\frac{j\sqrt{D}}{D_2},0)\,.$$

Thus, we have (modulo a change in the order of summation that we could justify) and writing $\psi = \psi(\tau,z_1,z_2)$,

$$(\theta_L,\psi)$$
$$= \sum_{\substack{D_1|D\\ \gamma\in\Gamma_0(D)/S(D_1)}} (2D_2)^{-1} \sum_{\substack{\lambda\in\mathcal{O}\\ j\in\mathbb{Z}}} (R_{34}(A_{(D_1)})R_{34}(\gamma)^{-1}.\psi)\begin{pmatrix}\lambda & \frac{j\sqrt{D}}{D_2}\\ 0 & \lambda'\end{pmatrix}.$$

Let us write $R(\sigma) = R_{12}(\sigma)R_{34}(\sigma)$, i.e. we write

$$R_{34}(A_{(D_1)})R_{34}(\gamma)^{-1} = R_{12}(\gamma)R_{12}(A^{-1}_{(D_1)})R(A_{(D_1)}\gamma^{-1})\,.$$

If α is a function on the space E_{12}, we have:

$$\sum_{\xi\in\Lambda} (R_{12}(\gamma)\cdot\alpha)(\xi) = \varepsilon(\gamma) \sum_{\xi\in\Lambda} \alpha(\xi)\,, \quad \text{for } \gamma\in\Gamma_0(D)$$

(the level of the even lattice Λ is D, and its discriminant $-D$)

$$\sum (R_{12}(A^{-1}_{(D_1)})\cdot\alpha)(\xi) = D_2^{-1}\psi(D_2) \sum_{\xi\in D_1\Lambda^*+\Lambda} \alpha(\xi)$$

from the Lemma 2.8.18, to be proven.

We denote for $\xi\in E_{12}$, $(x_3,x_4)\in\mathbb{R}\times\mathbb{R}$ the element

$\xi + x_3e_3 + x_4e_4$ of E by $(\xi; x_3, x_4)$.

Using the preceding remarks we have:

$$\sum_{\lambda\in\mathcal{O}} (R_{34}(A_{(D_1)})R_{34}(\gamma)^{-1}.\psi(\tau,z_1,z_2))\begin{pmatrix}\lambda & \frac{j\sqrt{D}}{D_2} \\ 0 & \lambda'\end{pmatrix} =$$

$$\sum_{\lambda\in\mathcal{O}} R_{12}(\gamma)R_{12}(A_{D_1}^{-1})(R(A_{(D_1)}\gamma^{-1}).\psi(\tau,z_1,z_2))\begin{pmatrix}\lambda & \frac{j\sqrt{D}}{D_2} \\ 0 & \lambda'\end{pmatrix} =$$

$$D_2^{-1}\psi(D_2)\varepsilon(\gamma) \sum_{\xi\in D_1\Lambda^*+\Lambda} (R(A_{(D_1)}\gamma^{-1}).\psi(\tau,z_1,z_2))(\xi; \frac{j\sqrt{D}}{D_2}, 0).$$

We recall

$$R(A_{(D_1)}\gamma^{-1})\psi(\tau,z_1,z_2) = j(A_{(D_1)}\gamma^{-1},\tau)^{-k}\psi(A_{(D_1)}\gamma^{-1}.\tau,z_1,z_2). \quad (2.8.6)$$

Let $\xi\in D_1\Lambda^*+\Lambda$, then $\xi \in \Lambda/D_2$ and $S(\xi) = \frac{2n}{D_2}$ (2.8.17). We write

$$\xi = \frac{\sqrt{D}}{D_2}\begin{pmatrix}\mu & 0 \\ 0 & -\mu'\end{pmatrix}, \text{ with } \mu \in \frac{\mathcal{O}}{\sqrt{D}} \text{ and } S(\xi) = \frac{2n}{D_2}.$$

Now $\psi(\tau,z_1,z_2)(\xi; \frac{j\sqrt{D}}{D_2}, 0) = (\frac{1}{\mu z_1+\mu' z_2-j})^k(\frac{\sqrt{D}}{D_2})^{-k}(\frac{2n}{D_2})^{k-1}e^{2i\pi n\tau/D_2}$

Thus, it is not difficult to see that

$$\sum_{\substack{D_1|D \\ \gamma\in\Gamma_0(D)/S_{(D_1)} \\ j\in\mathbb{Z} \\ \xi\in D_1\Lambda^*+\Lambda}} (D_2^{-1}\psi(D_2)\varepsilon(\gamma)(R(A_{(D_1)}\gamma^{-1}).\psi)(\xi; \frac{j\sqrt{D}}{D_2}, 0))$$

is absolutely convergent. Let us consider the partial sum Σ' over the (ξ, j) such that $\xi = \frac{\sqrt{D}}{D_2}\begin{pmatrix}\mu & 0 \\ 0 & -\mu'\end{pmatrix}$, with $S(\xi) = \frac{2n}{D_2}$ (i.e. $N(\mu) = \frac{D_2 n}{D}$) with n fixed. (We only have to consider $n > 0$, as $\psi(\tau,z_1,z_2)$ is supported on $S(\xi) > 0$). Then, by definition of ω_n^0, we have the equality:

$$(\sqrt{D})^k\, 2^{1-k} \sum_{\substack{\xi\in D_1\Lambda^*+\Lambda \\ S(\xi)=\frac{2n}{D_2} \\ j\in\mathbb{Z}}} D_2^{-1}\psi(D_2)(R(A_{(D_1)}\gamma^{-1}).\psi)(\xi;\tfrac{j\sqrt{D}}{D_2},0)$$

$$= D_2\omega^0_{D_2n}(z_1,z_2)n^{k-1}j(A_{(D_1)}\gamma^{-1},\tau)^{-k}e^{2i\pi n(A_{(D_1)}\gamma^{-1}.\tau)/D_2} .$$

Now, summing over γ in $\Gamma_0(D)/S_{(D_1)}$ the expression

$$\mathcal{E}(\gamma)j(A_{(D_1)}\gamma^{-1},\tau)^{-k}e^{2i\pi n(A_{(D_1)}\gamma^{-1}.\tau)/D_2}$$

we obtain finally (!)

$$(\sqrt{D})^k\, 2^{1-k}(\theta_L,\psi(\tau,z_1,z_2))$$

$$= \sum_{\substack{D_1|D \\ n=1}}^{\infty} \psi(D_2)D_2^{-1}\omega^0_{D_2n}(z_1,z_2)n^{k-1}G^n_{(D_1)}(\tau).$$

Reassembling together all the terms (D_2,n) such that D_2n is fixed, we obtain:

$$(\sqrt{D})^k\, 2^{1-k}\rho_k(\tau,z_1,z_2) = \sum_{n=1}^{\infty}\omega^0_n(z_1,z_2) \sum_{\substack{D_2|D \\ D_1D_2=D \\ D_2|n}} \psi(D_2)D_2^{-1}\left(\frac{n}{D_2}\right)^{k-1}G^{n/D_2}_{(D_1)}(\tau)$$

$$= \sum_{n=1}^{\infty} \omega^0_n(z_1,z_2)n^{k-1}G^{n/D_2}_{(D_1)}(\tau)$$

and our proposition is proven, modulo the two lemmas: 2.8.17, 2.8.18 that we prove now:

<u>Proof of 2.8.17:</u> It is immediate to see that the first member is contained in the second member. Now let $\xi_0 \in \Lambda$ such that $S_{12}(\xi_0,\xi_0) \in 2D_2\mathbb{Z}$, we have:

$$\xi_0 = a\begin{pmatrix}1 & 0\\ 0 & 1\end{pmatrix} + b\begin{pmatrix}\frac{1+\sqrt{D}}{2} & 0\\ 0 & \frac{1-\sqrt{D}}{2}\end{pmatrix},$$

$$S(\xi_0,\xi_0) = 2((a+\tfrac{b}{2})^2 - b^2\tfrac{D}{4}).$$

As D_2 divides D, this implies $2a + b = D_2u$, thus $b = D_2u - 2a$ and

$$\xi_0 = D_2u\begin{pmatrix}\frac{1+\sqrt{D}}{2} & 0\\ 0 & \frac{1-\sqrt{D}}{2}\end{pmatrix} - a\begin{pmatrix}\sqrt{D} & 0\\ 0 & -\sqrt{D}\end{pmatrix}.$$

The first term belongs to $D_2\Lambda$, the second to $D\Lambda^*$.

Proof of 2.8.18: We first prove the following abstract result:

2.8.28. Lemma: Let (E,S) an orthogonal space, with $\dim E = k$. Let Λ be an even lattice in (E,S) of level n_Λ and discriminant D. Let $\gamma = \begin{pmatrix}a & b\\ c & d\end{pmatrix}$ be an element of $SL(2,\mathbb{Z})$ such that $ac \equiv 0 \mod n_\Lambda$ and $acS(\xi,\xi) \in 2\mathbb{Z}$ for every $\xi \in \Lambda^*$, then

$$R_S(\gamma)\cdot\theta_\Lambda = s(\gamma)\theta_{c\Lambda^*+\Lambda}$$

with $s(\gamma) = D^{-1/2}c^{-k/2} \sum_{y\in\Lambda/c\Lambda} e^{-i\pi\frac{d}{c}S(y)}$, if $c \neq 0$.

(Remark. If the level n_Λ of Λ is odd, the condition $ac \equiv 0 \mod n_\Lambda$ implies $acS(\xi,\xi) \in 2\mathbb{Z}$, for $\xi \in \Lambda^*$.)

Proof: Let us first see by abstract consideration that $R_S(\gamma)\cdot\theta_\Lambda$ is proportional to $\theta_{c\Lambda^*+\Lambda}$. By definition $(R_S(\gamma)\cdot\theta_\Lambda,\varphi) = (\theta_\Lambda, R_S(\gamma)^{-1}.\varphi)$. We can compute $R_S(\gamma)^{-1}$ in the following way:

We consider our self-dual lattice $r = \Lambda^* \otimes P \oplus \Lambda \otimes Q$ with its character $\chi(\exp(\ell^* \otimes P + \ell \otimes Q)) = (-1)^{S(\ell,\ell^*)}$. The operator $(A_r(\gamma^{-1})\cdot\varphi)(n) = \varphi(\gamma\cdot n)$ transforms the space $H(r,\chi)$ into $H(\gamma^{-1}r,\gamma^{-1}\chi)$. We notice that the hypothesis of 2.8.28 implies that χ coincides with $\gamma^{-1}\cdot\chi$ on $\gamma^{-1}r \cap r$. Thus there is a natural intertwining operator $I_{r,\gamma^{-1}r}$ between the model $W(\gamma^{-1}r,\gamma^{-1}\chi)$ of W in $H(\gamma^{-1}r,\gamma^{-1}\chi)$ and the model $W(r,\chi)$ in $H(r,\chi)$, namely we average φ in

$$(I_{r,\gamma^{-1}r}\varphi)(n) = \sum_{\xi\in r/r\cap\gamma^{-1}r} \chi(\exp \xi)\, \varphi(n \exp \xi) .$$

The operator $R'(\gamma^{-1}) = I_{r,\gamma^{-1}r} \circ A_r(\gamma)$ satisfies the fundamental property $R'(\gamma^{-1})\, W(r,\chi)(n)\, R'(\gamma^{-1})^{-1} = W(r,\chi)(\gamma^{-1}\cdot n)$. Hence there exists a scalar $s(\gamma)$ such that the following diagram is commutative:

$$\begin{array}{ccc} H(\ell) & \xrightarrow{\quad R_s(\gamma^{-1})\quad} & H(\ell) \\ \Big\downarrow \theta^{\chi}_{r,\ell} & & \Big\downarrow \theta^{\chi}_{r,\ell} \\ H(r,\chi) & \xrightarrow{\quad s(\gamma)R'(\gamma^{-1})\quad} & H(r,\chi) \end{array} .$$

We now remark that $r/\gamma^{-1}r\cap r \simeq \Lambda^*/\Lambda^*\cap c^{-1}\Lambda$: as $\Lambda \otimes Q \subset r \cap \gamma^{-1}r$, we have

$r/\gamma^{-1}r\cap r \simeq \{\xi \otimes P\ ;\ \xi \in \Lambda^*\}$ modulo

$\{\xi \otimes P\ ,\ \xi \in \Lambda^*$, such that $\gamma(\xi \otimes P) \in r\}$

i.e. $r/\gamma^{-1}r\cap r \simeq \Lambda^*/\Lambda^*\cap c^{-1}\Lambda$.

Thus:

$$(\theta^{\chi}_{r,\ell}\ R_S(\gamma^{-1})\varphi)(0) = (\theta_\Lambda, R_S(\gamma^{-1})\cdot\varphi)$$

$$= s(\gamma)(I_{r,\gamma^{-1}r} \circ A_r(\gamma^{-1})\cdot\theta^{\chi}_{r,\ell}\varphi)(0)$$

$$= s(\gamma) \sum_{\xi\in\Lambda^*/\Lambda^*\cap c^{-1}\Lambda} (A_r(\gamma^{-1})\cdot\theta^{\chi}_{r,\ell}\varphi)(\exp \xi \otimes P)$$

$$= s(\gamma) \sum_{\xi\in\Lambda^*/\Lambda^*\cap c^{-1}\Lambda} (\theta^{\chi}_{r,\ell}\varphi)(\exp \xi \otimes \gamma P)$$

$$= s(\gamma) \sum_{\xi\in\Lambda^*/\Lambda^*\cap c^{-1}\Lambda} (\theta^{\chi}_{r,\ell}\varphi)(\exp \xi \otimes (aP + cQ)) .$$

We write: $\exp(\xi \otimes (aP + cQ)) = \exp c\xi \otimes Q \exp a\xi \otimes P \exp \frac{ac}{2} S(\xi,\xi)E$

As $\theta^{\chi}_{r,\ell}\varphi \in H(r,\chi)$, $\xi \in \Lambda^*$, $a\xi \otimes P \in r$ and $acS(\xi,\xi) \in 2\mathbb{Z}$, by our hypothesis,

$$(\theta^{\chi}_{r,\ell}\varphi)(\exp \xi \otimes aP + cQ) = (\theta^{\chi}_{r,\ell}\varphi)(\exp(c\xi \otimes Q))$$

and we obtain:

$$\sum_{\xi\in\Lambda^*/\Lambda^*\cap c^{-1}\Lambda} (\theta^{\chi}_{r,\ell}\varphi)(\exp \xi \otimes (aP + cQ))$$

$$= \sum_{\xi\in\Lambda^*/\Lambda^*\cap c^{-1}\Lambda} (\theta^{\chi}_{r,\ell}\varphi)(\exp c\xi \otimes Q)$$

$$= \sum_{\xi\in\Lambda^*/\Lambda^*\cap c^{-1}\Lambda} \sum_{u\in\Lambda} \varphi(\exp(u + c\xi) \otimes Q)$$

$$= s(\gamma)\ (\theta_{c\Lambda^*+\Lambda}, \varphi) ,$$

which is the first part of our assertion. Comparing with the Proposition 2.6.11, we clearly have: $s(\gamma) = D^{-1/2}c^{-k/2}c(0,0)$, which ends the proof of our Lemma 2.8.28.

2.8.29. It thus remains to compute $s(A_{(D_1)})$ for

$$A_{(D_1)} = \begin{pmatrix} D_2 & -p \\ D_1 & q \end{pmatrix} .$$

Let us consider the basis $e_1 = \sqrt{D}$, $e_2 = \frac{1+\sqrt{D}}{2}$ of $\mathcal{O}$ over $\mathbb{Z}$.
For $\lambda = ae_1 + be_2$, we have

$$s\begin{pmatrix} \lambda & 0 \\ 0 & \lambda' \end{pmatrix} = 2(a^2D + abD + b^2(\tfrac{D-1}{4})) .$$

Thus

$$e^{-2i\pi \frac{q}{D_1}(a^2D+abD+b^2(\frac{D-1}{4}))} = e^{2i\pi \frac{q(\frac{1-D}{4})}{D_1} b^2} .$$

Now

$$c(0,0) = \sum_{\substack{a\in\mathbb{Z}/D_1\mathbb{Z} \\ b\in\mathbb{Z}/D_1\mathbb{Z}}} e^{2i\pi \frac{q(\frac{1-D}{4})\, b^2}{D_1}}$$

$$= D_1\, D_1^{1/2} \left(\frac{\frac{1-D}{4}}{D_1}\right)\left(\frac{q}{D_1}\right) \varepsilon_{D_1} .$$

But

$$\left(\frac{\frac{1-D}{4}}{D_1}\right) = \left(\frac{1-D}{D_1}\right) = 1$$

$$\left(\frac{q}{D_1}\right) = \left(\frac{D_2}{D_1}\right) \quad \text{as } qD_2 \equiv 1 \text{ Mod } D_1$$

and

$$\varepsilon_{D_1}\left(\frac{D_2}{D_1}\right) = \left(\frac{D_1}{D_2}\right) \quad \text{if } D_1 \equiv 1 \text{ Mod } 4$$

$$\varepsilon_{D_1}\left(\frac{D_2}{D_1}\right) = -\left(\frac{D_1}{D_2}\right)i \quad \text{if } D_2 \equiv 3 \text{ Mod } 4.$$

Thus we obtain: $s(A_{(D_1)}) = D_2^{-1/2}\, \varepsilon_{D_1}(\frac{D_2}{D_1})$, Q.E.D.

2.8.23. Similarly that in the Section 2.7, we now consider the correspondence between cusp forms of weight k and character $(\frac{\cdot}{D})$ with respect to $\Gamma_0(D)$ and Hilbert modular forms of weight (k,k) given by:

$$(\bar{Z}f)(z_1,z_2) = \int \overline{\Omega(\tau,z_1,z_2)}\, f(\tau)\, d\mu_k(\tau)$$

$$(Zf)(z_1,z_2) = (Zf)(-\bar{z}_1,-\bar{z}_2).$$

We first need to express the development of $\omega_n^0(z_1,z_2)$ in Fourier series.

2.8.24. <u>Lemma</u>: $\omega_n^0(z_1,z_2) = \frac{(2\pi i)^k}{(k-1)!} \sum_{r=1}^{\infty} \sum_{\substack{\mu\in\delta^{-1} \\ \mu\gg 0 \\ N(\mu)=\frac{n}{D}}} r^{k-1} e^{2i\pi r(\mu z_1+\mu' z_2)}$.

<u>Proof</u>: As $N(\mu) > 0$, we have $\mu\mu' > 0$. Thus either $\mu > 0$ and $\mu' > 0$, or both are negative. As k is even, we can then sum only on $\mu > 0$, $\mu' > 0$, i.e. $\mu \gg 0$ and multiply the result by 2. Then for $z_1 \in P^+$, $z_2 \in P^+$, $\mu z_1 + \mu' z_2 \in P^+$, and our result is proven as in 2.7.24.

From the characteristic property of Poincare series we then obtain (up to a multiplicative constant):

$$(Zf)(z_1,z_2) = \sum_{\substack{n \\ \mu\gg 0 \\ \mu\in\delta^{-1} \\ N(\mu)=\frac{n}{D}}} \sum_{r=1}^{\infty} r^{k-1} e^{2i\pi r(\mu z_1+\mu' z_2)} \sum_{\substack{D_2|D \\ D_2|n}} \psi(D_2) D_2^{(k-1)} a_{(D_1)}^{n/D_2}(f)).$$

We put together all the term having the same coefficient. Let $\lambda \in \delta^{-1} = \frac{\mathcal{O}}{\sqrt{D}}$, then $\lambda\delta = (\lambda\sqrt{D})\mathcal{O}$ is an integral ideal, with norm $N(\lambda\delta) = N(\lambda)D$. We sum over all the cuples (r,μ) such that $r\mu = \lambda$. Thus r divides the integral ideal $(\lambda\delta) = \mathfrak{A}$, and $DN(\mu) = N(\mathfrak{A})/r^2$. Thus we obtain:

2.8.25. __Theorem:__ Let $f \in S_k(\Gamma_0(D),(\frac{\cdot}{D}))$ a cusp form of weight k. For each integral ideal $\mathfrak{A}$ of $\mathcal{O}$ we define:

$$c(\mathfrak{A}) = \sum_{r|\mathfrak{A}} r^{k-1} \sum_{\substack{D_2|D \\ D_2|N(\mathfrak{A})/r^2}} \psi(D_2)\, D_2^{k-1}\, a_{(D_1)}^{N(\mathfrak{A})/r^2 D_2}(f) .$$

(The first sum is over the natural numbers r dividing $\mathfrak{A}$, the second sum over the integers dividing D and $N(\mathfrak{A})/r^2$.) Then the series:

$$(Zf)(z_1,z_2) = \sum_{\substack{\nu\in\delta^{-1} \\ \nu\gg 0}} c(\nu\delta)\, e^{2i\pi(\nu z_1+\nu' z_2)}$$

is a cusp form of weight (k,k) for the Hilbert modular group.

__Remark:__ As proven by Zagier [37], this map coincides with the map of Doi Naganuma defined for the eigen functions of the Hecke operators.

2.9. Cohen lifting of modular forms.

Let K be a real quadratic field. We have discussed in 2.8 the kernel $\Omega(\tau, z_1, z_2)$ of the Doi-Naganuma correspondence constructed by Zagier. We will modify this construction in order to obtain the correspondence, conjectured by H. Cohen ([5]), between modular forms in one variable with respect to any congruence subgroup $\Gamma_0(N)$ and Hilbert modular forms in two variables. We consider $K = \mathbb{Q}(\sqrt{D})$, with $D \equiv 1 \bmod 4$ and we keep the notations of Section 2.8.

Let N be an integer. We denote by $\Gamma_0(N, \mathcal{O})$ the subgroup of $SL(2, \mathcal{O})$ defined by:

$$\Gamma_0(N, \mathcal{O}) = \{\gamma = \begin{pmatrix} a & b \\ c & d \end{pmatrix};\ \gamma \in SL(2, \mathcal{O}),\ c \in N\mathcal{O}\}.$$

Let χ be a character mod N. The map $\gamma = \begin{pmatrix} a & b \\ c & d \end{pmatrix} \to \chi(dd')$ is a character of $\Gamma_0(N, \mathcal{O})$ denoted by $\chi \circ N_{K/\mathbb{Q}}$. In this section we will prove the:

2.9.1. Theorem: Let k be an integer greater or equal to 3. Let

$$f(\tau) = \sum_{n=1}^{\infty} a(n)\, e^{2i\pi n\tau} \in S_k(\Gamma_0(N), \chi)$$

be a cusp form of weight k and character χ with respect to $\Gamma_0(N)$.

Let $K = \mathbb{Q}(\sqrt{D})$, with $D \equiv 1 \bmod 4$. We define for an integral ideal

$$c(\mathfrak{A}) = \sum_{\substack{r \in \mathbb{N}^+ \\ r \mid \mathfrak{A}}} r^{k-1} \left(\frac{r}{D}\right) \chi(r)\, a\left(\frac{N(\mathfrak{A})}{r^2}\right).$$

Then

$$C_f^K(z_1,z_2) = \sum_{\substack{\nu\in\delta^{-1}\\ \nu\gg 0}} c(\nu\delta)\, e^{2i\pi(\nu z_1+\nu' z_2)}$$

is a Hilbert modular form, with respect to the congruence subgroup $\Gamma_0(N,\mathcal{O})$ and character $\chi\cdot N_{K/\mathbb{Q}}$.

To prove this theorem we will reinterpret Cohen correspondence $f \to C_f^K$ as given by the Petersson inner product with a kernel

$$\Omega_\chi(\tau;z_1,z_2).$$

Let us introduce N' the smallest common multiple of N and D. We have $N' = ND'$, $N = N_0D''$ with $D'D'' = D$. We consider on $\mathbb{Z}/N'\mathbb{Z}$ the function $r \to \overline{\chi(r)}(\frac{r}{D})$. We define

2.9.2. $$u(h) = \sum_{h\in\mathbb{Z}/N'\mathbb{Z}} e^{-\frac{2i\pi hr}{N'}}\, \overline{\chi(r)}(\tfrac{r}{D}).$$

The function u is a function on $\mathbb{Z}/N'\mathbb{Z}$ such that $u(hx) = \chi(h)(\frac{h}{D})\, u_0(x)$, for $h \in (\mathbb{Z}/N'\mathbb{Z})^*$. The function u is an even function if χ is even, odd if χ is odd. We have the inversion formula:

2.9.3. $$\sum_{u\in\mathbb{Z}/N'\mathbb{Z}} u(h)\, e^{\frac{2i\pi hr}{N'}} = N'(\overline{\chi(r)}(\tfrac{r}{D})).$$

We define, for k and ψ of the same parity, $(z_1,z_2) \in P^+ \times P^+$:

2.9.4. $$\omega^0_{\chi,n}(z_1,z_2) = \sum_{\substack{j\in\mathbb{Z}\\ \mu\in\delta^{-1}\\ N(\mu)=\frac{n}{D}}} u(j)\left(\frac{1}{uz_1+u'z_2+(j/N')}\right)^k$$

2.9.5. <u>Lemma</u>: Let $(z_1,z_2) \in P^+ \times P^+$, then:

$$\omega^0_{\chi,n}(z_1,z_2) = c \sum_{\substack{\mu\in\delta^{-1} \\ \mu>>0 \\ N(\mu)=\frac{n}{D}}} \sum_{r=1}^{\infty} r^{k-1}(\tfrac{r}{D})\overline{\chi(r)}\, e^{2i\pi r(\mu z_1+\mu' z_2)}$$

where c is a nonzero constant.

<u>Proof</u>: If $N(\mu) = \frac{n}{D}$, $\mu\mu' > 0$. As the function

$$u(j)\ (\frac{1}{\mu z_1+\mu' z_2+(j/N')})^k$$

is unchanged under $(j,\mu,\mu') \to (-j,-\mu,-\mu')$, we can restrict the sum defining $\omega^0_{\chi,n}$ to be only over $\mu > 0$, $\mu' > 0$, i.e. $\mu>>0$.

Now writing

$$\sum_{j\in\mathbb{Z}} u(j) \left(\frac{1}{\mu z_1+\mu' z_2+(j/N')}\right)^k$$

$$= \sum_{h\in\mathbb{Z}/N'\mathbb{Z}} u(h) \sum_{j\in\mathbb{Z}} \left(\frac{1}{\mu z_1+\mu' z_2+(h/N')+j}\right)^k$$

and applying Poisson summation formula to the function

$$x \to (\frac{1}{z+x})^k = c \int_{\xi\geq 0} e^{2i\pi\xi x}\, e^{2i\pi\xi z}\, \xi^{k-1}\, d\xi\ ,$$

we obtain

$$\omega^0_{\chi,n}(z_1,z_2) = c \sum_{h\in\mathbb{Z}/N'\mathbb{Z}} u(h) \sum_{r=1}^{\infty} r^{k-1}\, e^{2i\pi\frac{rh}{N'}}\, e^{2i\pi r(\mu z_1+\mu' z_2)}$$

$$= c' \sum_{r=1}^{\infty} \overline{\chi(r)}(\tfrac{r}{D})\, r^{k-1}\, e^{2i\pi r(\mu z_1+\mu' z_2)}. \qquad (2.9.3)$$

We set $\Gamma_\infty = \begin{pmatrix} 1 & \mathbb{Z} \\ 0 & 1 \end{pmatrix}$. Let us consider the Poincare series G^q of weight k on $\Gamma_0(N)$ associated to the cusp at ∞:

$$G^q(\tau) = \frac{1}{2} \sum_{\gamma \in \Gamma_\infty \backslash \Gamma_0(N)} \chi(\gamma)^{-1} (c\tau+d)^{-k} e^{2i\pi q \gamma\cdot\tau} .$$

We now define:

2.9.6. $$\Omega_\chi(\tau;z_1,z_2) = \sum_{n=1}^{\infty} \omega^0_{\chi,n}(z_1,z_2)\, n^{k-1}\, G^n(\tau) .$$

The function $\Omega_\chi(\tau;z_1,z_2)$ is constructed in order that:

$$C_f^K(z_1,z_2) = \int_{\mathcal{F}} \overline{\Omega_\chi(\tau;-\overline{z}_1,-\overline{z}_2)}\, f(\tau)\, du_k(\tau)$$

with $du_k(\tau) = (\mathrm{Im}\,\tau)^{k-2}\, |d\tau\, d\overline{\tau}|$ and $\mathcal{F}$ is a fundamental domain for $\Gamma_0(N)$, i.e. Ω_χ is the kernel of the Cohen correspondence.

Hence the Theorem 2.9.1 will be a consequence of the:

2.9.7. <u>Theorem</u>: The function $\Omega_\chi(\tau;z_1,z_2)$ is

1) modular in τ with respect to $\Gamma_0(N)$ and character χ,

2) modular in (z_1,z_2) with respect to $\Gamma_0(N,\mathcal{O})$ and character $(\chi \circ N_{K/\mathbb{Q}})^{-1}$.

The first assertion is obvious by construction. To prove the second assertion, we will reexpress $\Omega_\chi(\tau;z_1,z_2)$ as a coefficient $\langle V_\chi, \phi^k(\tau;z_1,z_2)\rangle$, with $\phi^k(\tau;z_1,z_2)$ the function defined in 2.8.6, and V_χ a $\Gamma_0(N,\mathcal{O})$ semi-invariant distribution on E, and prove an identity analogous to the Zagier identity (2.8.16)

Let D_1 be a divisor of D, we consider the lattice

$$L_{D_1} = \{ \begin{pmatrix} \lambda & \sqrt{D}\, x_3 \\ \sqrt{D}\, x_4 & \lambda' \end{pmatrix} ; \lambda \in D_1 \mathcal{O} + \mathcal{O}\sqrt{D} \ ; \ x_3, x_4 \in \mathbb{Z} \} .$$

Then L_{D_1} is invariant under the action of $SL(2, \mathcal{O})$. (2.8.10)

2.9.8. We will now define an invariant function v under $SL(2, \mathcal{O})$, which will be crucial in our study of the map of Cohen. We need first the:

2.9.9. Lemma: a) $\lambda \in D_1 \mathcal{O} + \mathcal{O}\sqrt{D}$ if and only if $\lambda \in \mathcal{O}$ and $\lambda\lambda' \in D_1\mathbb{Z}$.

b) If $\lambda \in D_1\mathcal{O} + \mathcal{O}\sqrt{D}$, $\lambda + \lambda' \in D_1\mathbb{Z}$.

Proof: It is clear that if $\lambda \in D_1\mathcal{O} + \mathcal{O}\sqrt{D}$, then $\lambda\lambda' \in D_1\mathbb{Z}$. Now $\sqrt{D}, \frac{1+\sqrt{D}}{2}$ is a basis of $\mathcal{O}$ over $\mathbb{Z}$. Let $\lambda = a\sqrt{D} + b(\frac{1+\sqrt{D}}{2}) \in \mathcal{O}$, with $a,b \in \mathbb{Z}$; we have $\lambda\lambda' = \frac{b^2}{4} - D(a + \frac{b}{2})^2$. If $\lambda\lambda' \in D_1\mathbb{Z}$, we obtain that D_1 divides D and $\lambda \in \mathcal{O}\sqrt{D} + D_1\mathcal{O}$. b) is immediate.

We now consider the $SL(2, \mathcal{O})$-invariant lattice $L_0 = \frac{L_D}{D}$, i.e.

$$L_0 = \{ \begin{pmatrix} \lambda & \frac{j}{\sqrt{D}} \\ \frac{k}{\sqrt{D}} & \lambda' \end{pmatrix} ; \lambda \in \frac{\mathcal{O}}{\sqrt{D}} = \delta^{-1}, \ j,k \in \mathbb{Z} \} .$$

This is the lattice considered by Kudla [17].

Let D' be a fixed divisor of D. On the lattice L_0, we define the function $v_{D'}$ by the formulas:

a) $v_{D'}\begin{pmatrix} \lambda & \frac{j}{\sqrt{D}} \\ \frac{k}{\sqrt{D}} & \lambda' \end{pmatrix} = 0$, if $\lambda\lambda' - \frac{jk}{D} \notin \mathbb{Z}$.

b) Let $\lambda\lambda' - \frac{jk}{D} \in \mathbb{Z}$. Let $(k,D') = D'_1$ and $D'_2 = D'/D'_1$, then we define $v_{D'}$ by:

$$v_{D'}\begin{pmatrix} \lambda & \frac{j}{\sqrt{D}} \\ \frac{k}{\sqrt{D}} & \lambda' \end{pmatrix} = \left(\frac{-1}{D'_1}\right)\left(\frac{k}{D'_2}\right)\left(\frac{j}{D'_1}\right).$$

2.9.10. Proposition: The function $v_{D'}$ is invariant by the action of $SL(2,\mathcal{O})$.

Proof: Let us consider the matrices $g(\varepsilon,\alpha) = \begin{pmatrix} \varepsilon & \alpha \\ 0 & \varepsilon' \end{pmatrix}$ ($\varepsilon \in$ Units, $\alpha \in \mathcal{O}$) and $\sigma = \begin{pmatrix} 0 & 1 \\ -1 & 0 \end{pmatrix}$. By a theorem of Vaserstein [34], these matrices generate $SL(2,\mathcal{O})$.

We compute:

$$g(\varepsilon,\alpha)\begin{pmatrix} \lambda & \frac{j_1}{\sqrt{D}} \\ \frac{k_1}{\sqrt{D}} & \lambda' \end{pmatrix}(g(\varepsilon,\alpha)')^{-1} = \begin{pmatrix} u & \frac{j_2}{\sqrt{D}} \\ \frac{k_2}{\sqrt{D}} & u' \end{pmatrix},$$

with $k_2 = k_1$, $j_2 = j_1 - (\varepsilon\alpha'\lambda - \varepsilon'\alpha\lambda')\sqrt{D} - (\alpha\alpha')k_1$. For

$$p = \begin{pmatrix} \lambda & \frac{j}{\sqrt{D}} \\ \frac{k}{\sqrt{D}} & \lambda' \end{pmatrix}, \ \det p = \lambda\lambda' - \frac{jk}{D}.$$

Hence the set of points p such that $\det p \in \mathbb{Z}$ is invariant under $O(2,2)$, in particular under $SL(2,\mathcal{O})$. Let p be such that $\det p \in \mathbb{Z}$. We write $\lambda = \frac{u}{\sqrt{D}}$ with $u \in \mathcal{O}$, $k_1 = D_1'k'$, with $(k',D_1') = 1$. The relation $\det p \in \mathbb{Z}$ is:

$- \frac{uu'}{D} - \frac{D_1'k'j}{D} \in \mathbb{Z}$, hence $uu' \in D_1'\mathbb{Z}$ and $u \in \mathcal{O}D_1' + \mathcal{O}\sqrt{D}$.

Then $(\varepsilon\alpha'\lambda\sqrt{D}) = \varepsilon\alpha'u \in \mathcal{O}D_1' + \mathcal{O}\sqrt{D}$ and

$\varepsilon\alpha'\lambda\sqrt{D} - \varepsilon'\alpha\lambda'\sqrt{D} = (\varepsilon\alpha'u) + (\varepsilon\alpha'u)' \in D_1'\mathbb{Z}$. Thus $j_2 \equiv j_1 \bmod D_1'$,

$(\frac{j_2}{D_1'}) = (\frac{j_1}{D_1'})$, and clearly

$$v_{D'}(g(\varepsilon,\alpha)\, pg(\varepsilon,\alpha)'^{-1}) = v_{D'}(p).$$

Let us now compute the action of σ. We have:

$$\sigma\begin{pmatrix} \lambda & \frac{j}{\sqrt{D}} \\ \frac{k}{\sqrt{D}} & \lambda' \end{pmatrix}\sigma^{-1} = \begin{pmatrix} \lambda' & -\frac{k}{\sqrt{D}} \\ -\frac{j}{\sqrt{D}} & \lambda \end{pmatrix}.$$

Let

$$p = \begin{pmatrix} \frac{u}{\sqrt{D}} & \frac{j}{\sqrt{D}} \\ \frac{k}{\sqrt{D}} & \frac{-u'}{\sqrt{D}} \end{pmatrix},$$

such that $uu' + jk \in D\mathbb{Z}$. Let $k = D_1'k'$, $j = D_1''j'$, with $(k',D') = (j',D') = 1$, $D' = D_1'D_2' = D_1''D_2''$. The equality to be proven is

$$(\frac{-j}{D_2''})(\frac{-k}{D_1''})(\frac{-1}{D_1'}) = (\frac{j}{D_1'})(\frac{k}{D_2'})(\frac{-1}{D_1'}).$$

If $(D_1',D_1'') \neq 1$, both members of the equality are zero. Thus we consider the case where $D' = D_1'D_1''M$, i.e. $D_2' = D_1''M$, $D_2'' = D_1'M$.

The first member of the equality is:

$$(\frac{-1}{D'})(\frac{j}{D_1'})(\frac{j}{M})(\frac{k}{D_1''})(\frac{-1}{D_1''}),$$

while the second member is

$$\left(\frac{j}{D_1'}\right)\left(\frac{k}{M}\right)\left(\frac{k}{D_1''}\right)\left(\frac{-1}{D_1'}\right) .$$

We hence have to prove:

$$\left(\frac{-1}{D'}\right)\left(\frac{-1}{D_1''}\right)\left(\frac{j}{M}\right) = \left(\frac{k}{M}\right)\left(\frac{-1}{D_1'}\right) .$$

Let us write $4uu' = a^2 - Db^2$, with $a,b \in \mathbb{Z}$. The equality $4uu' + 4jk = a^2 - Db^2 + 4D_1'D_1''j'k' \in D\mathbb{Z}$ implies $a^2 \in D_1'D_1''\mathbb{Z}$. As D is square-free, $a = D_1'D_1''x$, with $x \in \mathbb{Z}$ and we obtain the relation

$$D_1'D_1''x^2 + 4j'k' \in \frac{D}{D_1'D_1''}\mathbb{Z} \subset M\mathbb{Z} .$$

This implies:

$$\left(\frac{D_1'}{M}\right)\left(\frac{D_1''}{M}\right) = \left(\frac{-1}{M}\right)\left(\frac{j'}{M}\right)\left(\frac{k'}{M}\right), \text{ i.e. } \left(\frac{j}{M}\right) = \left(\frac{-1}{M}\right)\left(\frac{k}{M}\right) .$$

As

$$\left(\frac{-1}{D'}\right)\left(\frac{-1}{D_1''}\right)\left(\frac{-1}{M}\right)\left(\frac{-1}{D_1'}\right) = \left(\frac{-1}{D'}\right)^2 = 1 ,$$

this completes the proof.

We come back to the study of the kernel $\Omega_\chi(\tau;z_1,z_2)$.

2.9.11. <u>Lemma</u>: We have:

$$\Omega_\chi(\tau;z_1,z_2) = c \sum_{\substack{\gamma\in\Gamma_\infty\backslash\Gamma_0(N) \\ j\in\mathbb{Z} \\ \lambda\in\mathcal{O}}} \chi(\gamma)^{-1}(R_S(\gamma)\cdot\psi^k(\tau;z_1,z_2)\begin{pmatrix} \lambda & \frac{\sqrt{D}j}{N} \\ 0 & \lambda' \end{pmatrix}$$

when c is a nonzero constant.

<u>Proof</u>: By 2.8.6, $R_S(\gamma)\cdot\phi(\tau;z_1,z_2) = (c\tau+d)^{-k}\,\phi(\gamma\cdot\tau;z_1,z_2)$

Let us write

$$\begin{pmatrix} \lambda & \frac{\sqrt{D}j}{N'} \\ 0 & \lambda' \end{pmatrix} = \sqrt{D}\begin{pmatrix} u & \frac{j}{N'} \\ 0 & -u' \end{pmatrix},$$

with $u \in \delta^{-1}$. Now

$$S\left(\begin{pmatrix} u & \frac{j}{N'} \\ 0 & -u' \end{pmatrix}, Q(z_1,z_2)\right) = -\left(\frac{j}{N'} + uz_2 + u'z_1\right).$$

Thus

$$(R_S(\gamma)\cdot\phi(\tau,z_1,z_2))\begin{pmatrix} \sqrt{D}u & \frac{\sqrt{D}j}{N'} \\ 0 & (\sqrt{D}u)' \end{pmatrix}$$

$$= 0\,,\quad \text{if} \quad uu' < 0$$

$$= c\left(uz_2+u'z_1+\frac{j}{N'}\right)^{-k}(Duu')^{k-1}(c\tau+d)^{-k}e^{2i\pi(\gamma\cdot\tau)(Duu')}$$

$$\text{if} \quad uu' > 0.$$

It is not difficult to see that the corresponding series is absolutely convergent. Rearranging the terms, we obtain 2.9.11.

2.9.12. Let us consider the decomposition of $E = E_{12} \oplus E_{34}$ in orthogonal subspaces, with

$$E_{12} = \begin{pmatrix} x_1 & 0 \\ 0 & x_2 \end{pmatrix}, \quad E_{34} = \begin{pmatrix} 0 & x_3 \\ x_4 & 0 \end{pmatrix}$$

and $S = S_{12} + S_{34}$. The forms S_{12} and S_{34} are both of signature (1,1). In the decomposition $L^2(E) = L^2(E_{12}) \otimes L^2(E_{34})$, R_S is written as $R_{12} \otimes R_{34}$.

We consider the lattice

$$\Lambda = \{\begin{pmatrix} \lambda & 0 \\ 0 & \lambda' \end{pmatrix}, \lambda \in \mathcal{O}\} \text{ in } E_{12}.$$

$S_{12}(*)$ is even-valued on Λ. With respect to S_{12},

$$\Lambda^* = \{\begin{pmatrix} u & 0 \\ 0 & u' \end{pmatrix}; u \in \delta^{-1}\}.$$

Hence $D\Lambda^* \subset \Lambda$. It follows from (2.6.9) that the θ-distribution $(\theta_\Lambda, \varphi) = \sum_{x \in \Lambda} \varphi(x)$ on E_{12} is semi-invariant for R_{12} under the group $\Gamma_0(D)$ with character $\mathcal{E}\begin{pmatrix} a & b \\ c & d \end{pmatrix} = (\frac{d}{D})$. As N' is a multiple of D, θ_Λ is semi-invariant under $\Gamma_0(N')$. Writing $\gamma = g\gamma_1$, with $g \in \Gamma_\infty \backslash \Gamma_0(N')$, $\gamma_1 \in \Gamma_0(N') \backslash \Gamma_0(N)$, we obtain

$$\Omega_\chi(\tau; z_1, z_2) =$$

$$\sum_{\gamma_1} \chi(\gamma_1)^{-1} \sum_{\substack{j \in \mathbb{Z} \\ \lambda \in \mathcal{O} \\ g \in \Gamma_\infty \backslash \Gamma_0(N')}} u(j)\chi(g)^{-1}(R_S(g)R_S(\gamma_1)\cdot \psi^k(\tau; z_1, z_2))\begin{pmatrix} \lambda & \frac{\sqrt{D}j}{N'} \\ 0 & \lambda' \end{pmatrix}.$$

Writing $R_S(g) = R_{12}(g) \otimes R_{34}(g)$, this is:

$$\sum_{\gamma_1} \chi(\gamma_1)^{-1} \sum_{\substack{j \in \mathbb{Z} \\ \lambda \in \mathcal{O} \\ g \in \Gamma_\infty \backslash \Gamma_0(N')}} u(j)\mathcal{E}(g)\chi(g)^{-1}(R_{34}(g)R_S(\gamma_1)\cdot \psi^k(\tau; z_1, z_2))\begin{pmatrix} \lambda & \frac{\sqrt{D}j}{N'} \\ 0 & \lambda' \end{pmatrix}.$$

2.9.13. Let us write (x_3, x_4) for the point $\begin{pmatrix} 0 & x_3 \\ x_4 & 0 \end{pmatrix}$ of E_{34}.

We now calculate, for a function φ on E_{34}, the distribution:

$$(\delta_\chi, \varphi) =$$

$$\sum_{g \in \Gamma_\infty \backslash \Gamma_0(N')} \mathcal{E}(g)\chi(g)^{-1} \sum_{j \in \mathbb{Z}} u(j)(R_{34}(g)\cdot\varphi)(\tfrac{\sqrt{D}j}{N'}, 0).$$

(We will see in the proof that indeed δ_ψ is a distribution.)

Let us use the operator $\mathcal{F}_{\ell_1,\ell}$ of 2.5.6 which transforms the representation R_{34} to the natural representation U of $SL(2,\mathbb{R})$ in $\mathbb{R}^2$. $\mathcal{F}_{\ell_1,\ell}$ is given by the Partial Fourier Transform in x_3.

Writing (x,y) or $\binom{x}{y}$ for $xP + yQ$ and using Poisson summation formula, we get:

$$\sum_{j\in\mathbb{Z}} u(j)(R_{34}(g)\cdot\varphi)\left(\frac{\sqrt{D}\,j}{N'},0\right)$$

$$= \sum_{h\in\mathbb{Z}/N'\mathbb{Z}} u(h) \sum_{j\in\mathbb{Z}} (R_{34}(g)\cdot\varphi)\left(\frac{\sqrt{D}\,h}{N'} + \sqrt{D}\,j,0\right)$$

$$= (\sqrt{D})^{-1} \sum \overline{\chi(j)}\left(\frac{j}{D}\right) U(g)\cdot(\mathcal{F}_{\ell_1,\ell}\varphi)\left(\frac{j}{\sqrt{D}},0\right)$$

$$= (\sqrt{D})^{-1} \sum \overline{\chi(j)}\left(\frac{j}{D}\right)(\mathcal{F}_{\ell_1,\ell}\varphi)\left(\frac{1}{\sqrt{D}}\, g^{-1}\cdot\binom{j}{0}\right)$$

$$= (\sqrt{D})^{-1} \sum_{\substack{j \\ (j,N')=1}} \overline{\chi(j)}\left(\frac{j}{D}\right) (\mathcal{F}_{\ell_1,\ell}\varphi)\left(\frac{1}{\sqrt{D}}\, g^{-1}\cdot\binom{j}{0}\right)$$

as $\overline{\chi(j)}\left(\frac{j}{D}\right)$ is 0 if j is not invertible mod N'.

Let us consider the action of $\Gamma_0(N')$ on $\mathbb{Z}P \oplus \mathbb{Z}Q$. The orbit of the point $\binom{1}{0}$ is the set $\{\binom{a}{N'c};$ with $(a,N'c) = 1\}$. Any element $\binom{m}{N'n}$, with $(m,N') = 1$, is thus of the form $jg\binom{1}{0}$, with $(j,N') = 1$ and $g \in \Gamma_0(N')$. Thus we obtain:

$$\sum_{g\in\Gamma_\infty\backslash\Gamma_0(N')} \varepsilon(g)(\chi(g))^{-1}\left(\sum_{j\in\mathbb{Z}} u(j)(R_{34}(g)\cdot\varphi)\left(\frac{\sqrt{D}\,j}{N'},0\right)\right)$$

$$= (\sqrt{D})^{-1} \sum_{m,n} \overline{\psi(m)}\left(\frac{m}{D}\right)(\mathcal{F}_{\ell_1,\ell}\varphi)\left(\frac{1}{\sqrt{D}}\,m, \frac{1}{\sqrt{D}}\,N'n\right)$$

$$= \frac{1}{N'} \sum_{j,k\in\mathbb{Z}} u(j)\, \varphi\left(\frac{\sqrt{D}\,j}{N'}, \frac{N'k}{\sqrt{D}}\right),$$

by using "again" Poisson summation formula. We have proved:

2.9.14. $(\delta_\chi,\varphi) = \frac{1}{N'} \sum_{j,k\in\mathbb{Z}} u(j)\, \varphi(\frac{\sqrt{D}\,j}{N'}, \frac{N'k}{\sqrt{D}})$.

Thus up to a multiplicative constant, we obtain the identity:

2.9.15. $\Omega_\chi(\tau;z_1,z_2) =$

$$\sum_{\gamma_i} \chi(\gamma_i) \sum_{\substack{j,k\in\mathbb{Z}\\ \lambda\in\mathcal{O}}} u(j)(R_S(\gamma_i)\cdot\psi^k(\tau;z_1,z_2))\begin{pmatrix} \lambda & \frac{\sqrt{D}\,j}{N'} \\ \frac{N'k}{\sqrt{D}} & \lambda' \end{pmatrix}.$$

Let us define on E the distribution C_χ, where

$$(C_\chi,\varphi) =$$

$$\sum_{\gamma_i\in\Gamma_0(N')\backslash\Gamma_0(N)} \chi(\gamma_i)^{-1} \sum_{\substack{j,k\in\mathbb{Z}\\ \lambda\in\mathcal{O}}} u(j)(R_S(\gamma_i)\cdot\varphi)\begin{pmatrix} \lambda & \frac{\sqrt{D}\,j}{N'} \\ \frac{N'k}{\sqrt{D}} & \lambda' \end{pmatrix}$$

i.e. $C_\chi = \sum_{\gamma_i\in\Gamma_0(N')\backslash\Gamma_0(N)} \chi(\gamma_i)^{-1} R_S(\gamma_i)^{-1}\cdot(\theta_\Lambda \otimes \delta_\chi)$

the equality 2.9.15 is then

$$\Omega_\chi(\tau;z_1,z_2) = (C_\chi,\psi^k(\tau;z_1,z_2)).$$

We now determine a system of representatives of $\Gamma_0(N')\backslash\Gamma_0(N)$ Let us consider D_1' a divisor of D'. We write $D' = D_1'D_2'$. As D is square free, $(D_1',D_2') = 1$. As $(ND_1',D_2') = 1$, we can find (p,q) such that $\alpha_{D_1'} = \begin{pmatrix} D_2' & p \\ ND_1' & q \end{pmatrix} \in \Gamma_0(N)$.

2.9.16. <u>Lemma</u>: A set of representatives of $\Gamma_0(N')\backslash\Gamma_0(N)$ is $\{\alpha_{D_1'}^{-1}\begin{pmatrix} 1 & x \\ 0 & 1 \end{pmatrix}\}$ where D_1' varies over the divisors of D' and x varies in $\mathbb{Z}/D_2'\mathbb{Z}$.

<u>Proof</u>: Let us consider the natural action of $\Gamma_0(N)$ on $\mathbb{Z}P + \mathbb{Z}Q$

We have, for $\gamma = \begin{pmatrix} a & b \\ Nc & d \end{pmatrix}$, $\gamma \cdot \binom{1}{0} = \binom{a}{Nc}$. Let $(c,D') = D_1'$, $D_2' = D'/D_1'$. Then $(Nc,D_2') = 1$ and $(Nc,a) = 1$. We thus can find (p',q') such that

$$\gamma' = \begin{pmatrix} a & p' \\ Nc & D_2'q' \end{pmatrix} \in \Gamma_0(N) .$$

Now $\gamma' \cdot \binom{1}{0} = \gamma\binom{1}{0} = (\gamma' \cdot \alpha_{D_1'})\, \alpha_{D_1'}^{-1}\binom{1}{0}$. But remark that $\gamma'\alpha_{D_1'} \in \Gamma_0(N')$. Hence we obtain $\gamma = \gamma'' \cdot \alpha_{D_1'}^{-1}\begin{pmatrix} 1 & x \\ 0 & 1 \end{pmatrix}$, with $\gamma'' \in \Gamma_0(N')$. The lemma follows, as $\alpha_{D_1'}^{-1}\begin{pmatrix} 1 & x \\ 0 & 1 \end{pmatrix} \alpha_{D_1'} \in \Gamma_0(N')$ if and only if $x \in D_2'\mathbb{Z}$.

2.9.17. <u>Lemma</u>:

$$R_{12}(\alpha_{D_1'}) \cdot \theta_\Lambda = c(D,N)\; D_1'^{1/2}\left(\frac{D_2'N_0}{D''D_1'}\right)\mathcal{E}_{D''D_1'}\; \theta_{D''D_1'\Lambda^*+\Lambda}$$

when $c(D,N)$ is a constant depending only of D and N.

<u>Proof</u>: We apply 2.6.11 Let $k \in \Lambda^*$, we need to calculate:

$$c(0,k) = \sum_{y \in \Lambda/ND_1'\Lambda} e^{-i\pi\frac{q}{ND_1'}S(y)}\; e^{\frac{2i\pi}{ND_1'}S(k,y)}\; e^{\frac{-i\pi D_2'}{ND_1'}S(k)} .$$

Let $y_0 \in \Lambda \cap ND_1'\Lambda^*$, then $S(y_0) \in 2ND_1'\mathbb{Z}$. If we apply the translation $y \to y + y_0$, we obtain the equality:

$$c(0,k) = e^{\frac{2i\pi}{ND_1'}S(k,y_0)}\; c(0,k) .$$

Hence $c(0,k) = 0$, unless

$$\frac{k}{ND_1'} \in (\Lambda \cap ND_1'\Lambda^*)^*, \text{ i.e. } k \in ND_1'\Lambda^* + \Lambda .$$

As $N = N_0 D''$ with $(N_0, D) = 1$ and $D\Lambda^* \subset \Lambda$,

$$D''D_1'\Lambda^* = D''D_1'(N_0\Lambda^* + D\Lambda^*) \subset ND_1'\Lambda^* + \Lambda, \text{ i.e.}$$
$$ND_1'\Lambda^* + \Lambda = D''D_1'\Lambda^* + \Lambda .$$

We suppose now $k \in D''D_1'\Lambda^* + \Lambda$. Remark that $D_2'k \in \Lambda$. Applying the translation $y \to y + D_2'k$, we then have:

$$c(0,k) = \sum_{y \in \Lambda/ND_1'\Lambda} e^{-i\pi \frac{q}{ND_1'} S(y)} = c(0,0) . \quad \text{(Recall that } qD_2' - pND_1' = 1\text{)}$$

Let us write $ND_1' = N_0 D'' D_1'$. As $(N_0, D''D_1') = 1$, we can write every $\lambda \in \Lambda$ on the form $\lambda = N_0\lambda_1 + D''D_1'\lambda_2$, where λ_1 ranges over $\Lambda/D''D_1'\Lambda$ and λ_2 ranges over $\Lambda/N_0\Lambda$. We obtain:

$$c(0,0) = c_1 c_2, \text{ with } c_1 = \sum_{y \in \Lambda/D''D_1'\Lambda} e^{\frac{-i\pi q N_0}{D''D_1'} S(y)} \text{ and}$$

$$c_2 = \sum_{y \in \Lambda/N_0\Lambda} e^{-i\pi q \frac{D''D_1'}{N_0} S(y)} .$$

The number c_2 is standard to calculate: N_0 is prime to D. Decomposing N_0 in prime n_s', we can diagonalize S over $\mathbb{Z}/n\mathbb{Z}$. As $\dim E_{12} = 2$, we see from 4.3 that c_2 depends only of the denominator N_0.

We now calculate c_1. Choosing $\sqrt{D}, \frac{1+\sqrt{D}}{2}$ as basis of $\mathcal{O}$ we obtain:

$$c_1 = \sum_{\substack{a\in\mathbb{Z}/D''D_1'\mathbb{Z} \\ b\in\mathbb{Z}/D''D_1'\mathbb{Z}}} e^{-2i\pi \frac{qN_0}{D''D_1'}(a^2D+abD+b^2((D-1)/4))},$$

i.e. as $D''D_1'$ divides D

$$c_1 = (D''D_1') \sum_{b\in\mathbb{Z}/D''D_1'\mathbb{Z}} e^{2i\pi \frac{qN_0}{D''D_1'}((1-D)/4)b^2}$$

$$= (D''D_1')(D''D_1')^{1/2} \left(\frac{qN_0}{D''D_1'}\right) \mathcal{E}_{D''D_1'} .$$

As $qD_2' \equiv 1 \bmod D''D_1'$, this concludes the proof of the Lemma 2.9.17.

2.9.18. We now analyze $R_{34}(\alpha_{D_1'})\cdot\delta_\chi$. Using the operator $\mathcal{F}_{\ell,\ell_1}$, we have:

$$(R_{34}(\alpha_{D_1'})\cdot\delta_\chi, \mathcal{F}_{\ell,\ell_1}\varphi)$$
$$= (\sqrt{D})^{-1} \sum_{j,k} \left(\frac{j}{D}\right) \overline{\chi(j)}\, (U(\alpha_{D_1'})^{-1}\cdot\varphi)\left(\frac{j}{\sqrt{D}}, \frac{N'k}{\sqrt{D}}\right).$$

The set $\{jP + N'kQ;\ j,k \in \mathbb{Z}\}$ is transformed by $\alpha_{D_1'}$ into the set $\{D_2'j'P + ND_1'k'Q;\ j',k' \in \mathbb{Z}\}$. The formula $\alpha_{D_1'}\binom{j}{N'k} = \binom{D_2'\ j'}{ND_1'\ k'}$ determines $j = qD_2'j' - pND_1'k'$. As χ is a character mod N and $qD_2' \equiv 1 \bmod N$, $\chi(j) = \chi(j')$. We write $D = D''D_1'D_2'$. We have $j \equiv j' \bmod D''D_1'$, $j \equiv k' \bmod D_2'$. Thus $\left(\frac{j}{D}\right) = \left(\frac{j}{D''D_1'}\right)\left(\frac{j}{D_2'}\right) = \left(\frac{j'}{D''D_1'}\right)\left(\frac{k'}{D_2'}\right)$.

We obtain thus:

$$(R_{34}(\alpha_{D_1'})\cdot\delta_\chi, \mathcal{F}_{\ell,\ell_1}\varphi)$$
$$= (\sqrt{D})^{-1} \sum_{j,k\in\mathbb{Z}} \left(\frac{j}{D''D_1'}\right) \overline{\chi(j)} \left(\frac{k}{D_2'}\right) \varphi\left(\frac{D_2'j}{\sqrt{D}}, \frac{ND_1'k}{\sqrt{D}}\right) .$$

Introducing the classes of $j \bmod ND_1'$, we need to calculate:

$$u_{D_1'}(h) = \sum_{j\in\mathbb{Z}/D_1'N\mathbb{Z}} \left(\frac{j}{D''}\right) \overline{\chi(j)} \left(\frac{j}{D_1'}\right) e^{-\frac{2i\pi hj}{ND_1'}} .$$

As $(D_1',N) = 1$, writing $j = D_1'j_1 + Nj_2$, we obtain

$$u_{D_1'}(h) = \left(\frac{D_1'}{D''}\right) \overline{\chi(D_1')} \left(\frac{N}{D_1'}\right) u_0(h)\, v_{D_1'}(h)$$

with

$$u_0(h) = \sum_{j\in\mathbb{Z}/N\mathbb{Z}} \left(\frac{j}{D''}\right) \overline{\chi(j)}\, e^{-\frac{2i\pi hj}{N}}$$

and

$$v_{D_1'}(h) = \sum_{j\in\mathbb{Z}/D_1'\mathbb{Z}} \left(\frac{j}{D_1'}\right) e^{-\frac{2i\pi jh}{D_1'}} = (D_1')^{1/2} \left(\frac{h}{D_1'}\right)\overline{\mathcal{E}_{D_1'}} .$$

Hence:

$$(R_{34}(\alpha_{D_1'})\cdot\delta_{\chi},\varphi)$$

$$= \frac{1}{N'} \left(\frac{D_1'}{D''}\right) \overline{\chi(D_1')} \left(\frac{N}{D_1'}\right)(D_1')^{1/2}\overline{\mathcal{E}_{D_1'}} \sum_{j,k} u_0(j)\left(\frac{j}{D_1'}\right)\left(\frac{k}{D_2'}\right)\varphi\left(\frac{\sqrt{D}\,j}{N'},\frac{ND_1'k}{\sqrt{D}}\right) .$$

We have:

$$\left(\frac{D_1'}{D''}\right)\left(\frac{N}{D_1'}\right)\overline{\mathcal{E}_{D_1'}}\left(\frac{D_2'N_0}{D''D_1'}\right)\mathcal{E}_{D''D_1'} = \left(\frac{N'}{D''}\right)\left(\frac{D''D_2'}{D_1'}\right)\overline{\mathcal{E}_{D_1'}}\ \mathcal{E}_{D''D_1'}$$

$$= \left(\frac{N'}{D''}\right)\left(\frac{D_2'}{D_1'}\right)\left(\frac{D_1'}{D''}\right)\mathcal{E}_{D''} \text{ using the relation } \mathcal{E}_{pq} = \mathcal{E}_p\mathcal{E}_q\left(\frac{p}{q}\right)\left(\frac{q}{p}\right).$$

Thus

$$(R_S(\alpha_{D_1'})\cdot(\theta_\Lambda \otimes \delta_\chi), \varpi)$$

$$= c'' \overline{\chi(D_1')}\, D_1' \left(\frac{D_2'}{D_1'}\right)\left(\frac{D_1'}{D''}\right) \sum_{\substack{j\in\mathbb{Z}\\ k\in\mathbb{Z}\\ \lambda\in D''D_1' \frac{\mathcal{O}}{\sqrt{D}}+\mathcal{O}}} u_0(j)\left(\frac{j}{D_1'}\right)\left(\frac{k}{D_2'}\right) \varphi\begin{pmatrix} \lambda & \frac{\sqrt{D}\,j}{N'} \\ \frac{ND_1'k}{\sqrt{D}} & \lambda' \end{pmatrix}$$

We now calculate: $C_{D_1',\chi} =$

$$\sum_{x\in\mathbb{Z}/D_2'\mathbb{Z}} \chi\left(\alpha_{D_1'}^{-1}\begin{pmatrix}1 & x\\ 0 & 1\end{pmatrix}\right)^{-1} R_S\begin{pmatrix}1 & x\\ 0 & 1\end{pmatrix}\cdot R_S(\alpha_{D_1'})\cdot(\theta_\Lambda \otimes \delta_\chi).$$

If

$$\lambda \in D''D_1' \frac{\mathcal{O}}{\sqrt{D}} + \mathcal{O}, \quad S\begin{pmatrix} \lambda & \frac{\sqrt{D}\,j}{N'} \\ \frac{ND_1'k}{\sqrt{D}} & \lambda' \end{pmatrix} = -2\left(\lambda\lambda' - \frac{jk}{D_2'}\right) \in 2\,\frac{\mathbb{Z}}{D_2'}.$$

Hence

$$\sum_{x\in\mathbb{Z}/D_2'\mathbb{Z}} \left(R_S\begin{pmatrix}1 & x\\ 0 & 1\end{pmatrix}\cdot\varphi\right)\begin{pmatrix} \lambda & \frac{\sqrt{D}\,j}{N'} \\ \frac{ND_1'k}{\sqrt{D}} & \lambda' \end{pmatrix}$$

$$= \left(\sum_{x\in\mathbb{Z}/D_2'\mathbb{Z}} e^{-2i\pi x(\lambda\lambda' - \frac{jk}{D_2'})}\right)\cdot\varphi\begin{pmatrix} \lambda & \frac{\sqrt{D}\,j}{N'} \\ \frac{ND_1'k}{\sqrt{D}} & \lambda' \end{pmatrix} \neq 0,$$

only if $\lambda\lambda' - \frac{jk}{D_2'} \in \mathbb{Z}$. In this case this is $D_2'\,\varphi\begin{pmatrix} \lambda & \frac{\sqrt{D}\,j}{N'} \\ \frac{ND_1'k}{\sqrt{D}} & \lambda' \end{pmatrix}$.

We remark that if $\lambda \in \frac{\mathcal{O}}{\sqrt{D}}$ and $\lambda\lambda' \in \mathbb{Z}/D_2'$, then $\lambda \in D''D_1' \frac{\mathcal{O}}{\sqrt{D}} + \mathcal{O}$. As $\chi(\alpha_{D_1'}^{-1}) = \chi(D_2')$, we obtain

$$(C_{D_1',\chi},\varphi) = c'''\left(\frac{D_2'}{D_1'}\right)\left(\frac{D_1'}{D''}\right) \sum_{\substack{j,k\in\mathbb{Z}\\ \lambda\in\delta^{-1}\\ \lambda\lambda'-\frac{jk}{D_2'}\in\mathbb{Z}}} u_0(j)\left(\frac{j}{D_1'}\right)\left(\frac{k}{D_2'}\right) \varphi\begin{pmatrix} \lambda & \frac{\sqrt{D}\,j}{N'} \\ \frac{ND_1'k}{\sqrt{D}} & \lambda' \end{pmatrix}.$$

We write $\frac{\sqrt{D}\,j}{N'} = \frac{D''j}{N/\sqrt{D}}$. For $D''D_1'D_2' = D \equiv 1 \bmod 4$, we have:

$$\left(\frac{D_2'}{D_1'}\right)\left(\frac{D_1'}{D''}\right)\left(\frac{D''}{D_1'}\right)\left(\frac{D_1'}{D_2'}\right) = \left(\frac{D_2'D''}{D_1'}\right)\left(\frac{D_1'}{D''D_2'}\right) = \mathcal{E}_{D_2'D''}\cdot\mathcal{E}_{D_1'} = (\mathcal{E}_{D_1'})^2 = \left(\frac{-1}{D_1'}\right).$$

Hence, we rewrite:

$$(C_{D_1',\chi},\varphi) = c''' \sum_{\substack{\lambda\in\delta^{-1}\\ j\in D''\mathbb{Z}\\ k\in D_1'\mathbb{Z}\\ \lambda\lambda'-\frac{jk}{D}\in\mathbb{Z}}} \left(\frac{-1}{D_1'}\right)u_0\left(\frac{j}{D''}\right)\left(\frac{j}{D_1'}\right)\left(\frac{k}{D_2'}\right) \varphi\begin{pmatrix} \lambda & \frac{j}{N/\sqrt{D}} \\ \frac{Nk}{\sqrt{D}} & \lambda' \end{pmatrix}$$

and the distribution C_χ is proportional to the distribution V_χ where

$$(V_\chi,\varphi) = \sum_{D_1'|D'} \sum_{\substack{\lambda\in\delta^{-1}\\ j\in D''\mathbb{Z}\\ k\in D_1'\mathbb{Z}\\ \lambda\lambda'-\frac{jk}{D}\in\mathbb{Z}}} \left(\frac{-1}{D_1'}\right)\left(\frac{j}{D_1'}\right)\left(\frac{k}{D_2'}\right)u_0\left(\frac{j}{D''}\right) \varphi\begin{pmatrix} \lambda & \frac{j}{N/\sqrt{D}} \\ \frac{Nk}{\sqrt{D}} & \lambda' \end{pmatrix}.$$

We obtain the identity $\Omega_\chi(\tau;z_1,z_2) = (V_\chi,\psi^k(\tau;z_1,z_2))$. Let

$$\omega_{\chi,n}(z_1,z_2) =$$

$$\sum_{D_1'|D'} \sum_{\substack{\lambda\in\mathcal{O}\\ u\in D''\mathbb{Z}\\ v\in D_1'\mathbb{Z}\\ \lambda\lambda'-uv=Dn}} u_0\left(\frac{u}{D''}\right)\left(\frac{u}{D_2'}\right)\left(\frac{v}{D_1'}\right)\left(\frac{-1}{D_1'}\right)\left(\lambda z_1+\lambda' z_2+\frac{u}{N}-Nvz_1z_2\right)^{-k}.$$

Then we obtain an identity analogous to Zagier identity [25].

2.9.19. Theorem: We have the identity:

$$\Omega_\chi(\tau;z_1,z_2) = \sum_{n=1}^{\infty} \omega^0_{\chi,n}(z_1,z_2)\, n^{k-1}\, G^n(\tau)$$

$$= \sum_{n=1}^{\infty} \omega_{\chi,n}(z_1,z_2)\, n^{k-1}\, e^{2i\pi n\tau}.$$

It remains to see that the forms $\omega_{\chi,n}(z_1,z_2)$ are Hilbert modular forms for the congruence subgroup $\Gamma_0(N,\theta)$. This will follows from the:

2.9.20. Proposition: The distribution V_χ is semi-invariant under $\Gamma_0(N,\theta)$.

Proof: Let us consider the automorphism of E given by

$$x \to \begin{pmatrix}1 & 0\\ 0 & N\end{pmatrix} x \begin{pmatrix}1 & 0\\ 0 & N\end{pmatrix}^{-1}.$$

The transformed distribution V'_χ under this automorphism is:

$$(V'_\chi,\varphi) = \sum_{D'_1|D} \sum_{\substack{\lambda\in\delta^{-1}\\ j\in D''\mathbb{Z}\\ k\in D'_1\mathbb{Z}\\ \lambda\lambda'-\frac{jk}{D}\in\mathbb{Z}}} \left(\frac{-1}{D'_1}\right)u_0\left(\frac{j}{D''}\right)\left(\frac{j}{D'_1}\right)\left(\frac{k}{D'_2}\right)\varphi\begin{pmatrix}\lambda & \frac{j}{\sqrt{D}}\\ \frac{k}{\sqrt{D}} & \lambda'\end{pmatrix}$$

and we need to prove that V'_χ is semi-invariant under the action of the subgroup $\Gamma'_N(\theta) = \{\begin{pmatrix}a & b\\ c & d\end{pmatrix} \in SL(2,\theta);\ b \in N\theta\}$.

Let us consider the distribution

$$(D_\chi,\varphi) = \sum_{\substack{\lambda\in\delta^{-1}\\ j\in D''\mathbb{Z}\\ k\in\mathbb{Z}\\ \lambda\lambda'-\frac{jk}{D}\in\mathbb{Z}}} u_0\left(\frac{j}{D''}\right)\varphi\begin{pmatrix}\lambda & \frac{j}{\sqrt{D}}\\ \frac{k}{\sqrt{D}} & \lambda'\end{pmatrix}.$$

Recall that we have defined the function $v_{D'}$ on L_0 in 2.9.10. We have $V'_\chi = v_{D'} D_\chi$. Thus, as $v_{D'}$ is invariant under the full group $SL(2,\mathcal{O})$, we only have to prove that D_χ is semi-invariant under $\Gamma'_N(\mathcal{O})$. We have:

$$\begin{pmatrix} a & Nb \\ c & d \end{pmatrix} \begin{pmatrix} \lambda & \frac{j_1}{\sqrt{D}} \\ \frac{k_1}{\sqrt{D}} & \lambda' \end{pmatrix} \begin{pmatrix} d' & -Nb' \\ c' & a' \end{pmatrix} = \begin{pmatrix} u & \frac{j_2}{\sqrt{D}} \\ \frac{k_2}{\sqrt{D}} & u' \end{pmatrix}$$

with $j_2 = aa'j_1 - N^2(bb')k_1 + N(a'b\lambda' - ab'\lambda)\sqrt{D}$.

Now if $j_1 \in D''\mathbb{Z}$ and $\lambda\lambda' - \frac{j_1 k_1}{D} \in \mathbb{Z}$, this implies $D\lambda\lambda' \in D''\mathbb{Z}$, hence $(\sqrt{D}\,\lambda) \in D''\mathcal{O} + \mathcal{O}\sqrt{D}$ and $(a'b\lambda' - ab'\lambda)\sqrt{D} \in D''\mathbb{Z}$

As D'' divides N, we see that $j_2 \in D''\mathbb{Z}$ and that $\frac{j_2}{D''} \equiv aa'\frac{j_1}{D''} \bmod N$; the proposition is then clear.

Bibliographical Notes.

Section 2.1: The model of the representation W as an induced representation by a cocompact subgroup of N is due to P. Cartier [4] in view of applications to theta functions. A further reference is also L. Auslander and R. Tolimieri [2].

Section 2.2: The results of this section are classical.

We have benefitted from notes of Harold Stark [32] and of the articles of G. Shimura [30] and T. Shintani [31].

The Theorem 2.2.10 is a symmetric way to express reciprocity relations between these Gauss sums $b(\ell_1,\ell_2)$ (Formula 2.2.8).

We follow in the proof of 2.2.18 the text of S. Lang [18] and refer to it for basic definitions and properties of the quadratic residue symbol (we have adopted the convention of Shimura [30]).

The computation of $k(g)$ is due to Igusa [13]. The proof given here relies on the Theorem (1.7.6) of Part I.

The function v_Z (2.2.33) appears as the "lowest energy state" in many contexts. We didn't discuss here the Fock model for the representation W of the Heisenberg group, to which this vector is intimately related. We refer to Bargmann [3] as the basic reference, for this realization.

The article of Weil [36] is an underlying reference.

Section 2.3: More information on the group $SL(2,\mathbb{R})$ and on the K-decomposition of the representation $T_{k,0}$ can be found in the book [19] of S. Lang.

For Poincare series and the Petersson inner product, we refer to Lehner J. ([20], Chapter VIII).

<u>Section 2.4</u>: The Theorem 2.4.9 in this precise form is due to G. Shimura [30]. The fact that the representation R is a sum of two irreducible representations of the symplectic group in even and odd functions goes back to D. Shale [29].

<u>Section 2.5</u>: The main references for this section are Rallis-Schiffmann [22]-[23], R. Howe [11] and M. F. Vigneras [35].

The Proposition 2.5.6 is due to I. Segal [27].

The decomposition of the representation R_S of $SL(2,\mathbb{R}) \times O(p)$ is given by M. Saito [26].

The isomorphism of the symplectic Lie aglebra $sp(B)$ with the Lie algebra of differential operators $\{p_i q_j + q_j p_i, p_i p_j, q_i q_j\}$ with $q_i = \frac{\partial}{\partial x_i}$, $p_j = x_j$, is at the basis of the infinitesimal formula 2.5.13. We refer to A. A. Kirillov [16] for the infinitesimal corresponding formulas.

The Theorem 2.5.24 is basically due to M. F. Vigneras [35].

We explicit from 2.5.25 to 2.5.31 the following idea of R. Howe [12]: the explicit decomposition of the tensor product of representation T_k of the holomorphic discrete series with the representation $T_{k'}$ of the antiholomorphic discrete series, as determined by E. Gutkin [9] and J. Repka [25] should lead us to the results of Rallis and Schiffmann. Following this line, we obtained the Theorem 2.5.28, which gives a simple proof of the needed results of ([22]-[23]).

We are indebted to Victor Kac for pointing out to us the modular forms constructed by Hecke in [10], and to Dale Petersson for discussing the proof of Hecke [10] that we reproduce here (2.5.34).

The paragraphs 2.5.37-2.5.44 summarize results of [15]: We determined in this article all possible covariant holomorphic maps from the Siegel upper-half plane D to the space of R_S transforming according to the automorphy factor $\tau(CZ+D)$, for τ a finite dimensional representation of $GL(n,\mathbb{C})$.

Section 2.6: The Theorem 2.6.9 in this precise form is due to G. Shimura [30].

The Proposition 2.6.11 in this form is due to T. Shintani [31].

The Theorem 2.6.16 is due to Hecke [10].

The Theorem 2.6.20 is due to A. Andrianov [1]. Our proof is based on the Theorem 1.7.6 of Part I.

Section 2.7: The main reference of this section is G. Shimura [30].

The idea of using the Weil representation associated to a form of signature (2,1) is due to Niwa [21] and Shintani [31].

The proof of 2.7.25 is based on the general approach of Rallis-Schiffmann [24] to Zagier identity.

Section 2.8: The main reference is D. Zagier [37]. Our approach is based on Rallis-Schiffmann [24].

The Theorem 2.8.16 is due to Zagier. Our proof follows the general line of Rallis-Schiffmann [24].

<u>Section 2.9:</u> The main result (Theorem 2.9.1) of this section is due to the author. This theorem was conjectured by H. Cohen [5] who also determined transformation properties of C_f^K on some subgroup of $\Gamma_0(N,\mathcal{O})$ [6].

Our approach is based on the explicit computation of the kernel of this correspondence in the form of a q-series. This approach was suggested to us by the article of Kudla [17].

References

[1] A. N. Andrianov: "Symmetric squares of zeta-functions of Siegel's modular forms of genus 2," Trudy Mat. Inst. Steklov, vol. 142, (1976), pp. 22-45.

[2] L. Auslander and R. Tolimieri: Abelian harmonic analysis, theta-functions and functions algebras on a nilmanifold, Lecture Notes in Mathematics, No. 436, Springer-Verlag, (1975), Berlin-Heidelberg-New York.

[3] V. Bargmann: "On a Hilbert space of analytic functions and an associated integral transform," Part I, Comm. on Pure and Appl. Math., vol. 14, (1961), pp. 187-214.

[4] P. Cartier: "Quantum mechanical commutation relations and theta functions," Proc. Sympos. Pure Math., vol. 9, (1966), pp. 361-383.

[5] H. Cohen: "Formes modulaires à deux variables associées a une forme à une variable," C. R. Acad. Sci. Paris, t. 281, (1975), pp. 753-755.

[6] H. Cohen: "A lifting of modular forms in one variable to Hilbert modular forms in two variables," Modular functions of one variable VI, Proceedings, Bonn, (1976), Lecture Notes, no. 627, Springer-Verlag; Berlin, Heidelberg, New York.

[7] K. Doi and H. Naganuma: "On the functional equation of certain Dirichlet series," Inventiones Math. vol. 9, (1969), pp. 1-14.

[8] M. Eichler: "Introduction to the theory of algebraic numbers and functions," Academic Press, (1966), New York-London.

[9] E. Gutkin: "Overcomplete systems of subspaces and symbols of operators," Functional Analysis and its Applications, vol. 9, (1975), pp. 89-90; English translation, pp. 260-262.

[10] E. Hecke: "Über einen neuen Zusammenhang zwischen elliptschen Modulfunktionen und indefiniten quadratischen Formen," Mathematische Werke, pp. 418-428, Vandenhoeck and Ruprecht, Göttingen (1970).

[11] R. Howe: "θ-series and invariant theory," Automorphic forms, representations and L- functions, Proceedings of symposia in pure mathematics, vol. XXXIII, part 2, pp. 275-286, A.M.S., Providence, R. I.

[12] R. Howe: "On some results of Strichartz and of Rallis and Schiffmann," Journal of Functional Analysis, vol. 32, (1979), pp. 297-303.

[13] J. I. Igusa: "On the graded ring of theta-constants," (II), American Journal of Mathematics, vol. 88, (1966), pp. 221-236.

[14] J. I. Igusa: "Theta functions," Springer-Verlag, (1972), Berlin-New York.

[15] M. Kashiwara and M. Vergne: "On the Segal-Shale-Weil representation and harmonic polynomials," Inventiones Math., vol. 44, (1978), pp. 1-47.

[16] A. A. Kirillov: "The characters of unitary representations of Lie groups," Functional Analysis and its Applications, vol. 3, (1976), (Translation by Consultant Bureau).

[17] S. Kudla: "Theta functions and Hilbert modular forms," Nagoya Math. J., vol. 69, (1978), pp. 97-106.

[18] S. Lang: Algebraic number theory, Addison-Wesley, (1970), Reading, Mass.

[19] S. Lang: SL(2,R), Addison-Wesley, (1975), Reading, Mass.

[20] J. Lehner: "Discontinuous groups and automorphic functions," A.M.S., Providence, R. I., (1964).

[21] S. Niwa: "Modular forms of half-integral weight and the integral of certain theta-functions," Nagoya Math. J., vol. 56, pp. 147-161, (1975).

[22] S. Rallis and G. Schiffmann,: "Discrete spectrum of the Weil representation," Bull. Amer. Math. Soc., vol. 83, pp. 267-270.

[23] S. Rallis and G. Schiffmann: "Automorphic forms constructed from the Weil representation, holomorphic case," Amer. Journal of Math. 100, (1978), pp. 1049-1122.

[24] S. Rallis and G. Schiffmann: "On a relation between SL_2 cusp forms and cusp forms on tube domains associated to orthogonal groups," preprint.

[25] R. Repka: "Tensor products of unitary representations of SL(2,R)," American Journal of Mathematics, vol. 100, (1978), pp. 747-774.

[26] M. Saito: "Representations unitaires des groupes symplectiques," J. Math. Soc. Japan, vol. 24, (1972), pp. 232-251.

[27] I. Segal: "Transforms for operators and symplectic automorphisms over a locally compact group," Math. Scand., vol. 13, (1963), pp. 31-43.

[28] J. P. Serre and H. Stark: "Modular forms of weight 1/2," Modular functions of one variable, VI, Proceedings, Bonn, (1976), Lecture Notes in Mathematics No. 627, Springer-Verlag, Berlin, Heidelberg, New York.

[29] D. Shale: "Linear symmetrics of free boson fields," Trans. Amer. Math. Soc., vol. 103, (1962), pp. 149-167.

[30] G. Shimura: "On modular forms of half integral weight," Ann. of Math., vol. 97, (1973), pp. 440-481.

[31] T. Shintani: "On construction of holomorphic cusp forms of half integral weight," Nagoya Math. J., vol. 58, (1975), pp. 83-126.

[32] H. Stark: "On modular forms from L-functions in number theory," to appear.

[33] S. Sternberg: "Some recent results on the metaplectic representation," Group theoretical methods in Physics, Proceedings, Tübingen (1977), Lecture Notes in Physics, Springer-Verlag, Berlin, Heidelberg, New York.

[34] L. W. Vaserstein: "On the group SL_2 over Dedekind rings of arithmetic type," Math. U.S.S.R. Sbornik, vol. 18 (1972), pp. 321-332.

[35] M. F. Vigneras: "Series theta des formes quadratiques indefinies," Modular functions of one variable VI, Bonn, (1976), Lecture Notes, No. 627, Springer-Verlag Berlin, Heidelberg, New York.

[36] A. Weil: Sur certains groupes d'operateurs unitaires, Acta Math., vol. 111 (1976), pp. 143-211.

[37] D. Zagier: Modular forms associated to real quadratic fields, Inventiones Math., vol. 30, (1975), pp. 1-46.

Printed in the United States
By Bookmasters